Acoustics of Studios and Auditoria

LIBRARY OF IMAGE AND SOUND TECHNOLOGY

ACOUSTICS OF STUDIOS AND AUDITORIA

V. S. MANKOVSKY

BASIC MOTION PICTURE TECHNOLOGY

L. BERNARD HAPPÉ

WIDE-SCREEN CINEMA AND STEREOPHONIC
SOUND

MICHAEL Z. WYSOTSKY

COLOR FILM FOR COLOR TELEVISION

RODGER J. ROSS

ACOUSTICS OF STUDIOS AND AUDITORIA

V. S. MANKOVSKY

Edited and introduced by
Dr Christopher Gilford
*Reader in Acoustics, Department of Building,
University of Aston in Birmingham*

FOCAL PRESS

LONDON & NEW YORK

ISBN 0 240 50711 8

Translated from the Russian by Gordon Clough

Printed and bound in Great Britain at
the Pitman Press, Bath

Contents

Foreword to English Edition

V. S. Mankovsky is a lecturer in the Leningrad Institute of Cinema Technicians, and his book is written as a textbook for students in this and other similar institutes. In the main, he has succeeded in presenting the whole of acoustics, sound insulation, and noise control as a neat and logical sequence, with sufficient basic theory to give the student a thorough understanding of the subject and enable him to apply his knowledge to new situations. At the same time, there is a mass of design information which will make the book serve also as a work of reference.

Many of the Soviet texts in applied science which are now becoming available in English illustrate the differences in outlook and methods which have arisen from the virtual independence of the work in the USSR from that in progress in the West. Although knowledge and discovery advance in parallel over a broad front, the way in which they do so often differs markedly in the two regions and each has its own areas of relative advance or backwardness. This book is no exception; for instance, the section on stereophony is devoted entirely to spaced-microphone techniques, no mention being made of the use of crossed-microphones which have yielded consistently better results for over ten years, yet there is little doubt that the methods of analysis of public address systems in large enclosures are ahead of those used in Britain. These differences add greatly to the interest of the book.

A very attractive feature is the fact that all the essential theory is dealt with by the use of mathematical methods which are within the range of the average applied science undergraduate or even of the bright A-level student. The treatment of studio acoustics in Chapter 2 to 4 by the statistical, geometric, and wave methods in parallel is most elegantly carried out and, I feel, by itself makes the translation worthwhile. The sections on electrically and acoustically coupled rooms are particularly well treated.

In editing Mr A. G. Clough's translation I have taken the liberty, in the interests of readability, of doing a good deal of pruning of many sentences which, in the Russian original, were unnecessarily repetitive or circumlocutory. In doing so I have been careful to preserve any emphasis which the author may have wished to make and to avoid any alterations to his logical development or style of discussion. Where there were differences in terminology from those in general use in English speaking countries, they are pointed out in footnotes. For technical reasons connected with the production of the book, it was decided to retain the continental forms of the symbols for logarithms and for trigonometric and hyperbolic functions. We think that most readers will be sufficiently familiar with these forms not to be inconvenienced. The references at the end of the book presented some difficulty because a large proportion are to Russian language sources which will not be readily accessible. It was decided, however, to retain them, in transliterated form, as their absence would be an obstacle to any reader who wished to follow the author's ideas outside the covers of the book. For the same reason, design tables are printed in straight translation, although western equivalents of the equipment or materials listed may not be immediately recognizable.

It is to be hoped that the reader will find the book as a whole as useful and stimulating as I have found the work of editing it.

DR CHRISTOPHER GILFORD,

Reader in Acoustics,
Department of Building,
University of Aston in
Birmingham.

Preface

This book deals with effects produced in enclosed spaces by sound, with the physical laws that give rise to these effects, and with methods used to analyse them. It also considers questions of the influence of these effects on the quality of sound transmission, how to determine criteria for the evaluation of acoustics within enclosed spaces, and means by which good acoustic conditions can be achieved. These questions are discussed in the light of existing theory and with the help of experimental data which have greatly increased in the last ten or twelve years.

The book is used as a textbook for the course on Acoustics of Film Studios and Cinemas for students specializing in sound techniques in the Leningrad Institute of Cinema Technicians. It includes material on the practical aspects of setting up optimum acoustic conditions in enclosed spaces such as recording and dubbing studios, radio and television studios, amd rooms for sound reproduction.

At present, multi-purpose halls are being built. These need acoustics suitable for sound reproduction via amplifiers and also for direct listening to dramatic and musical performances. This book therefore includes information on the acoustics of such halls, and also necessary data on sound amplification.

The book also includes some data on acoustic conditions in enclosed spaces intended for stereophonic reproduction and on partially enclosed spaces.

Analysis of the physical reality of the acoustic processes in given enclosures cannot alone serve as a basis for the qualitative evaluation of the acoustics of these enclosures, as this kind of evaluation depends also on the subjective judgment of the listener, which in turn is based on the psycho-physiology of hearing. However, despite the great amount of work which has been done in this field, there are still many questions to which a definitive answer has not yet been found.

11

For this reason the author has been obliged in some cases (e.g. in his consideration of the theory of coupled enclosures) to base himself on physical propositions and on their mathematical analysis without any detailed analysis of the psychological side of the matter, and in yet other cases (e.g. in considering the theory of optimum reverberation) to restrict himself merely to setting out a number of rather contradictory experimental results.

By approaching the question in this way the reader will be able to make a properly oriented study of the effects and form a critical evaluation of the qualitative characteristics of the processes under consideration.

Many basic problems are dealt with in this book with the help of mathematical formulae which will be familiar to students of higher mathematics, and it is assumed that students of sound techniques will already have studied the bases of physical and physiological acoustics and are familiar with electro-acoustical equipment. In cases where formulae used for practical calculations and acoustic design are quoted, worked examples are given.

Bearing in mind that this is a textbook, the subject is dealt with systematically, and mathematical analysis is given at each stage although the same analyses can be found in other books. For the same reason, short summaries of conclusions are given under some headings, so that the student can revise the basic points related to the given subjects.

In order that the results of theoretical calculations should correspond with the acoustic characteristics of real rooms, methods of acoustic measurements in enclosures are set out on pages 342–72.

The book will be useful to acoustic designers, to sound engineers, sound mixers, and to specialists working on sound transmission in the cinema and in broadcasting.

The author would like to express his gratitude to V. A. Burgov, Ya. Sh. Vakhitov, A. N. Kacherovich, and A. D. Khokhlov for their many valuable comments on the manuscript, and also to the members of the Acoustics Department of the Leningrad Institute for film technicians, for their active discussion of the text.

1 Acoustic Processes in Enclosures

1.1 The sound field in enclosed rooms

The vibrations of a sound source placed in unlimited space cause disturbances in the surrounding atmosphere. These disturbances, in the form of variations in pressure, spread out in all directions from the sound source. A sound field is created around the sound source. This field can be defined by laws relating sound pressure with space-time coordinates.

The presence of walls creating a partly or wholly enclosed space around the source changes the character of the sound field. The enclosed volume of air becomes a system which is not only excited while the sound source is emitting, but which can continue in a state of vibration after the source has ceased to emit.

The character of a sound field which arises in a partly or wholly enclosed space is closely connected with the linear dimensions of the space. Where the linear dimensions of the enclosed space are small in comparison with the wave length, as, for example, in a Helmholz resonator or in a tube, normal modes of vibration in the space arise only at a few frequencies. For systems of this kind the number of normal modes is small.

Enclosures, and particularly those which are used as large auditoria, have linear dimensions which are significantly larger than the wavelength even for low sound frequencies. Such enclosures, or rather the volumes of air which they contain, are vibratory systems with a wide spectrum of characteristic frequencies. Every normal mode of vibration of these systems gradually dies away.

With such complex and irregular signals as speech or music, characteristic vibrations arise in the air volume of the room at frequencies corresponding to the component frequencies of this complex signal. As the make-up of the spectrum of the signal is constantly changing, each change is accompanied by the appearance of new characteristic vibrations which superimpose themselves on the previously excited vibrations which have not yet died away.

Thus each of the many impulses which make up a speech or musical phrase is accompanied by a gradually decaying reverberation which has an influence on the basic signal produced in the room and changes its time-structure.

The decay of the reverberation takes a shorter or longer time depending on the rate of dispersal of the sound energy, which in turn depends primarily on losses at the boundaries of the room. Moreover, individual vibrations, which make up the reverberation, may decay at different speeds. The dissimilar speeds at which the normal modes of vibration of the enclosure decay result from the fact that the sound absorptive properties of various areas of the surfaces forming the boundaries of the enclosure may be very different one from another. As a result, the sound field becomes uneven.

Thus the differences in the speeds at which the normal modes of vibration excited in the enclosure decay can lead to a major change in the sound field, i.e. to a variation in signal level from one part of the room to another.

The response/frequency characteristic of an air space, like that of any other vibratory system, depends on the dimensions of the system, or, in this case, on the dimensions of the enclosure. This characteristic may be very different from the frequency spectrum of the basic signal. In this case, those components of the signal which coincide with the characteristic frequencies of the room stand out, as a result of resonance, from the others which have no equivalent in the normal frequency spectrum of the enclosure. A greater or lesser emphasis of frequency components of the basic signal may also be caused by differences in the speed at which various normal modes decay. All this may result in a change in the frequency structure of the basic signal, or, in other words, lead to an alteration in its tone colour.

A consideration of the physical aspect of the sound processes in enclosures is not enough to judge the acoustic characteristics of the enclosures. A full judgment may be made only after a subjective evaluation of these processes, carried out with the help of listening apparatus.

From the point of view of aural perception of phenomena typical of enclosures, most interesting are those particular characteristics of hearing, such as the ability of the ear to receive sequences of sound impulses and sum their energy, provided that the interval between the impulses does not exceed a specific time. This time interval is known as the minimum delay time.

Haas's experiments[88] have shown that the minimum delay time for speech depends on the speech tempo, the intensity of the signal, its tone colour and the general delay time of reverberation characteristic of the given enclosure. He established in particular that if the speed of speech is reduced from 7·4 to 3·5 syllables per second, the minimum delay increases from

14

40 to 92 msec, while an increase in the difference between signal levels of from 0 to 6 dB results in an increase in the minimum delay of from 68 to 172 msec. Suppression of low or high frequencies leads to a change in the minimum delay of from 68 to 80 or 105 msec respectively.

The minimum delay time for music is significantly greater than that for speech and, depending on the nature of the work, reaches from 150 to 250 msec.

Thus if the time structure of the basic signal and the normal modes of vibration of the air space within the enclosure which accompany the basic signal are such that the interval between it and a reflection from the enclosure does not exceed the minimum delay time, they are received by the ear in blended form, as reverberation. But if the intervals are greater than the

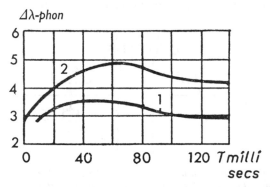

Fig. 1.1. Graphs of increase of loudness of a speech signal when combined with a reflected signal of equal loudness. 1: at a level of 55 phon. 2: at a level of 70 phon.

minimum delay, there is a gap between the signals and the ear identifies the reflection as a distinct echo, which is particularly noticeable when the level of the reflection is comparable with the level of the basic signal.

If the normal vibration modes aroused by the basic signal are at a comparatively high level and follow the basic signal at a delay no greater than the minimum delay time, they blend with this signal and intensify it. Consequently, the normal modes of vibration give sound signals in a room a certain amplification, the degree of which depends on the absorption of energy at the boundaries of the enclosure.

This can be demonstrated in practice by the fact that a change in the distance between the sound source and the receiver results in a significantly smaller change in signal level in an enclosure than it does in the open air. Figure 1.1 gives some idea of the magnitude of the increase in loudness which takes place when a speech signal is reinforced by an equally loud delayed reflection. This figure shows the graph of the dependence of this increase $\Delta\lambda$ on the delay time τ.

Changes in conditions at the boundaries of the enclosure result in alteration of the level of delayed reflections, and also in an alteration in the increase of loudness caused by the normal modes of the airspace of the enclosure, which makes the sound field in the enclosure uneven.

Thus the air space of the enclosure has a significant influence on the signal made within it by:

1. Accompanying it with a characteristic reflection which may take the form of a separate distinct echo, i.e. introducing time changes into its structure.

2. Altering its tone colour, i.e. by introducing changes into its frequency spectrum.

3. Increasing its level by the energy derived from the normal modes of vibration of the enclosure.

4. Creating dissimilar reception conditions in different parts of the enclosure, i.e. by introducing a spatial change into the signal.

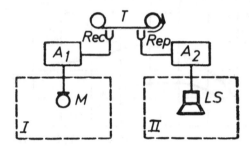

Fig. 1.2. Block diagram of a sound transmission channel. M: microphone. A_1 and A_2 amplifiers. T: magnetic tape. Rec. and Rep.: recording and reproducing heads. LS: loudspeaker. I and II: first and second enclosures.

It follows that the enclosure is a complex element playing a part in the process of sound radiation and reception, and with a significant influence on the quality of the sound. In the case of the transmission of sound via a recording and reproducing chain, or via a low frequency broadcasting chain, the enclosure can be considered as one of the links in the chain together with such other links as the microphones, amplifiers and loudspeakers. A recording and reproducing chain as used, for instance, in the cinema, can be schematically represented as in Fig. 1.2. The first and second enclosures shown in the figure are the enclosures for the recording and reproduction of sound respectively.

1.2 Methods of analysing the sound field in an enclosure

The changes a sound field undergoes if the sound source is not placed in unrestricted space but in an enclosure can be both useful and harmful from the point of view of aural perception.

The reverberation which accompanies every impulse of an irregular signal (e.g. speech or music) can be useful if the duration is limited. Under

16

these conditions speech can sound more lively, and music can become more full-sounding by the establishment of the necessary link between succeeding musical tones. A noticeably long reverberation renders speech less intelligible, and makes musical notes overlap, particularly in rapid passages. In this case, without even taking into consideration the possibility of the appearance of a gap exceeding the minimum delay time, the quality of sound becomes worse.

The overstressing of some characteristic frequency or another of the enclosure, resulting in a change in the timbre of the basic signal to a degree where it no longer corresponds with accepted impressions of natural sound sources, must also have an adverse effect on the reception of speech and music.

The fact that the loudness of a sound in certain conditions is increased by virtue of the normal modes of vibration of the air contained in the enclosure is useful because, contrary to experience in unconfined space, the increase in loudness is particularly noticeable by listeners fairly remote from the sound source. But the possible unevenness of the increase in loudness at different parts of the enclosure, resulting from variable rates of energy decay at the boundaries of the enclosure, is undesirable because it disturbs the uniformity of hearing conditions for listeners who are situated at various parts of the enclosure.

A careful analysis of the sound field created in an enclosure by specific sound sources is needed to determine under what conditions certain changes in the signal produced by an enclosure are useful or harmful, and to determine what factors have an influence on the quality of speech and music. Such an analysis can be carried out on the basis of any of the theories of room acoustics.

One theory—the wave theory—is based on the fact that the air space contained within an enclosure, acts as a vibratory system when it is excited by a sound signal. Unlike a string or a membrane this system is three dimensional and has a complex spectrum of characteristic frequencies.

A strict analysis of the wave processes going on in the enclosure allows the physical nature of these processes to be revealed, and leads to an understanding of the way in which they influence the acoustical properties of the enclosure.

The wave theory of room acoustics, developed by Morse and Bolt from one-dimensional experiments first suggested by Rayleigh in 1878, is capable of strict mathematical proof but the mathematics are complex and cumbersome.

A second theory, which was developed almost in parallel with the wave theory, idealizes the physical processes in the enclosure, and leaves aside any consideration of the wave nature of sound. In this theory the sound field in an enclosure is regarded as a field through every point of which are passing simultaneously a large number of waves reflected from the bounding surfaces of the enclosure.

Just as the energy from the sound source is radiated in all directions, so the reflected waves from any point within the enclosure also travel in all possible directions. The phases of waves passing through each such point

17

may therefore be considered to be randomly distributed, bearing in mind that natural sounds (speech and music) produced within the enclosure may be described as random and irregular signals. This allows us to determine the energy at any given point in the enclosure, without taking into account phase shifts between waves, as the sum of the mean energy values of reflections reaching the given point.

The study of combinations of random phenomena which have common properties, such as are the combinations of reflections reaching each point within an enclosure, is carried out by means of mathematical statistics based on the theory of probability. The statistical method cannot disclose any detailed physical picture of the phenomenon; its advantage, however, lies in the fact that with simple mathematics based on data of the results of the process it allows us to reach an objective conclusion about the quantitative aspects of the process and about its possible shortcomings.

These last remarks make it clear why the statistical method can be used in the study of the acoustical processes within an enclosure, and how the approach to the analysis of the sound field within an enclosure outlined above can serve as a sufficient basis for the currently existing statistical theory of room acoustics.

The method of applying mathematical statistics to random events may give a far from objective picture of the process if for any reason the phenomena cease to be random. For instance, if some surfaces within the room have focusing properties, the random nature of the direction of arrival of reflected waves is lost, and the result of any statistical analysis of a problem of this type is inaccurate. In this case another method can be used in which the sound field is considered as a combination of rays, constructed according to the laws of geometric optics. Thus the sound wave is replaced by a ray along which sound energy is transferred.

The geometrical construction of the pattern of rays makes it possible to determine their point of impact on the surfaces bounding the enclosure, and to calculate the energy losses resulting from the sound absorbing properties of the actual materials which cover these surfaces. This is particularly important when the walls of the enclosure are covered with a variety of materials with differing properties. Under these conditions the sound field in an enclosure can be specified by the magnitude of sound energy, determined for any point by calculating the loss from reflection incurred by all rays passing through that point.

1.3 The acoustic peculiarities of different types of enclosures

The degree of usefulness or harmfulness of signal variations caused by the properties of the enclosure, and which define the acoustic conditions within it, depends not only on the psycho-physiological peculiarities of hearing, but also on the character of natural sounds (speech or music) and on the properties of the secondary sources and receivers of sounds (loudspeakers, microphones) used for sound transmission in enclosures.

18

Bearing this in mind, enclosures may be divided into three basic groups: those for direct listening (lecture halls, theatres and concert halls); those for sound transmission via an electro-acoustic system or by radio (film-studios, cinema auditoria and studios); and those where sound amplifying systems are to be used (stadia, lecture halls, etc.).

In rooms of the first category, those for direct listening, natural signals are used and are picked up directly by the ear of the listener. The radiation and reception processes take place in the same enclosure, i.e. these processes are adjacent in space and time.

A principal characteristic of enclosures of this nature is that the power of the sound sources employed in them is comparatively small, and is confined by the limitations of the human voice or of musical instruments. The amplifying effect of the characteristic vibrations is very important in these enclosures, particularly when they are large in size. It is also important, however, that the reverberation engendered by the characteristic frequencies should not be too prolonged and should not interfere with the clarity of the sound.

In transmitting sound via an electro-acoustic system, two enclosures are used: the primary, which houses the natural sound sources and the microphone, which are together involved in the sound recording process, and the secondary enclosure, in which are placed the loudspeaker and the listeners taking part in the process of sound reception. In this case the primary sound source is separated from the listener, and connection between them is accomplished by means of an electro-acoustic system or by a radio channel.

The processes of radiation and reception are in some cases separated in space and in time (e.g. the recording and reproduction of a film sound track) or in others merely in space (e.g. a radio broadcast). As a result of this the sound signals transmitted by the loudspeaker in the secondary enclosure contain within themselves the reverberation characteristic of the primary enclosure. Moreover, the transmitted signals undergo further changes because of the effect on them of the secondary enclosure.

If it is borne in mind that sound signals in enclosures designed for listening (e.g. cinema auditoria) are emitted at levels amplified by 10 to 15 dB, it is clear that the amplification of the signals involves also the amplification of the reverberation. The overall distortion in conditions of sound reproduction is greater than in direct listening conditions.

Halls with sound amplification systems fall into a special category. They have the properties of enclosures for direct listening, inasmuch as the listener is directly linked with the prime source of sound, and also the properties of enclosures for sound transmission, in view of the parallel action of the system with a secondary receiver and radiator (the microphone and loudspeakers).

The presence of two means of transferring information from the sound sources to the listener results in the overlapping of echoes resulting from normal vibration modes induced by both primary and secondary sources. As a result of this the influence of the acoustics of the enclosure on the quality of the sound transmission is more obvious than in an enclosure for direct listening.

Some enclosures equipped with sound amplification systems show a significant difference in the distances separating the listener from the primary and secondary sources. This situation can disturb the time-relations both between the two basic signals, coming to the listener by two different paths, and also between their echoes.

As the acoustic conditions in enclosures are connected with the character of the transmitted signal, and even with the type of radiator or receiver of sound, the first and second types of enclosure can be further divided into sub-groups. The first group of enclosures can be divided into enclosures intended for speech and for music, and the second group into primary enclosures (for sound recording) and secondary enclosures (for sound reproduction).

Practice shows that in classifying enclosures one should proceed not only from the character of the sound sources and receivers involved in the process of sound transmission. Enclosures may be classified by the use to which they are put, employing a number of technical aspects of the sound transmission process. For example, enclosures used for sound transmission in the film industry can be divided into the following groups depending on the demands which are made on them in terms of acoustic conditions:

1. Synchronized shooting stages.
2. Music recording studios (dubbing stages).
3. Re-dubbing stages.
4. Speech recording studios and speech dubbing stages.
5. Sound effects studios.
6. Viewing rooms.
7. Cinema auditoria.

The variety of enclosures intended for different uses in sound transmission shows that particular demands must be made of them in regard to acoustic conditions. These demands can be formulated only as a result of a detailed study of the sound processes in the given enclosure, the factors which influence them, and also of the conditions which ensure the best reception of sound.

2 The Statistical Theory of Room Acoustics

2.1 Acoustic processes in an enclosure as presented by statistical theory

The study of the sound field in an enclosure is best first approached from the point of view of statistical theory, because it is simpler both in its basic conceptions of acoustic processes in enclosures and also in the mathematics it uses.

The basic proposition of this theory is its conception of the sound field as a field consisting of a large number of waves reflected from the bounding surfaces of the enclosure and dispersing themselves in different directions. It is along these directions that sound energy is transferred. On this basis, sound processes in an enclosure can be set out in the following way.

If a sound source is switched on in an enclosure, the sound waves spreading out from it are reflected from the bounding surfaces of the enclosure, and having lost part of their energy, are re-distributed until they again come up against a bounding surface. This impact produces a further reflection and a further energy loss and so on.

The energy losses are caused by the fact that when the sound wave penetrates the frontier between two media, part of the energy it carries is absorbed by the new medium. The absorption capacity of media of various kinds can be characterized by the coefficient of sound absorption, set out in the form

$$\alpha = \frac{\varepsilon_{absorbed}}{\varepsilon_{incident}} \qquad 2.1$$

where $\varepsilon_{incident}$ represents the sound energy falling on the surface of the medium and $\varepsilon_{absorbed}$ the energy absorbed by this medium.

21

The difference between α and unity, i.e. $1 - \alpha = \beta$, is called the energy coefficient of reflection. It is clear that

$$\beta = \frac{\varepsilon_{reflected}}{\varepsilon_{incident}}, \qquad\qquad 2.2$$

where $\varepsilon_{reflected}$ is the sound energy reflected from the surface.

In the first instant after a sound source is switched on, a listener in the enclosure receives the energy of the direct sound wave leaving the source. To this, fractionally later, is added the energy of the reflected wave whose path from the source to the listener is shortest. Then the listener receives second and third reflections and so on. The order in which these reach the listener and the intervals between them may not be constant.

For instance a wave twice reflected from surfaces close to the sound source might reach the listener earlier than a wave which has been reflected only once from a very remote surface in the enclosure. The addition of energy from each successive wave is not always progressively smaller, the more so as it is also affected by the sound absorbing capacity of the material from the surface of which any given wave has been reflected.

A similar unevenness may also be observed as a result of the fact that sound waves reaching the listener after their first, second or more reflections, may have different phase shifts. Consequently, while there is a general tendency towards a gradual growth of energy, variations in the number of reflections experienced by sound arriving within a short interval, variations in the coefficients of absorption of the inner surfaces of the enclosure, and also phase differences, may introduce a noticeable unevenness into the process of the growth of energy in an enclosure.

As a result of the gradual, although uneven decrease in energy of delayed reflections, the growth of energy ceases after a certain time and the general energy in the enclosure reaches a constant level. This takes place when the addition of energy from the sound source is equal to the loss of energy caused by absorption at the boundaries of the enclosure.

If, after a stationary level of energy has been reached, the action of the sound source is ended, the sound perceived by the listener does not disappear immediately. A short time after the source has ceased to function, the direct sound wave stops and the listener receives the energy of those first reflected waves which have come by a longer path. Then the second and third reflected waves which have travelled even longer paths reach the listener. The energy of these waves is smaller. After a certain interval of time the energy of arriving waves decreases so much that the ear can no longer perceive it and sound disappears.

Thus after the sound source has been switched off, the general energy in the enclosure gradually loses those of its components connected with the various reflections, and this process of loss takes place in the same sequence in which the reflections were built up. The energy in the enclosure decays and, if the intervals between the succeeding reflections is greater than the minimum delay time, there is an echo accompanying each signal as it is interrupted.

22

Sound processes in an enclosure as set out above can still not be examined by mathematical statistics because the combinations of reflected waves which play a part in the processes do not have the necessary common factors. These common factors can be achieved only by a simplified view of certain properties of the sound field in an enclosure.

The first such simplified view of the properties of the sound field relates to the manner of propagation of sound waves.

It has already been pointed out that these waves move in different directions and, if the losses at each reflection are small, a large number of waves passes simultaneously through every point in space within the enclosure. In order that this property of reflected waves may become a constant factor in the process, the arrival of reflected waves at every point in the room from different directions should be equally probable.

The second simplification relates to the proposition that the sound energy at any point in the enclosure is the sum of the mean values of the energy of all reflections passing through that point. In this proposition the common factor is that there is an equal probability of all possible phase shifts for all of the many waves which are being added.

It can be shown that this proposition is true for a number of signals for various values for their delay time τ.

If to the signal $f(t_1)$ is added a delayed reflection $f(t_1 - \tau)$, the mean power of the total signal can be represented as

$$P_\Sigma(t) = \frac{1}{T'} \int_{t-T'}^{t} [f(t_1) + f(t_1 - \tau)]^2 \, dt = P_1 + P_2 + 2r(t) \qquad 2.3$$

where

$$P_1 = \frac{1}{T'} \int_{t-T'}^{t} f^2(t_1) \, dt, \quad P_2 = \frac{1}{T'} \int_{t-T'}^{t} f^2(t_1 - \tau) \, dt \qquad 2.4$$

are the mean powers of each signal for the time T', and

$$r(t) = \frac{1}{T'} \int_{t-T'}^{t} f(t_1) \cdot f(t_1 - \tau) \, dt \qquad 2.5$$

is the value known as the auto-correlation function of the signal $f(t)$.

If we reckon that $P_1 = P_2 = P$, the equation 2.3 can be written as

$$P_\Sigma(t) = 2P + 2r(t) = 2P \left[1 + \frac{r(t)}{P} \right] = 2P[1 + \rho(t)] \qquad 2.6$$

where $\rho(t) = \dfrac{r(t)}{P}$ represents the value known as the coefficient of coherence.

If the delay of the reflection is zero, then, as follows from the equation 2.5

$$r(t) = P_1 \quad \text{and} \quad \rho(t) = 1$$

23

and the mean total power of the signals according to equation 2.6 will be $P_\Sigma = 4P$. This case suits the addition of two so-called coherent signals.

As the delay time increases, the functions $r(t)$ and $\rho(t)$ gradually decrease and when the coefficient of coherence approaches zero, the mutual connection between the signals is lost. In this case, as follows from the formula 2.6, the total power is found by adding arithmetically the mean powers of the signals. In these conditions the signals are known as non-coherent.

The experimentally determined dependence of the coefficient of coherence $\rho(t)$ on the delay time τ for a speech signal is shown in Fig. 2.1. It follows from the graph that for speech with a delay of about 10 msec the coefficient

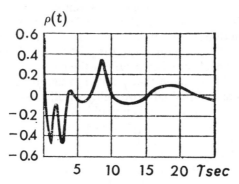

Fig. 2.1. The dependence of the coefficient of coherence on the delay time of a reflected speech signal.

of coherence is almost equal to zero, and in this case it is sufficiently justifiable to allow the addition of the mean values of the energies brought by the reflected waves to each point. This is also permissible for any other irregular signal, e.g. for music.

Finally, to use the statistical method of analysis we must make one more assumption which simplifies the physical picture of the sound field in an enclosure. It must be assumed that the enclosure is large in relation to the wavelength and that its surfaces have a low and on the whole similar sound absorbing capacity. Under these conditions the number of waves superimposed on one another and reaching the given point of the enclosure from different directions will be great enough to ensure an even distribution of sound energy throughout the enclosure.

Considering the relation of the total sound energy within the enclosure to its volume, which is known as the volume density of sound energy, it can be said that at any point in the interior space of the enclosure at any given moment the volume density of sound energy is equal.

Thus, in order to be able to undertake a statistical analysis of the sound field in an enclosure, we have to a certain extent to idealize the field, allowing the following assumptions:

1. Reflected waves reach all points within the enclosure from different directions, and all these directions are equally probable.

24

2. If the mean energy values of all reflections passing through a given point are arithmetically added, they will define the general sound energy at that point in space.

3. The density of sound energy at any given moment is the same throughout the enclosure.

A sound field with the properties outlined above is known as a diffuse sound field.

When we speak of a diffuse sound field we are referring to a field of such a nature that the reflected waves within it are mutually oriented in as disorderly and chaotic a manner as their amplitudes and phases are distributed in space.

A diffuse sound field can be achieved on the condition that really non-uniform surfaces of the enclosure are replaced by others, ideally uniform, the coefficient of absorption of which is equal to the mean coefficient of absorption of the non-uniform surfaces. Moreover, the form of the idealized enclosure should be such that any section of the path followed by a wave between successive reflections should be equal to the mean length of such sections of path in the enclosure which actually exists.

The mean length of path sections between two successive reflections is known as the mean free path of a sound wave.

2.2 The mean coefficient of sound absorption

The coefficient of sound absorption α as expressed in the equation 2.1 can be interpreted as a ratio expressing the proportion of the energy carried by the wave absorbed in a single reflection. This interpretation follows from this way of expressing formula 2.1.

$$\varepsilon_{absorbed} = \alpha \varepsilon_{incident}$$

However, if we take into account the fact that the reflection from each new surface may take place at an angle different from the preceding one, and that the sound absorbing properties of any material may depend on the angle of incidence and reflection of the wave, such an interpretation makes the coefficient of sound absorption a random and varying quantity, and consequently one which cannot define the absorption capacity of one material as a whole. The sound absorbing capacity of a material can be defined as the average of the values found for waves striking it from all angles.

Certain conditions have to be met to allow the establishment of a diffuse sound field. One such condition is that reflected waves should have an equal probability of distributing themselves in all directions. From this follows the condition that there should also be an equal probability that these waves should make impact with the surfaces of the enclosure at any angle. Bearing this in mind, we can speak of the mean coefficient of absorption as the ratio of energy absorbed by a certain uniform surface to the amount of energy falling diffusely upon that surface. In this case the mean coefficient of sound absorption is the most probable of the values which can be determined for each angle of incidence.

The mean coefficient of sound absorption, defined by taking into account the great variety of angles of incidence of waves distributed in the enclosure, can characterize the enclosure only if the surfaces which compose the enclosure are sufficiently uniform in their physical properties. If the interior surfaces of an enclosure consist of N physically dissimilar units, which are, however, equal in area one to another, the mean coefficient of absorption has to be calculated taking into account the variety of non-uniform units expressed as a simple mean value arrived at by the formula

$$\alpha_{mean} = \frac{\alpha_1 + \alpha_2 + \alpha_3 + \ldots + \alpha_N}{N} = \frac{1}{N} \sum_i \alpha_i, \qquad 2.7$$

where $\alpha_1, \alpha_2 \ldots \alpha_i$ are diffuse coefficients of absorption for each unit.

As under the conditions of real enclosures, absorbers of different physical characteristics occupy different areas, the coefficient of absorption in these cases should be found by calculating the relative value of the area of each unit as a mean value thus:

$$\alpha_{mean} = \alpha_1 \frac{S_1}{S} + \alpha_2 \frac{S_2}{S} + \ldots + \alpha_N \frac{S_N}{S} = \frac{1}{S} \sum_i \alpha_i S_i \qquad 2.8$$

where $S_1, S_2 \ldots S_N$ are areas of the various non uniform units, $\alpha_1 S_1, \alpha_2 S_2 \ldots \alpha_N S_N$ the general absorptions of each of them, S is the total area of all the interior surfaces of the enclosure, and $\sum \alpha_i S_i = \alpha_{mean} S = A_1$ is the total absorption of the whole enclosure.

If the enclosure contains various objects and people, then in calculating the mean coefficient of sound absorption we have to take into account the sound absorbing capacity of both the surfaces bounding the enclosure and also that of the objects within the enclosure.

To find the total absorption of a number of objects of the same type we multiply the equivalent absorption of one object by the total number of objects in the enclosure. By equivalent absorption in this case is understood the total absorption of one square metre of surface having a diffuse co-efficient of absorption numerically equal to the absorption produced by one object. It is clear that the total absorption of all groups of objects can be expressed by the equation

$$A_2 = \alpha_1 N_1 + \alpha_2 N_2 + \ldots + \alpha_k N_k = \sum_m \alpha_m N_m$$

where N_m is the number of objects of the same type and α_m is the equivalent absorption of each of them.

Finally, the total absorption of sound energy in an enclosure which contains people and non-uniform objects is expressed by the equation

$$A = A_1 + A_2 = \sum_i \alpha_i S_i + \sum_m \alpha_m N_m$$

and the mean coefficient of sound absorption

$$\alpha_{mean} = \frac{A_1 + A_2}{S} = \frac{1}{S}(\sum_i \alpha_i S_i + \sum_m \alpha_m N_m) \qquad 2.9$$

This approach to the calculation of the mean coefficient of sound absorption is equivalent to replacing the actual enclosure under consideration with an ideal room in which the total absorption is evenly spread throughout all its surfaces.

2.3 The mean free time and mean free path in an enclosure

Every sound wave which undergoes repeated reflections from the surfaces of an enclosure can be represented as a broken line whose sections have lengths l_1, l_2, l_3, etc. The length of each section is equivalent to the length of a free run of the wave, and as they are different from one another we have to follow a procedure similar to that used in determining the coefficient of absorption and to find a mean length of the free run of a wave for the given enclosure. This can be done by considering the combination of all the sections making up the reflected wave and defining the arithmetical mean:

$$l_{mean} = \frac{l_1 + l_2 + l_3 + \dots + l_k}{k} = \frac{1}{k}\sum_k l_k$$

The mean free path can be established in another way, by taking the combinations of path sections between two successive reflections for a large number of reflected waves. The results obtained by these two methods cannot differ significantly from one another, but both methods involve measuring the length of a great many sections representing the paths between successive reflections of a single wave, or between two reflections of a large number of waves. The determination of the lengths of these sections is a laborious process, and so, as the mean length of the free run of a wave must depend on the linear dimensions of the enclosure, let us try to find a shorter way of establishing it.

In a room of volume V, let us isolate on one of its surfaces a certain area dS (Fig. 2.2). Let a sound wave impinge on this area at an angle θ. Then the energy falling on this area during the time dt will be proportional to the volume of the cylinder on the base dS and with a height h set out in Fig. 2.2, as

$$h = c_0\, dt \cos \theta$$

where $c_0 . dt$ is the path travelled by the wave during the time dt.

The volume of the cylinder is

$$dV = dS\, c_0\, dt \cos \theta$$

Let us find out what is the probability that of all the waves distributed in the enclosure at an angle of θ whether any will fall on the area dS.

According to the theory of probability, the simple probability of an event taking place is a fraction, of which the numerator is the number of cases favourable to the occurrence of the event, and the denominator is the total number of cases.

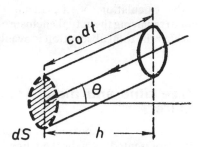

Fig. 2.2. Diagram for the determination of probability ω_1.

Thus the probability of a wave falling on the area dS is

$$\omega_1 = \frac{1}{2}\frac{dV}{V} = \frac{dS\, c_0\, dt \cos \theta}{2V} \qquad\qquad 2.10$$

(The coefficient $\frac{1}{2}$ is introduced because the distribution of waves towards the area and moving away from it are equally probable.)

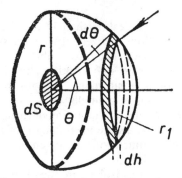

Fig. 2.3. Diagram for the determination of probability ω_2.

Now let us define the probability that, of all the waves having different directions, a wave will impinge on the given area at angle θ. For this purpose we enclose the area dS with a hemisphere of arbitrary radius r. (Fig. 2.3.)

As in the first case, the probability of the impact of a wave at angle θ or at an angle close to it $\theta + d\theta$ can be expressed by a fraction, the numerator of which is the area of a belt of radius r_1, and the denominator the area of the hemisphere.

It follows from Fig. 2.3 that the height of the belt

$$dh = r \,.\, \sin d\theta$$

or, as the angle $d\theta$ is small:

$$dh = r \cdot d\theta$$

The area of the belt, taking into account the relationships following from Fig. 2.3, is defined as

$$S_b = 2\pi r_1 \, dh = 2\pi r^2 \sin \theta \, d\theta$$

The probability we are seeking, then is:

$$\omega_2 = \frac{S_b}{S_{\text{hemisphere}}} = \frac{2\pi r^2 \sin \theta \, d\theta}{2\pi r^2} = \sin \theta \, d\theta \qquad 2.11$$

The probability of the simultaneous happening of a number of events is equal to the product of the simple probabilities of each of those events. On this basis the probability of a sound wave at angle θ impinging on the area dS during the time dt, according to equations 2.10 and 2.11, will be

$$\omega_3 = \omega_1 \cdot \omega_2 = \frac{c_0}{4V} \sin 2\theta \, d\theta \, dt \, dS \qquad 2.12$$

As the probability of a general event is equal to the sum of the probabilities of individual events, the probability of a sound wave impinging on any area at any angle between 0 and $\dfrac{\pi}{2}$ will be:

$$\omega_4 = \int\limits_0^S \int\limits_0^{\pi/2} \frac{c_0}{4V} \sin 2\theta \, d\theta \, dt \, dS$$

or, after integration and the insertion of these limits:

$$\omega_4 = \frac{c_0 S}{4V} \cdot dt \left(-\frac{\cos 2\theta}{2} \right) \Big|_0^{\pi/2} = \frac{c_0 S}{4V} \, dt \qquad 2.13$$

If the time interval dt is extended up to the point (up to τ) where the impact of a sound wave on any given section of the interior surfaces of the enclosure ceases to be a probability and becomes a certainty, then we can say that

$$1 = \frac{c_0 S}{4V} \tau$$

(The probability of a certainty is equal to 1.)
The indicated interval of time which is the mean time between two reflections of sound will be defined as:

$$\tau = \frac{4V}{c_0 S} \qquad 2.14$$

From this, the mean number of reflections for a unit of time:

$$n = \frac{1}{\tau} = \frac{c_0 S}{4V} \qquad 2.15$$

29

and the mean free path of a sound wave

$$l_{\mathrm{mean}} = c_0\tau = \frac{4V}{S} \qquad\qquad 2.16$$

The results of the calculations of τ, n, and l_{mean} according to the formulae 2.14, 2.15 and 2.16 are well enough in accord with the data obtained by using one of the methods which were considered at the beginning of this section. Some deviations observed in comparison of the results show themselves in variations in the numerical coefficient when enclosures of varying forms are being analysed. But these deviations connected with the dependence of the numerical coefficient on the shape of the enclosure have no practical significance, as they do not normally exceed 3–5 per cent.

2.4 The growth and decline of sound energy in an enclosure

The conditions under which the statistical method can be applied, and which we have considered above, allow the derivation of a mathematical expression for the relation between the magnitude of the sound energy in an enclosure and its parameters.

If we assume that the absorption of sound energy at the limits of an enclosure goes on just as continuously as the energy keeps arriving from the sound source, we can write, on the basis of the law of conservation of energy that the rate of growth of energy in an enclosure is the difference between the energy P_a radiated by the source in unit time and the energy P_p absorbed in the same time, i.e. that

$$V\frac{\mathrm{d}E}{\mathrm{d}t} = P_a - P_p \qquad\qquad 2.17$$

where V is the volume of the enclosure, and E is the density of sound energy.

It is clear that the amount of sound energy in an enclosure can be expressed as the product of its density multiplied by the volume of the enclosure, i.e. as EV. The energy absorbed after each reflection will in this case be $EV\alpha_{\mathrm{mean}}$, and the total loss of energy in a unit of time will be increased by the number of times that similar reflections take place in that unit of time, i.e. it will be $EV\alpha_{\mathrm{mean}}n$.

If we insert the average number of reflections taking place in one second, according to equation 2.15, the loss of energy in unit time can be written thus

$$P_p = EV\alpha_{\mathrm{mean}}\frac{c_0 S}{4V} = \frac{c_0\alpha_{\mathrm{mean}}S}{4}E$$

By replacing $\dfrac{c_0\alpha_{\mathrm{mean}}S}{4V}$ with δ, this last expression can be written in the form:

$$P_p = \delta VE$$

and the differential equation (2.17) as

$$V\frac{\mathrm{d}E}{\mathrm{d}t} + \delta VE = P_a \qquad\qquad 2.18$$

To solve the equation 2.18 both halves must be multiplied by δ and only one variable factor left in each half, after which the equation appears in the form

$$\delta V \, dE = \delta(P_a - \delta VE) \, dt$$

or

$$\delta \, dt = \frac{\delta V}{P_a - \delta VE} dE$$

After integration of the equation, we get

$$\delta t = - \ln(P_a - \delta VE) + C \qquad\qquad 2.19$$

At the moment of starting ($t = 0$), i.e. at the moment when the sound source is switched on, $E = 0$. If we insert these values in the preceding equation, the constant of integration is seen to be:

$$C = \ln P_a$$

Inserting this value into equation 2.19, we then have:

$$\ln(P_a - \delta VE) - \ln P_a = - \delta t$$

or

$$\ln\left(1 - \frac{\delta VE}{P_a}\right) = - \delta t$$

whence

$$1 - \frac{\delta VE}{P_a} = e^{-\delta t}, \qquad \frac{\delta V}{P_a}E = 1 - e^{-\delta t}$$

If the so-called index of decay of sound energy δ is replaced by its value, then we get

$$E = \frac{4P_a}{c_0 \alpha_{\text{mean}} S}[1 - e^{-(c_0 \alpha_{\text{mean}} S/4V)t}] \qquad\qquad 2.20$$

The expression 2.20 characterizes the process of the growth of sound energy in an enclosure and is graphically represented by section A of the curve in Fig. 2.4a.

The second term in the bracketed factor is called an exponential function, and thus the process of the growth of sound energy in an enclosure under the conditions we have noted above is usually described as exponential. As time t is increased this term rapidly approaches zero, the density of sound energy in the enclosure becoming constant and equal to:

$$E = E_0 = \frac{4P_a}{c_0 \alpha_{\text{mean}} S} \qquad\qquad 2.21$$

Thus the expression 2.21 characterizes the equilibrium which occurs when the energy radiated by the source and that absorbed by the enclosure balance out and the net growth of energy ceases. This situation corresponds to section B of the curve in Fig. 2.4a.

If we assume that the decay of sound energy in the enclosure is continuous after the sound source is switched off, then the decay process is represented by the equation 2.17, with the first term of the right hand side made zero. Consequently the equation describing the process of decline of sound energy in an enclosure, after rearrangement, can be written in the form:

$$\frac{\mathrm{d}E}{E} = -\,\delta\,\mathrm{d}t \qquad\qquad 2.22$$

The solution of this equation will be:

$$\ln E + C = -\,\delta t$$

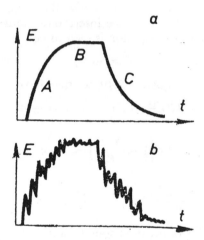

Fig. 2.4. Curves of growth A and decline C of sound energy: (a) where absorption is constant, (b) in actual conditions.

If we replace the constant of integration by its values reached from the initial conditions $t = 0$ and $E = E_0$, as $C = -\ln E_0$ the solution of equation 2.22 can be rewritten as

$$\ln E - \ln E_0 = -\,\delta t, \qquad \frac{E}{E_0} = \mathrm{e}^{-\delta t}$$

or, finally, as

$$E = E_0\,\mathrm{e}^{-\delta t} = E_0\,\mathrm{e}^{-(c_0\alpha_{mean}S/4V)t} \qquad\qquad 2.23$$

If the variation of energy density E with time t, as represented by the equation 2.23, is expressed in graphical form, it will be similar to section C of Fig. 2.4a. Consequently this graph taken as a whole expresses the process of change undergone by sound energy in an enclosure when the sound source in the enclosure is switched on and then switched off.

This graph gives an idealized representation of the process of growth and decay of energy in an enclosure. In actual enclosures, the graph has a rather irregular form as is shown in Fig. 2.4b, occasioned by the inevitable fluctuation of energy caused, for instance, by uneven absorption at the boundaries of the enclosure or by disturbances of the sequence of arrival of the sound waves which we have referred to earlier.

If we analyse the equations 2.21 and 2.23, we can establish that for the extreme case of total reflection of sound energy, i.e. where $\alpha_{mean} = 0$, the density of sound energy in the room, when the source is switched on, increases to an infinite extent, and when the source is switched off remains constant thereafter. This is fully consistent with our conception of the actual character of the sound field in an enclosure, if the situation could be achieved in practice.

For the other extreme case, the total absorption of sound energy, i.e. $\alpha_{mean} = 1$, equation 2.23 becomes:

$$E = \frac{4P_a}{c_0 S} e^{-c_0 St/4V}$$

indicating the existence of the process of sound energy decay, the rate of which is dependent on the relationship S/V. This is inconsistent with the real conditions in the sound field.

In reality total absorption is a property of a space which has boundaries whose density and elasticity are equal to that of the air which fills the space. Consequently in a case of total absorption of sound energy at the limits of the space involved we have a situation equivalent to the distribution of the wave in an open field, the decay in which is in no way connected with the dimensions of any enclosure. All this shows that the formula 2.23 is incompatible with the second limiting case, and that using this formula is not valid for values of α_{mean} close to 1.

These circumstances prompted Eyring to reconsider the method of solving the problem of the change in sound energy in an enclosure when the sound source is switched on and off. At the basis of his new method was the proposition that the absorption of energy, both during the period of growth and during that of decay, is not constant but proceeds by steps at intervals of time equal to the mean free time τ of a sound wave. All other assumptions remained the same.

Let a sound source of power P_a be switched on at the moment $t = 0$; in this case, in a period of time equivalent to the mean free time τ, the energy in the enclosure rises to

$$\varepsilon_0 = P_a \tau$$

If we assume that the coefficient of absorption of all surfaces in the enclosure is identical and equal to α, we can write that after the first reflection, and until time $t = 2\tau$, the energy remaining in the enclosure will be

$$\varepsilon'_{rem} = P_a \tau (1 - \alpha)$$

But during that same interval the source produces energy equal to $P_a \tau$ and consequently, when $t = 2\tau$, the total energy in the enclosure is:

$$\varepsilon_1 = \varepsilon_0 + \varepsilon'_{rem} = P_a \tau [1 + (1-\alpha)]$$

33

After the second reflection, (when $t = 3\tau$) there remains energy in the enclosure to the value:

$$\varepsilon''_{rem} = \varepsilon_1(1 - \alpha) = P_a\tau[(1 - \alpha) + (1 - \alpha)^2]$$

and further energy $\varepsilon_0 = Pv$ is added.

The total energy in the enclosure is then

$$\varepsilon_2 = \varepsilon_0 + \varepsilon''_{rem} = P_a\tau[1 + (1 - \alpha) + (1 - \alpha)^2]$$

Clearly, after $(k - 1)$ reflections, i.e. at the time

$$t = k\tau \qquad\qquad 2.24$$

the total energy in the enclosure is:

$$\varepsilon_{k-1} = \varepsilon_0 + \varepsilon_{rem}^{(k-1)} = P_a\tau[1 + (1 - \alpha) + (1 - \alpha)^2 + \ldots + (1 - \alpha)^{k-1}]$$

The factor in the square brackets is a geometrical progression, the sum of which is well known to be

$$\frac{(1 - \alpha)^k - 1}{(1 - \alpha) - 1}$$

Hence, the total energy in the enclosure is:

$$\varepsilon_{k-1} = P_a\tau\frac{1 - (1 - \alpha)^k}{\alpha}$$

and the volume density of the energy

$$E = \frac{\varepsilon_{k-1}}{V} = \frac{P_a\tau}{V\alpha}[1 - (1 - \alpha)^k]$$

If we insert the value τ from equation 2.14 into this formula, and the value of k from equation 2.24, the volume density of energy may be shown as:

$$E = \frac{4P_a}{c_0\alpha_{mean}S}\left[1 - (1 - \alpha)^{(c_0S/4V)t}\right] \qquad\qquad 2.25$$

Putting $(1 - \alpha) = e^x$, whence $x = \ln(1 - \alpha)$, we may write

$$(1 - \alpha)^{(c_0S/4V)t} = e^{[\ln(1-\alpha)c_0S/4V]t} \qquad\qquad 2.26$$

If we now substitute into equation 2.25 values of the factors outside and inside the brackets derived from equations 2.21 and 2.26, we get:

$$E = E_0\{1 - e^{[\ln(1-\alpha)c_0S/4V]t}\} \qquad\qquad 2.27$$

It is clear that as $\alpha < 1$ and the logarithm of a fraction has a negative value,

$$\lim_{t \to \infty} \left| e^{[\ln(1-\alpha)c_0S/4V]t} \right| = 0$$

34

and the density of energy in the enclosure will asymptote to a steady value. Then as $t = \propto 1$,

$$E = E_0 = \frac{4P_a}{C_0 \alpha S} \qquad 2.28$$

If the sound source is switched off, then after an interval of time equivalent to the mean free time, i.e. after the first reflection, the energy density in the enclosure falls to

$$E_\tau = E_0(1 - \alpha) \qquad 2.29$$

After two reflections, i.e. after an interval $t = 2_\tau$ this density is:

$$E_{2\tau} = E_0(1 - \alpha)^2 \qquad 2.30$$

After k reflections, (where $t = k\tau$)

$$E_{k\tau} = E_0(1 - \alpha)^k$$

If we replace k with its value derived from equation 2.14, the last expression may be written thus:

$$E_{k\tau} = E_0(1 - \alpha)^{\frac{c_0 S}{4V} t}$$

Finally, from equation 2.26 the density of energy in the enclosure at time t after the switching off of the sound source may be written in the form:

$$E = E_0 \, e^{[c_0 S \ln(1-\alpha)/4V]t} \qquad 2.31$$

We will consider that

$$\frac{c_0 S \ln (1 - \alpha)}{4V} = \delta_1 \qquad 2.32$$

and then the decay process is expressed in a more general formula similar to formula 2.23:

$$E = E_0 \, e^{\delta_1 t} \qquad 2.33$$

If we substitute into expression 2.31 the same limiting values of α which we substituted into expression 2.23, it is easily shown that where $\alpha = O$ the conditions in the sound field remain as before. Where $\alpha = 1$, because the exponent in the expression 2.31 becomes infinitely large and negative, the energy density falls instantaneously to zero. This is fully consistent with our conception of the physical reality of the decay process in conditions of total absorption and gives grounds to assume that Eyring's method provides a correct solution of the problem for all values of the coefficient of sound absorption.

The equation 2.31, which expresses the process of the decay of sound in an enclosure caused by energy losses at the boundaries of the enclosure, does not take into account energy losses in the air. These losses are proportionate

35

to the distances covered by the sound wave, and they can therefore be expressed by the equation

$$-\frac{dE_a}{E} = m_a \, dx$$

which after the distance x has been replaced by the product $c_0 t$, may be rewritten

$$\frac{dE_a}{E} = - m_a c_0 \, dt \qquad 2.34$$

where m_a is the coefficient of attenuation of sound energy in air.

The total losses of sound energy taking place both by losses at the boundaries of the enclosure and as a result of absorption in air are equal to the sum of the losses expressed by the equations 2.22 and 2.34, i.e.

$$\frac{dE}{E} = - (\delta + m_a c_0) \, dt \qquad 2.35$$

Because the equations 2.22 and 2.35 are of identical form, so, too, are their solutions, i.e. the solution of 2.35 is given by the equation

$$E = E_0 \, e^{-(\delta + m_a c_0)t} \qquad 2.36$$

As the expression 2.23, which is the solution of the equation 2.22, is similar to the expression 2.33, then we can write that

$$E = E_0 \, e^{(\delta_1 - m_a c_0)t} \qquad 2.37$$

or, following the equation 2.32

$$E = E_0 \, e^{\{[c_0 S \ln(1-\alpha)/4V] - m_a c_0\}t} = E_0 \, e^{[S \ln(1-\alpha) - 4 m_a V/4V]c_0 t} \qquad 2.38$$

The analysis of sound processes in enclosures by the statistical method, and a consideration of the resulting formula (2.20, 2.27, 2.23 and 2.31), and expressions (2.21 and 2.35), allow us to state the following conclusions.

1. The growth and decay of energy in a diffuse sound field follow exponential curve.

2. The steepness of this curve is determined by the exponent of this curve. The rates of growth and fall of energy increase in proportion to the increase of the coefficient of absorption α and of the ratio $\dfrac{S}{V}$.

3. As this ratio depends on the shape and dimensions of the enclosure, the rates of growth and decay of sound energy in the enclosure are also dependent on these factors.

4. In establishing the rates at which the energy in the enclosure grows and decays, we must take into account not only absorption at the boundaries of the enclosure but also absorption of energy in the air. Losses from absorption in the air increase with increases in the size of the enclosure.

5. During the time in which a source is radiating, when the energy absorbed in the enclosure becomes equal to the energy emitted by the

36

source, a condition of equilibrium is reached. At this time the density of sound energy in the enclosure becomes constant, and its magnitude is determined by the power of the sound source P_a and the total absorption αS of the surfaces bounding the enclosure.

6. Eyring's analysis of the build-up and decay of source in an enclosure reflects the physical reality of these processes more accurately than does the method based on the concept of the continuous growth or decay of energy.

2.5 Reverberation time

If the graph in Fig. 2.4a was drawn so that the density of sound energy was shown on a logarithmic scale along the vertical axis, the new graph would have the form shown in Fig. 2.5. This choice of scale shows the growth of sound in an enclosure as a comparatively rapid process, while the decay time is prolonged for a considerable period.

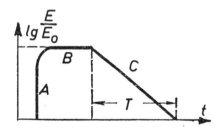

Fig. 2.5 Graph of aural perception of the process of the growth and decay of sound in an enclosure.

The graph of the relationship between the logarithm of energy density and time has a practical interest in that it expresses closely the actual way in which sound is heard. This results from the fact that the ear does not react to the intensity of a disturbance, but to a value close to the logarithm of this intensity. Thus what the ear hears can be better characterized by reference to the logarithm of sound energy density (usually termed the level) than to the value of the energy itself.

As speech, music, and other natural sounds, being irregular, are series of signals succeeding one another, the prolongation occasioned by the decay may cause each signal to overlap in time the succeeding signals, masking them to some degree. The masking effect, shown in Fig. 2.6, may produce interference, and, depending on the duration of the echo, impair the sound quality.

The process of the decay of the energy in the enclosure is known as reverberation, and the duration of the echo, i.e. the time the signal takes to be reduced to the threshold of hearing, is known as the reverberation time.

Having regard to the interference caused by the decaying sound, reverberation time has been used in the statistical theory as a factor enabling an evaluation of the acoustic properties of an enclosure to be made.

Reverberation time, as may be seen from the equation 2.31 is connected not only with the factors which characterize the enclosure (S, V and α) but also with the initial energy density E_0. The greater E_0 becomes, the longer will be the time needed for the sound to die away to the threshold of hearing.

In order that the reverberation process should be definable in terms of the acoustic properties of the enclosure only, and that it should not depend on the intensity of sound energy radiated by the sound source, a standard definition of reverberation time has been adopted.

By standard reverberation time T, as proposed by W. C. Sabine, who laid the basis of the statistical theory of room acoustics, is understood the time taken for sound energy in an enclosure to fall to 10^{-6} times its original

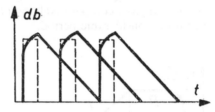

Fig. 2.6. Graphs of growth and decay of sound impulses in an enclosure

level (a fall in the level of energy of 60 dB). From the definition of standard reverberation time it follows that $t = T$ when

$$\frac{E}{E_0} = 10^{-6}$$

If the equation 2.37 is rewritten as:

$$\frac{E}{E_0} = e^{(\delta_1 - m_a c_0)t}$$

and the right hand sides of these two equations are equated, then we may write that

$$10^{-6} = e^{(\delta_1 - m_a c_0)T}$$

After converting this expression into logarithms to the base 10 we have:

$$-6 = (\delta_1 - m_a c_0)\, T \log_{10} e$$

and from this we derive standard reverberation time

$$T = \frac{6}{-\, (\delta_1 - m_a c_0) \log_{10} e} \qquad\qquad 2.39$$

or, by inserting the value of δ_1 into this formula in accordance with equation 2.32:

$$T = -\frac{24}{c_0 \log_{10} e} \cdot \frac{V}{S \ln (1 - \alpha) - 4m_a V}$$

If the numerical coefficient is calculated, and all factors in this last equation are stated in the metric system, this equation can be expressed in the form:

$$T = \frac{0{\cdot}164V*}{-S\ln(1-\alpha) + 4m_aV} \qquad\qquad 2.40$$

The resultant formula was first produced by Eyring, and bears his name.

If in the equation 2.39, m_a is given the value of $8{\cdot}98 \cdot 10^{-3}\,\dfrac{f^2}{\rho_0 c_0^2}$ and if it is reckoned that $\delta_1 = 0$, then one can determine how the time T_0, during which sound being distributed in unlimited space decays by 60 dB, is

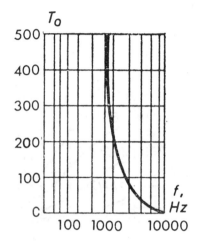

Fig. 2.7. The dependence of decay time in air, T_0, on frequency.

dependent on the frequency f. This dependence is shown as a graph in Fig. 2.7.

As can be seen from Fig. 2.7 losses of sound energy in the air have an influence on the decay process of sound and on the reverberation time only at high frequencies, starting from 2000–4000 Hz, where the influence becomes suddenly noticeable.

This conclusion is confirmed by the experimental graphs of the dependence of m_a on the relative humidity of the air. The graphs have been drawn for various frequencies and are shown in Fig. 2.8. In fact, as will be seen from the curves in this figure, the coefficient of attenuation of sound energy in air at a frequency of 1000 Hz or even 2000 Hz in a humidity of from 50–80 per cent is so small that the second term in the denominator of the fraction in expression 2.40 may become insignificant in relation to the first term of the denominator, particularly when we are dealing with small enclosures. Therefore, for such enclosures, and frequencies of less

* Editor's Note: This constant is usually quoted as $0{\cdot}161$, the figure for a temperature of 20°C.

39

than 4000Hz, we may ignore the term $4m_aV$ so that the formula 2.40 can be written:

$$T = \frac{0 \cdot 164V}{-S \ln (1 - \alpha)}$$ 2.41

or, converting to logarithms to base 10:

$$T = \frac{0 \cdot 071V}{-S \log_{10} (1 - \alpha)}$$ 2.42

The omission of the factor $4m_aV$ is not permissible for enclosures having a comparatively large volume at frequencies of 2000 to 4000 Hz, as, for example, where $V = 5000$ m³, the total absorption by the contained air reaches hundreds of units.

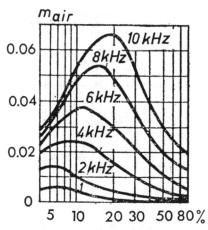

Fig. 2.8. Graphs of the dependence of the coefficient of attenuation of sound energy in air (m_a) on relative humidity at frequencies of 1, 2, 4, 6, 8, and 10 kHz.

For higher signal frequencies, as can be seen from Figs. 2.7 and 2.8, absorption by the air becomes very large, both in comparison with the first term of the fraction in the expression 2.40, and absolutely. Because of this, there is no possibility at these frequencies of controlling reverberation time by altering the overall absorption of the bounding surfaces of the enclosure; moreover, this time becomes so small that it no longer has any effect on sound quality, and so the need to control it diminishes. In practical conditions, therefore, the calculation of reverberation time is restricted to frequencies below 4000 Hz.

The formula 2.41 can be further transformed if it is borne in mind that the expression $- \ln (1 - \alpha)$ can be written as:

$$- \ln (1 - \alpha) = \alpha + \frac{\alpha^2}{2} + \frac{\alpha^3}{3} + \cdots$$

It can be seen from the calculations and graphs in Fig. 2.9 that if $\alpha \leqslant 0 \cdot 3$ the denominator of the fraction in expression 2.41 can be replaced, with a sufficiently close degree of approximation, by the first two terms of the series, and if $\alpha \leqslant 0.2$, one need use only the first. In this case the formula 2.41 takes on this form:

$$T = \frac{0.164V}{\alpha S} \qquad 2.43$$

It was in this form that W. C. Sabine first presented reverberation time in 1896, and for this reason formula 2.43 is often called Sabine's formula.

The formula 2.43 can be derived directly from the expression 2.23 if the ration $\dfrac{E}{E_0}$ is taken as equal to 10^{-6}, as in the determination of standard reverberation time from expression 2.37. Thus, it becomes clear than an analysis of sound processes in an enclosure which relies on the concept of

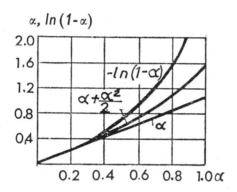

Fig. 2.9. The dependence of $-\ln(1-\alpha)$ on the coefficient of absorption α.

continuous absorption of energy at the boundaries of the enclosure (as examined at the beginning of paragraph 2.4), is valid only for enclosures with a mean coefficient of absorption not exceeding 0.2.

All the foregoing formulae are based on a series of assumptions which may not always be acceptable in practice. In fact, where an enclosure has a poor shape, or where it contains focusing surfaces, the directional distribution of waves ceases to be random. In large enclosures where the mean free path of a wave is large, the number of reflections before the decay process is complete, may be insufficiently large for the statistical approach to be fully justified, particularly if the enclosure is well deadened. Nor can one reckon that the probability of sound waves falling on particular parts of the absorbent surfaces remains the same wherever the surfaces are placed.

It is also clear that if the position of the sound source within the enclosure is changed, the order of reflections will change with it, and therefore the method of averaging ignores the actual interrelation of acoustic conditions in the enclosure and the position in the enclosure of the source of sound.

Finally, the statistical method of evaluating the acoustic qualities of an enclosure does not take into account the advantages and disadvantages which may result from the fact that the reverberation process may consist of several stages with different decay rates.

All this indicates a need for a critical approach to calculations made according to the formulae of the statistical theory, and also that reverberation time is not a sufficient criterion of the acoustic properties of an enclosure. In an evaluation of these properties a great part is played by the diffuseness of the field and by the time structure of the reflections, which is determined by their delay time and their relative levels.

The results of this chapter may be briefly summaried as follows:

1.   Reverberation time depends on the volume of the enclosure, the areas of its interior surfaces, and their mean coefficient of absorption; it also depends on the attenuation by the air contained in the enclosure, the shape of the enclosure, and the positions of the sound source and of sound absorbing materials.

2.   Reverberation time in one and the same enclosure is not constant at all frequencies, because sound absorption, both in the air and at the boundaries, is dependent on frequency.

3.   For practical purposes it is wise to carry out calculations of reverberation time at frequencies below 4000 Hz.

4.   In the calculation of reverberation time for small enclosures, losses of sound energy in the air may be ignored, and the simple formulae 2.41 and 2.42 may be used.

5.   The absorption of sound energy in the air must be taken into account in the calculation of the reverberation time of a large enclosure, but only at frequencies of from 2000 to 4000 Hz. Formula 2.40 should be used in this case.

6.   The calculation of reverberation time may be simplified if the mean coefficient of absorption of the given enclosure does not exceed 0·2. In this case, formula 2.43 may be used.

7.   As reverberation time is not an absolute or unique criterion for the evaluation of the acoustic properties of an enclosure, to exclude possible error, the correct formula must be applied for any particular case, with an understanding of the physical principles.

2.6   The decay of sound energy in acoustically coupled enclosures

In practice, there are cases where the quality of sound depends on the acoustic properties of not one, but several, enclosures. Such enclosures are described as 'acoustically coupled'. There are two kinds of coupling-acoustic, or two-way, and electro-acoustic, or one-way, enclosures.

The first kind of coupling envisages two or more enclosures connected to each other through openings of arbitrary size (Fig. 2.10a). When a sound source is switched on in one enclosure, the sound energy passes from the first enclosure to the second and back again, thus altering the character of the sound field in both rooms. Here, the process of decay of sound energy in each of the connected rooms is governed by the acoustic and geometric properties of both enclosures, and by the areas of the coupling aperture. An example of this kind of coupling is that between a sound shooting stage and a set erected on it, or the link between the main part of a cinema auditorium and that part of it which is situated below the balcony.

If one enclosure is acoustically coupled to another, it is possible to construct, on the basis of the law of conservation of energy, a differential equation similar to equation 2.17 for an isolated enclosure.

To determine the growth of energy in the primary enclosure which houses the source of sound, it is necessary to subtract from the energy radiated by

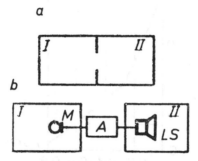

Fig. 2.10. Coupled enclosures, acoustic (a) and electroacoustic coupling (b).

the source in unit time both the energy absorbed in the primary enclosure P_{1_n}, and also the energy which leaves the enclosure and passes into the secondary enclosure P_{1_s}. For greater accuracy we should also take into account the energy which, after having been reflected from the surfaces of the secondary enclosure, comes back through the coupling aperture and returns to the primary enclosure, P_{2_s}.

So, for the primary enclosure, we can write:

$$V_1 \frac{dE_1}{dt} = P_a - P_{1_n} - P_{1_s} + P_{2_s} \qquad 2.44$$

When a situation is reached where the growth of energy in the enclosure ceases, i.e. when $\frac{dE_1}{dt} = 0$, the equation 2.44 takes the form

$$P_{1_n} + P_{1_s} = P_a + P_{2_s} \qquad 2.45$$

By analogy with equation 2.18, we can show that

$$P_{1_n} = \frac{c_0 \alpha_1 S_1}{4} E_{0_1}; \qquad P_{1_s} = \frac{c_0 S}{4} E_{0_1}, \qquad \text{and } P_{2_s} = \frac{c_0 S}{4} E_{0_2} \qquad 2.46$$

where α_1 and S_1 are respectively the mean coefficient of absorption and the area of the absorbing surfaces of the primary enclosure. By making the

43

appropriate substitution, the equation 2.45 can be written in the form:

$$\frac{c_0(\alpha_1 S_1 + s)}{4} E_{0_1} = P_a + \frac{c_0 s}{4} E_{0_2} \qquad 2.47$$

It is clear that the equation for the energy density in the secondary enclosure will be similar to equation 2.47, with the one exception that the term P_a will be missing, as the sound source is located only in the primary enclosure.

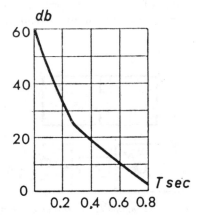

Fig. 2.11. The process of decay of sound energy in a coupled enclosure which contains the source and receiver of sound.

Thus, for the secondary enclosure

$$\frac{c_0(\alpha_2 S_2 + s)}{4} E_{0_2} = \frac{c_0 s}{4} E_{0_1} \qquad 2.48$$

From this equation it follows that

$$E_{0_2} = \frac{s}{\alpha_2 S_2 + s} \cdot E_{0_1} \qquad 2.49$$

After substituting the value of E_{02} in equation 2.47, we find that

$$\frac{c_0(\alpha_1 S_1 + s)}{4} E_{0_1} = P_a + \frac{c_0 s}{4} \cdot \frac{s}{\alpha_2 S_2 + s} \cdot E_{0_1}$$

whence, after some rearrangement,

$$E_{0_1} = \frac{4P_a}{c_0(\alpha_1 S_1 + s)\left[1 - \dfrac{s^2}{(\alpha_1 S_1 + s)(\alpha_2 S_2 + s)}\right]} \qquad 2.50$$

We can introduce the following symbols

$$Q_1 = \frac{s}{\alpha_1 S_1 + s} \quad \text{and} \quad Q_2 = \frac{s}{\alpha_2 S_2 + s} \qquad 2.51$$

44

where Q_1 and Q_2 are values which are called the coupling coefficients between the primary and secondary enclosures.

By substituting these coefficients into expressions 2.50 and 2.49 we can write equations for the energy density in each of the two coupled enclosures in a stable condition:

$$E_{0_1} = \frac{4P_a}{c_0(\alpha_1 S_1 + s)(1 - Q_1 Q_2)} \qquad 2.52$$

$$E_{0_2} = E_{0_1} \cdot Q_2 \qquad 2.53$$

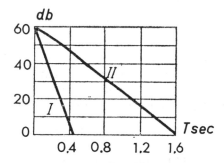

Fig. 2.12. Graphs of the decay of sound energy in an enclosure without (I) and with (II) acoustic coupling.

In order to determine the density of sound energy in coupled enclosures during the decay period, we have to solve equation 2.44 for the primary enclosure and a similar equation for the secondary. The energy densities in the two coupled enclosures is thus shown to be:

$$E_1 = \frac{E_{0_1}}{\frac{m_2}{m_1} - 1} \left[\left(\frac{m_2}{m_1} e^{-m_1 t} - e^{-m_2 t} \right) - \frac{m_2}{m_1} (1 - Q_1 Q_2)(e^{-m_1 t} - e^{-m_2 t}) \right]$$

$$2.54$$

$$E_2 = \frac{E_{0_2}}{\frac{m_2}{m_1} - 1} \left(\frac{m_2}{m_1} e^{-m_1 t} - e^{-m_2 t} \right) \qquad 2.55$$

where m_1 and m_2 are coefficients depending on the parameters of the enclosures (α_1, S_1, V_1, α_2, S_2, and V_2) and of the area of the connecting opening.

Now let us examine special cases of the expressions 2.52, 2.53, 2.54 and 2.55.

1. Let $s = 0$, i.e. there is no connection between the enclosures. In this case it is clear from equations 2.51 that $Q_1 = Q_2 = 0$ and that the energy

45

density in the primary and secondary enclosures in a stable condition will be, according to equations 2.52 and 2.53,

$$E_{0_1} = \frac{4P_a}{c_0 \alpha_1 S_1} \quad \text{and} \quad E_{0_2} = O$$

As was to be expected, the first expression is identical to the expression 2.21, for an isolated enclosure. The second expression reduces to zero, which is correct as where $s = 0$ no energy penetrates to the secondary enclosure.

Where $s = 0$, m_1 tends towards infinity, and the expressions 2.54 and 2.55 take on the simple form:

$$E_1 = E_{0_1} e^{-m_2 t} \quad \text{and} \quad E_2 = O$$

which corresponds to the case of the isolated enclosure (equation 2.23) and indicates the absence of acoustic link.

2. Let $Q_1 = Q_2 = 1$. The coupling coefficients can only take this value, as can be seen from equations 2.51, if $\alpha_1 S_1 = \alpha_2 S_2 = 0$. This is impossible in practice. Consequently neither of the coupling coefficients can be greater than or equal to unity.

3. Let $\alpha_1 S_1 \lll s \ggg \alpha_2 S_2$. This means that Q_1 and Q_2 are almost unity, and the second item of the equation 2.54 approaches zero. In this case the expressions 2.54 and 2.55 become identical. Consequently, where the connecting aperture is large in area the energy densities in the rooms become equal.

On the basis of this analysis we can form a number of conclusions.

1. If the connecting aperture is very small, each of the connected enclosures can be regarded as isolated.

2. The acoustic coupling coefficients are always smaller than unity.

3. Where the opening is large in area, the two coupled enclosures may be regarded as one large independent enclosure.

Formulae 2.54 and 2.55 show that the process of the decay of sound energy in linked enclosures is expressed not as a simple exponential function, but as the sum or the difference of such functions with different indices of decay.

This leads to the fact that the energy level decay curve in acoustically coupled enclosures is no longer a straight line, as is the case for independent enclosures. If the source and receiver of sound are in one of the linked enclosures, it can be shown experimentally that the process is represented by a bent line (Fig. 2.11). The section of the line up to the bend is determined by the decay of energy in the primary enclosure, while after the bend it is determined mainly by the decay of energy returning from the secondary enclosure.

Moreover, calculations made by means of the formulae quoted show that the duration of the decay process depends less on the volume of the enclosures than on the coefficients of absorption. This can be seen from Fig. 2.12, which shows decay curves for a large deadened room (curve I)

46

and for the same enclosure acoustically linked with a small, undeadened enclosure (curve II).

From all this we have three further conclusions:

1. The process of the decay of sound in linked enclosures is not subject to simple exponential law.

2. The character of the overall curve of the decay process depends both on the parameters of the linked enclosures and on whether the sound source and receiver are in one room or in different rooms.

3. The process of decay in linked enclosures is basically determined by the more reverberant enclosure, even in cases where the live enclosure is small by comparison with the second, more damped enclosure.

2.7 The decay of sound energy in enclosures coupled electro-acoustically

Electro-acoustic coupling supposes the presence of two enclosures coupled to each other by an electro-acoustic system. As follows from Fig. 2.10b, in this case the quality of the sound received in the secondary enclosure depends initially on the acoustic properties of the primary enclosure, which affects the signal reaching the microphone in a specific way. But in addition to this, the transmitted signal is subjected to further changes as a result of the acoustic properties of the secondary enclosure when the signal is reproduced by the loudspeaker.

Thus the process of the decay of sound signals in the secondary enclosure depends on the acoustic properties of both the primary and the secondary enclosures. It is clear that there is no reverse action, i.e. there is no effect of the secondary enclosure on the sound heard in the primary enclosure.

An example of an electro-acoustic coupling system is that which exists between a sound recording studio and a cinema auditorium.

Let there be two enclosures, connected by an electro-acoustic system. Then for the secondary enclosure we can construct a differential equation similar to the equation 2.18:

$$V_2 \frac{dE_2}{dt} + V_2 \delta_2 E_2 = P_{a_2} \qquad 2.56$$

If the sound source in the primary enclosure is switched off, the process of decay in that enclosure follows the law

$$E_1 = E_{0_1} e^{-\delta_1 t}$$

And the power radiated by the loudspeaker in the secondary enclosure diminishes according to the same law, i.e.

$$P_{a_2} = P_{0_2} e^{-\delta_1 t}$$

where P_{0_2} is the power of the loudspeaker in the steady state. This power, substituting the value of the index of decay δ as in equation 2.21, is given by:

$$P_{0_2} = E_{0_2} \cdot \frac{c_0 \alpha_2 S_2}{4} = E_{0_2} V_2 \delta_2$$

and consequently

$$P_{a_2} = P_{0_2} e^{-\delta_1 t} = E_{0_2} V_2 \delta_2 e^{-\delta_1 t} \qquad 2.57$$

By substituting the values of P_{a_2} taken from equation 2.57 into equation 2.56 we get the following differential equation:

$$\frac{dE_2}{dt} + \delta_2 E_2 - \delta_2 E_{0_2} e^{-\delta_1 t} = 0 \qquad 2.58$$

To solve this equation, we must first solve the complementary equation:

$$\frac{dE_2'}{dt} + \delta_2 E_2' = 0 \qquad 2.59$$

The solution of this equation is the same as that of equation 2.22, i.e.

$$E_2' = C e^{-\delta_2 t} \qquad 2.60$$

In the same form we will try to find the solution of the particular equation 2.58, i.e. we will consider that

$$E_2 = z e^{-\delta_2 t} \qquad 2.61$$

where z is an unknown function of t.

To find this function, which determines the solution of equation 2.58, we differentiate equation 2.61 and get

$$\frac{dE_2}{dt} = \frac{dz}{dt} e^{-\delta_2 t} - z \delta_2 e^{-\delta_2 t} \qquad 2.62$$

Now we substitute the values of E_2 and $\dfrac{dE_2}{dt}$ from equations 2.61 and 2.62 into equation 2.58, and then, after some simplifications, we get

$$\frac{dz}{dt} e^{-\delta_2 t} = \delta_2 E_{0_2} e^{-\delta_1 t}$$

Hence, after rearranging the variables

$$dz = \delta_2 E_{0_2} e^{(\delta_2 - \delta_1) t} dt$$

After both parts of the equation have been integrated, the required function z is found to be;

$$z = \delta_2 E_{0_2} \frac{e^{(\delta_2 - \delta_1) t}}{\delta_2 - \delta_1} + C_1 \qquad 2.63$$

We substitute this value of z into equation 2.61, and the solution of equation 2.58 becomes

$$E_2 = \frac{\delta_2}{\delta_2 - \delta_1} E_{0_2} e^{-\delta_1 t} + C_1 e^{-\delta_2 t} \qquad 2.64$$

To find the value of the arbitrary constant C_1 we consider the initial

48

conditions, i.e. where $t = 0$. In this case $E_2 = E_{0_2}$, and the expression 2.64 takes the form

$$E_{0_2} = E_{0_2} \frac{\delta_2}{\delta_2 - \delta_1} + C_1$$

whence

$$C_1 = E_{0_2} - \frac{\delta_2}{\delta_2 - \delta_1} E_{0_2} = E_{0_2} \left(1 - \frac{\delta_2}{\delta_2 - \delta_1} \right) = -E_{0_2} \frac{\delta_1}{\delta_2 - \delta_1} \qquad 2.65$$

After this value of the constant C_1 has been substituted into equation 2.64, the solution equation 2.58 is

$$E_2 = \frac{\delta_2}{\delta_2 - \delta_1} E_{0_2} e^{-\delta_1 t} - \frac{\delta_1}{\delta_2 - \delta_1} E_{0_2} e^{-\delta_2 t} \qquad 2.66$$

and finally

$$E_2 = \frac{E_{0_2}}{\delta_2 - \delta_1} (\delta_2 e^{-\delta_1 t} - \delta_1 e^{-\delta_2 t}) \qquad 2.67$$

If we consider the expression 2.67, the following conclusions emerge:

1. The decay of sound energy in the secondary enclosure is governed by the data of both coupled enclosures.

2. The decay of sound in the secondary enclosure is different from the decay in the primary enclosure. And as can be seen from a comparison of equations 2.67 and 2.60, the first of these does not follow the exponential law.

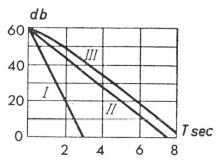

Fig. 2.13. Decay curves for a primary and secondary enclosure without any connection (I, II), and where there is electro-acoustic coupling between them (III).

This conclusion is confirmed by the graphs in Fig. 2.13, calculated by formulae 2.60 and 2.67 for two separate enclosures (graphs I and II) and for the same enclosures linked by means of an electro-acoustic system, (graph III).

In this latter case the process of the decay of energy level is not expressed by a straight line, but instead indicates a departure from exponential law.

49

3. The resultant decay curve (see Fig. 2.13) is always higher than the decay curve of either of the two interconnected enclosures, and nearer to that of the enclosure whose index of decay δ is the lower.

4. If the primary enclosure is heavily damped (δ_1 large) then the decay process, where there is electro-acoustic coupling, is governed mainly by the acoustic properties of the secondary enclosure, and vice versa. This follows from the fact that where $\delta_1 \to \infty$ the first term in the equation 2.67 tends towards zero and the equation takes the form

$$E_2 = E_0 \, e^{-\delta_2 t}$$

2.8 Resultant reverberation

The dependence of the overall reverberation time on the reverberation time of each of the enclosures coupled electro-acoustically was established by M. A. Sapozhkov[75] and L. D. Rozenberg[70].

It follows from the conclusion[4] of the previous section that if the decay index of the primary enclosure is significantly larger than the decay index of the secondary, or vice versa, the law of decay of sound energy is almost exponential and almost unaffected by the enclosure for which the value of δ is large. From this it becomes clear that an electro-acoustic coupling between two enclosures has the greatest effect on the overall process of decay when the decay indices of the primary and secondary enclosures are equal, i.e. where $\delta_1 = \delta_2 = \delta$ or where $T_1 = T_2 = T$.

If these values are substituted into equation 2.67, the result is indeterminate, being zero divided by zero. To find this indeterminate quantity, we can use Lopital's rule, i.e. we can establish the limit towards which the expression 2.67 is tending when $\delta_1 \to \delta_2$. Then,

$$E_2 = \lim_{\delta_1 \to \delta_2} f(\delta_1; \delta_2) = \lim_{\delta_1 \to \delta_2} \frac{[E_{0_2}(\delta_2 \, e^{-\delta_1 t} - \delta_1 \, e^{-\delta_2 t})]}{(\delta_2 - \delta_1)} =$$

$$= \underset{\delta_1 \to \delta_2}{-E_{0_2}(-\delta_2 t \, e^{-\delta_1 t} - e^{-\delta_2 t})}$$

or where $\delta_1 = \delta_2 = \delta$,

$$E_2 = E_{0_2} \, e^{-\delta t}(\delta t + 1) \qquad\qquad 2.68$$

whence we get the ratio

$$\frac{E_2}{E_{0_2}} = (\delta t + 1) \, e^{-\delta t} = k \, e^{-\delta t}$$

where

$$k = \delta t + 1$$

To obtain effective reverberation time, we proceed as for formula 2.43 and make the ratio equal to 10^{-6}, then,

$$k \, e^{-\delta T_p} = 10^{-6}$$

If we take the logarithm of this expression, we get:

$$\log k - \delta T_p \log e = -6$$

from whence

$$T_p = \frac{6 + \log k}{\delta \log e} \qquad\qquad 2.69$$

In order to express the effective reverberation in terms of the reverberation times of the coupled enclosures, we must find the connection between this time and the index of decay δ which was used in the last equation.

In connection with the formula for reverberation time of a single enclosure it was shown that

$$\frac{E}{E_0} = e^{-\delta T} = 10^{-6}, \text{ or } \delta T \log e = 6$$

whence

$$\delta = \frac{6}{\log eT} = \frac{6}{0.434T} = \frac{13.8}{T} \qquad\qquad 2.70$$

Let us substitute the value of δ into the expression for T_p, then,

$$T_p = \frac{6 + \log k}{13.8 \log e} T = \left(1 - \frac{1}{6} \log k\right) T \qquad\qquad 2.71$$

The value denoted in the last equation by the letter k can also be expressed in terms of the reverberation time of the coupled enclosures. In fact, if we substitute into the expression for k the value of δ found earlier, we can write that

$$k = 1 + \delta T_p = 1 + \frac{13.8}{T} T_p$$

Substitution of the value of k into equation 2.71 changes its form as follows:

$$T_p = \left[1 + \frac{1}{6} \log \left(1 + \frac{13.8}{T} T_p\right)\right] T \qquad\qquad 2.72$$

By solving this equation, we have the following simple relationship between overall reverberation time and the reverberation time of linked enclosures when both enclosures are equally damped:

$$T_p = 1.208T \qquad\qquad 2.73$$

This expression, 2.73, allows us to extend the last of the conclusions drawn from equation 2.67. It can now be formulated as follows.

If one of the enclosures linked by an electro-acoustic system, is heavily damped (T_1 or $T_2 \to 0$), then the resultant reverberation time is close to the reverberation time of the other, less damped enclosure. If both enclosures are equally damped ($T_1 = T_2$), the resultant reverberation differs by 20.8 per cent from that of either of the linked enclosures alone.

51

The theoretical conclusions reached above coincide closely with data obtained by L. D. Rozenberg.

From the experimental graphs in Fig. 2.14, we can easily establish that if the reverberation times in the linked enclosures are the same, with values of, for example, 0·5, 1 or 2 seconds, the values of the resultant reverberation closely correspond with those obtained by calculations based on formula 2.73.

When the reverberation time in the two linked enclosures is not the same, the resultant reverberation can be found from the graphs in Fig. 2.14 or from the empirical formula of M. A. Sapozhkov, according to which

$$T_p{}^3 = T_1{}^3 + T_2{}^3 \qquad\qquad 2.74$$

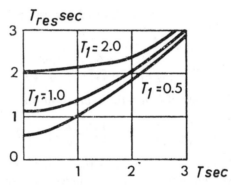

Fig. 2.14. The dependence of resultant reverberation time on the reverberation time of the secondary enclosure. T_1 is the reverberation time of the primary enclosure.

Formulae obtained from considerations of enclosures with unidirectional coupling must be approached with some care. This applies particularly to formulae 2.67, 2.72, 2.73 and 2.74. These formulae, however, are in ideal circumstances, and do not take into account all aspects of the physical conditions of the sound field and of the perception of the process of the decline of sound energy in linked enclosures. During the derivation of equation 2.57, it was stated that the sound energy radiated by the loud-speaker in the secondary field changes in exactly the same way as in the primary field where the microphone is placed. This statement, although correct from the quantitative point of view, is not always so from a qualitative point of view.

The reverberation process in the first enclosure is caused by the gradual diminution of energy in the progressively delayed reflections which reach the receiving point from all directions. The reverberation thus created is transmitted by way of a single channel and a loudspeaker to the secondary enclosure. The loudspeaker emits from one point only. Thus the field transmitted by the electro-acoustic system from one enclosure to another is not strictly identical with the actual field of the primary enclosure. As a result, the reverberation processes of the primary and secondary enclosures are not identical, and the listener observes a difference between the acoustics of the primary enclosure coming from the loudspeaker, and the reverberation process in the secondary enclosure. The acoustics of the primary enclosure are heard by the listener as a certain prolongation of sounds produced by natural sources, and this is heard against the background of the reverberation in the secondary enclosure.

To achieve a more organic connection between the primary and secondary fields, it is necessary that the field created by the electro-acoustic system should be identical with the field in the primary enclosure as regards the directions from which the sound waves arrive at the listener's ear. If this is done, the character and quality of the resultant field more closely correspond to their mathematical representation in the formulae quoted above. These demands are met to a certain extent by the field created by a multichannel electro-acoustic system of the kind used in stereophony.

The theory of linked enclosures can be used to analyse the mutual action of reverberant signals produced by a studio and an echo chamber when these signals are combined, as in film dubbing by an electro-acoustic system.

2.9 Optimum reverberation time

The duration of reverberation has a considerable effect on the quality of sounds received in an enclosure. The question arises as to what reverberation time gives the best possible quality of sound in an enclosure of any volume.

A consideration of the following example will help in replying to this question.

In an enclosure where $T = 2$ sec, the decay of the syllables of speech with an average duration of syllable of 0.2 sec is represented by the solid line curves in Fig. 2.15. If we reckon that the intensity of the first

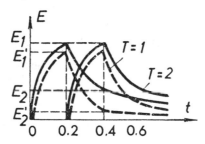

Fig. 2.15. Masking effect of attenuated signals, for reverberation times of 1 and 2 seconds.

syllable has declined to a quarter (from E_1 to E_2) by the moment of enunciation of the second syllable ($t = 0.4$ secs), the difference in levels between the first and second syllables at this instant is $10 \log \dfrac{E_1}{E_2} = 6\text{dB}$.

If the reverberation time in the enclosure is reduced to 1 sec, the decay of syllables is represented by the dotted line curves in Fig. 2.15. It will be seen

53

from this figure that because reflected energy is reduced, the intensity of the second syllable at the moment of its pronunciation is slightly reduced (from E_1 to E'_1). But the intensity of the first syllable at this moment has fallen even more (from E'_1 to E'_2). If we say that the ratio $\dfrac{E'_1}{E'_2} = 15$ at the instant of pronunciation of the second syllable, the difference in level between two adjacent syllables is 17 dB. It is clear that in the second case the reverberation of the first syllable will offer less interference to the perception of the second syllable.

Fig. 2.15 shows that the interfering action of the reverberation process can be avoided by considerably reducing reverberation time. However, by doing this, the increased clarity of speech is accompanied by an unpleasant jerkiness, as a result of which the quality of sound heard remains poor.

With music, a long reverberation time produces excessive confusion, while a short reverberation time impairs the tonal quality of the sound.

As both long and short reverberation times have a bad effect on the quality of the sound, it becomes clear that there must be some intermediate value of reverberation time which will give the best sound quality. This reverberation time is known as the optimum reverberation time. As it is connected with subjective perception, it can be determined only by means of experiment.

Experiments have shown that optimum reverberation time has different values depending on the use for which the enclosure is intended, its dimensions, the sound frequencies, the nature of the sound source, and the type of musical work to be played.

Let us examine the influence of each of these factors.

1. *The dependence of T_{opt} on the use to which the enclosure is put.* The requirements of the room acoustics, which must be the starting point of an evaluation of the sound quality of speech or music, cannot be identical for all enclosures. As high-quality speech transmission primarily demands clarity and intelligibility, different conditions are required for listening to music, conditions under which music would be heard with less clarity but with the necessary sonority and melodiousness. It follows that an enclosure designed for speech listening should have a shorter reverberation time than an enclosure for music. Moreover (see Chapter 1.3), because of the coupling between the primary and secondary enclosures involved in the transmission of sound by an electro-acoustic system, and also because of the higher level of sound reproduction in the latter enclosure, the optimum reverberation time in a room for sound reproduction should be less than in a room of similar volume intended for direct listening.

Finally it also follows from the analysis of coupled enclosures that the acoustic characteristics of the primary and secondary enclosures should differ from one another. If the secondary enclosure has a longer reverberation time than the primary enclosure, a more natural sounding quality of sound transmission can be achieved, by reason of the wider distribution of directions from which the reverberant sound comes.

54

The optimum reverberation time should be chosen in the first place therefore to suit the purpose of the given enclosure.

2. *The dependence of T_{opt} on the volume of the enclosure.* For faithful hearing of speech or music, it is essential that the signals should be sufficiently loud. However, as a result of the limited acoustic power of natural sound sources, the intensity of sound energy in the enclosure decreases as the volume of the enclosure is increased, so making listening conditions worse.

Equation 2.21 shows that the intensity of sound energy in an enclosure in a steady state is

$$E_0 = \frac{4P_a}{c_0 \alpha S} = 0 \cdot 012 \frac{P_a}{\alpha S}$$

As it follows from equation 2.43 that

$$\alpha S = \frac{0 \cdot 164 V}{T}$$

then substituting this value into the expression for E_0 we get

$$E_0 = 0 \cdot 012 \frac{P_a T}{0 \cdot 164 V} = 0 \cdot 073 \frac{P_a T}{V} \qquad 2.75$$

This equation 2.75 shows that the decrease in the intensity of sound energy due to an increase in the volume of the enclosure V, can to a certain extent be compensated by an increase in reverberation time T. It should however be pointed out that such an increase can be made only within small limits which allow the preservation of the necessary clarity of hearing, and which will not disturb people's normal conception of the relationship between the size of an enclosure and the duration of its reverberation time. Consequently, in enclosures for direct listening a certain increase in optimum reverberation time accompanying an increase in size is desirable. The character of the relationship between reverberation time and the size of an enclosure, then, is determined by the latter.

In auditoria for listening to speech, apart from meeting the demand for clarity, we must also try to meet another condition, that of correctly transmitting the tone colour of speech.

The intelligibility of speech is characterized by the articulation index, by which is understood the ratio of correctly understood syllables to the total number of syllables spoken. The articulation index A is measured in percentages, and percentages of 96 and above, 85 and above and 75 and above are considered, respectively, to be excellent, good, and satisfactory.

As was demonstrated by V. Knudsen[36], the articulation index depends on a number of factors, with which it is connected by the following equation

$$A\% = 96 \, k_L k_T k_N k_S \qquad 2.76$$

where the ks are factors representing the reductions caused to the coefficient of articulation on account of the loudness L, reverberation time T, level of interfering sound N, and the shape and dimensions of the enclosure S.

55

The dependence of each of the coefficients k_L, k_T, k_N, and k_S on the corresponding factors was determined experimentally. These relationships are shown as graphs in Figs. 2.16, 2.17 and 2.18.

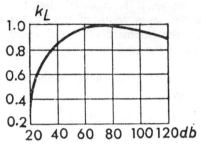

Fig. 2.16. Curve showing dependence of coefficient k_L on the loudness level of the sound.

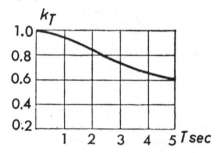

Fig. 2.17. Curve showing dependence of coefficent k_T on reverberation time.

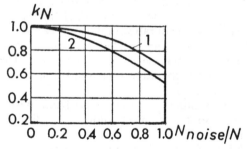

Fig. 2.18. Curves showing the dependence of coefficient k_N on the relation of noise level to signal level when the noise is audible (1) to the speaker, and when it is not (2).

If we consider that for normal rectangular enclosures with a normal relative noise level (e.g. corresponding with Fig. 2.18 where $\frac{N_N}{N} = 0.25$) the two last coefficients are close to unity, the equation for the articulation index can be stated:

$$A\% = 96k_Lk_T \qquad\qquad 2.77$$

From the equation 2.77 and from Figs. 2.16 and 2.17, it follows that to achieve an improvement in speech clarity we must reduce reverberation time and simultaneously increase the level of loudness, or at least to keep the level constant for every listener. However, for a fixed mean power from a speaker's voice a decrease in reverberation time results not in an increase, but in a reduction of loudness because of the reduction in reflected sound energy reaching the listener. It follows from this that the optimum reverberation time may be found by reducing the reverberation time to a point where the increase in clarity in speech associated with this reduction ceases because of the simultaneous reduction in the level of loudness.

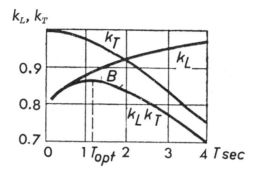

Fig. 2.19. Determining the optimum value of reverberation time to give maximum values for k_L and k_T.

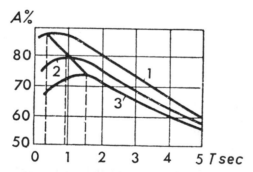

Fig. 2.20. Graphs of $A\% = f(T)$ for enclosure volumes of 700 (1), 11,200 (2), and 44,800 (3) cubic metres. The dotted lines show optimum reverberation times for these volumes.

As can be seen from the graphs in Fig. 2.19, which have been constructed for an enclosure with a volume of 11,200 m³, values of coefficients k_L and k_T can be selected for maximum articulation (Point B). It is clear that in this case reverberation time is optimal for an enclosure of the given volume.

If the values k_L and k_T are defined for enclosures of other volumes (they are different for different enclosures as a result of the fact that in large enclosures a speaker instinctively increases the loudness of his voice) and formula 2.77 is used to find the dependence $A\% = f(T)$ for each of these enclosures, we get a set of curves similar to those shown in Fig. 2.20.

57

Each curve in Fig. 2.20 has a clearly defined maximum which corresponds to the optimum value of reverberation time for an enclosure of the given volume. Having found the values of reverberation time corresponding to the maxima of the curves in Fig. 2.20, we can construct a curve showing the way in which optimum reverberation time is related to the volume of the enclosure for speech. This relation is shown in Fig. 2.21 by curve 1.

To meet the second condition set out above it is essential that in drama theatres, where a natural quality of voice timbre is extremely important, the optimum reverberation time is 10–50 per cent greater than that envisaged by curve 1 in Fig. 2.21 (see Fig. 2.21 curve 2).

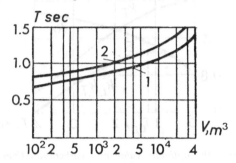

Fig. 2.21. The dependence of optimum reverberation time on the volume of auditoria (1) and of theatres (2).

The dependence of optimum reverberation time on the volume of an enclosure designed for music has been established by a number of workers both theoretically and experimentally.

The problem was first solved theoretically by S. Y. Lifshits[47] who, starting from the conclusion of the formula, based his calculations on the premises which it established that the product of the level of loudness L multiplied by the duration of the period of decline of sound t should be a constant value. He thus produced the following formula for optimum reverberation time:

$$T_{opt} = 0.41 \log V \quad (T \text{ in secs, } V \text{ in m}^3) \qquad 2.78$$

The relationship given by this formula between optimum reverberation time and the volume of an enclosure intended for listening to music is shown by curve 1 in Fig. 2.22.

The assumptions on which Lifshits rested his theory seem today to be rather dubious. The ear integrates the energy of a decaying sound only within the limits of a comparatively short period of time, so it can hardly be claimed that where the product Lt is constant the listener derives the same impression from a weak but slowly fading sound as from one which is intense but which fades rapidly. It is clear that this assumption is valid only for sounds which are similar in intensity.

Moreover, Lifshits, in his formula for optimum reverberation time, considering enclosures of various volumes, took the power of sources in them to be the same. This is not generally correct, because small enclosures are normally used for solo performances, while large halls are used for more powerful sound sources, such as choirs or orchestras.

V. Knudsen[36] considered that the mean power of probable sound sources in an enclosure increases in proportion to the $\frac{2}{3}$ power of the volume of the enclosure. This conclusion fits the data from a number of concert halls. However, introducing this correction to Lifshits' formula but retaining the doubtful premise about the constancy of the product of Lt, Knudsen arrived at results which are some 15–20 per cent higher than those of Lifshits but which in principle differ little from them.

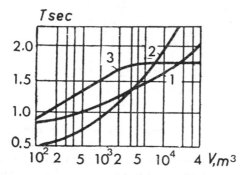

Fig. 2.22. The relationship of optimum reverberation time to the volume of concert halls. Curve 1 is from formula 2.78, curve 2 from formula 2.79, and curve 3 from formula 2.80.

These not wholly successful attempts to produce a theoretical definition of the dependence of optimum reverberation time on the size of the enclosure are not surprising in view of the fact that this dependence cannot be strictly defined and invariant. The choice of the optimum reverberation time for a concert hall or similar enclosure cannot be based solely on physical premises, as the evaluation of the quality of the reproduction of music depends on the aesthetic taste of the listeners. As the musical tastes of individuals differ, we can speak of optimum reverberation time only as a certain mean value acceptable to a large group of listeners.

This last remark leads us to a consideration of the experimental data on the relationship between optimum reverberation time and volume. Let us look at the acoustic parameters of a number of concert halls with good acoustic qualities. Fig. 2.23 shows the mean values of reverberation time for a number of European concert halls which are considered to have good acoustics. As can be seen from the figure, the reverberation times of these halls are contained within a belt which rises only gently.

A comparison of theoretical and experimental results can be made from Fig. 2.22; curve 1 corresponds to the theoretical formula 2.78, and curve 2 represents the empirical formula

$$T_{\text{opt}} = 0{\cdot}09 V^{1/3} \qquad\qquad 2.79$$

which is recommended by Mayer and Thiele on the basis of their analysis of acoustically satisfactory concert halls. Curves 1 and 2 of Fig. 2.22, show that the theoretical recommendations give rather higher mean values for optimum reverberation time for small concert halls and rather lower values for large halls, as compared with the experimental ones.

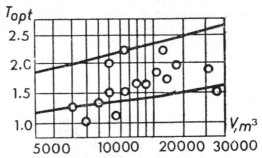

Fig. 2.23. Mean values of reverberation times in concert halls with good acoustic conditions.

This difference between theoretically and experimentally determined relationships $T_{opt} = f(V)$ is no chance one. A number of experiments carried out over the last twenty years have given results which are far from unanimous. As can be seen from Fig. 2.24, the recommendations for optimum reverberation time for concert halls of various dimensions which

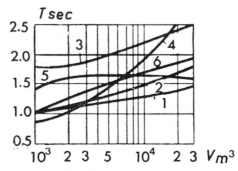

Fig. 2.24. The dependence of optimum reverberation time on the volume of music studios according to the data of Bettinger (1) Beranek (2) Bruel (3) Mayer and Thiele (4) Kuhl (5) and Lifshits (6).

have come from the work of Rettinger, Beranek, Bruel, Thiele and others (see curves 1, 2, 3, 4, 5, 6) are so different that it is extremely difficult to use them for practical design calculations.

The only thing that the majority of these curves have in common is their comparatively small slope. According to Fig. 2.24 a 10–30 times increase in volume is associated with an increase of only 30–50 per cent in optimum reverberation time, which is an indication of the comparatively

60

small dependence of the optimum reverberation time of an enclosure for music on the volume of the enclosure.

3. *The dependence of T_{opt} on the type of ensemble and the nature of the musical work.* If music enclosures of large volume are used for suitably powerful sound sources, the optimum reverberation time should be greater than if these enclosures are used for sound sources of comparatively low power. This indicates the need to choose differing values of optimum reverberation time for the best hearing of the sounds of a solo instrument, a small ensemble, and a large orchestra. Moreover, there should be a certain variation in the reverberation time according to the nature of the instruments in ensembles which are numerically of the same size, as wind instruments, for instance, are more powerful than strings.

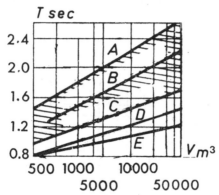

Fig. 2.25. *The dependence of optimum reverberation time on different types of performance.*

When slow, solemn and melodic music, as for instance an organ work, is being played, the optimum reverberation time should be longer than for quick, rhythmic dance music. A decrease in the time intervals between separate notes needs a decrease in reverberation time in order to ensure the purity and clarity of the musical reproduction.

The influence of these factors on optimum reverberation time is illustrated in Fig. 2.25 which shows how optimum reverberation time depends on the volume of the enclosure for various numbers and types of instruments and various kinds of musical works.

Thus the musical optimum of reverberation can have no single value, as it depends not merely on the size of the enclosure but also on the type of musical ensemble and on the character of the musical work.

The uncertainty arising from the great variety of recommendations about the choice of optimum reverberation time could not last long, as it excluded any possibility of carrying out correct design calculations of room acoustics. In 1951–54, W. Kuhl[45] carried out experiments in twenty music studios and concert halls with volumes ranging from 2000 to 14,000 m³, and with reverberation times ranging from 1·1 to 2·6 sec. In these enclosures a symphony orchestra recorded passages of classical, romantic and modern music. Then the recordings were reproduced in a well-damped room and were listened to by a large number of experts who gave their evaluation of the sound quality.

An analysis of the results of the experiment showed that for the large enclosures which were included in the experiment, the optimum reverberation time did not depend on the size of the enclosure. But the character of the music was found to have a significant influence on optimum reverberation time. The optimum time for modern music was found to be 1·48 sec, for classical music 1·54 sec, and for romantic music 2·07 sec.

As music studios and concert halls are used for music of all three types, it is suggested by way of a compromise that for enclosures larger than 2000 m³, the reverberation time should be chosen in accordance with

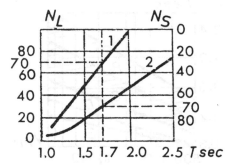

Fig. 2.26. Graph for the choice of optimum reverberation time for music studios: percentage of evaluations "reverberation high"; percentage of evaluations "reverberation low".

Fig. 2.26, where curve 1 responds to the evaluation 'high reverberation' for modern music and curve 2 to the evaluation 'low reverberation' for romantic music. Relying in the opinion of 70 per cent of the experts, the compromise value chosen was $T_{opt} = 1·7$ sec.

On the basis of experience of work in small music studios ($V = 200$–300 m³) it has been established that their reverberation time should be close to 1 sec. Consequently for studio volumes from 200 to 2000 m³, the optimum reverberation time should increase from 1·0 to 1·7 sec, which is expressed in the formula

$$\log T_{opt} = -\,0·374 + \tfrac{1}{6} \log V \qquad\qquad 2.80$$

This approach to the selection of optimum reverberation time is illustrated by curve 3 in Fig. 2.22.

Kuhl's rational decision in his consideration of the problem of determining optimum reverberation time should be noted. Despite the considerable difference in this time for music of different styles, Kuhl, basing his decision on the fact that music enclosures are used for listening to music of any style and character, has chosen a compromise solution, and has settled for a mean value of optimum reverberation time.

It is clear that a similar approach should be recommended in regard to the other factors which influence the choice of optimum reverberation time, in particular in regard to musical taste, and variations in the type and composition of the ensemble.

This recommendation seems particularly well founded in relation to all-purpose music enclosures, which are used for solo recitals and for orchestral concerts with wide ranges of programme. A serious factor in favour of the recommendation is the fact that, as follows from Fig. 2.23, good acoustic conditions in enclosures which are nearly or exactly of the same size, remain good despite a fairly wide diversity in their reverberation times.

It would be no great error if at the present time curve 3 in Fig. 2.22 were taken as a mean not merely for symphonic music, but for music in

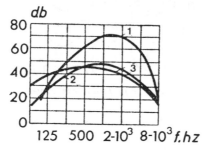

Fig. 2.27. The frequency distribution of the mean loudness level of an orchestra (1), female speech (2), male speech (3).

general. This is justified by the fact that when curve 3 is transferred to the graph in Fig. 2.24 (curve 5) it occupies an almost average position relative to the other curves on the graph.

Where there is some specialization in the use of a music enclosure, e.g. if it is intended for large orchestras with a choir, or for individual solo performances, or for dance music, the optimum reverberation time should be chosen to have a greater or smaller value than that found from curve 3 in Fig. 2.22. It should correspond to the purpose of the enclosure and lie within the limits of the shaded zone in Fig. 2.25.

4. *The dependence of T_{opt} on frequency*. The choice of an optimum mean reverberation time for an enclosure, according to the volume and the character of the musical work to be performed, is not enough to consider the problem solved. It is found that a reverberation time which is optimum for one frequency may not be so for other frequencies. It changes, in the first place, because of the variable sound absorption by the surfaces of the enclosure at different frequencies, and secondly, because the spectra of natural sounds are not the same (Fig. 2.27).

The influence of the first of these factors may be removed by a suitable choice of sound absorbing materials used in the treatment of the enclosure. The influence of the second factor, however, cannot be eliminated, and should be taken into account in determining the dependence of optimum reverberation time on frequency.

It is clear that the optimum frequency characteristic of reverberation time should be such that individual components of a complex sound should not be suppressed or stressed relative to one another by the action of reverberation.

S. Lifshits considered that this requirement could be met by preserving his previously stated condition about the constancy of the product of the

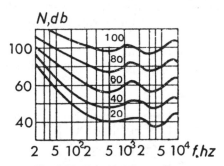

Fig. 2.28. Equal loudness curves.

level of loudness of a signal multiplied by its decay time for all frequency components of speech and music.

According to the equal loudness curves shown in Fig. 2.28 we can see that with a constant value of the intensity level N the loudness level L varies with variations in frequency. The connection between these values can be expressed by a formula,

$$L = \frac{N}{A} \qquad 2.81$$

where A is a coefficient calculated from the graphs Fig. 2.28, which is dependent on frequency as shown in Fig. 2.29. It follows from the curves

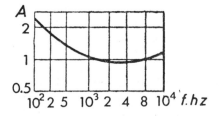

Fig. 2.29. The dependence of the mean value of coefficient A on frequency.

in Fig. 2.27 that the intensity level N of natural sounds is also dependent on frequency. Consequently, to keep the product Lt constant at all frequencies it would be necessary for reverberation time t to be inversely proportional to the relation 2.81. By finding the values of A and N for each of the

64

frequencies in Figs. 2.29 and 2.27, we can define the dependence of optimum reverberation time on frequency. This dependence for music is shown by curve 1 in Fig. 2.30.

V. Knudsen, in his researches on the dependence of optimum reverberation time on frequency, based himself on a principle somewhat different from that considered above. He argued that to ensure good sound quality, all the frequency components of the sound should decay to the threshold of hearing at the same instant.

It can be seen from the curves in Fig. 2.27 that the levels of low- and high-frequency components of speech and music are significantly lower than those of middle frequency components. Consequently, in order to satisfy Knudsen's condition, the reverberation time needed to give optimum results at high and low frequencies should be significantly greater than the optimum reverberation time at middle frequencies.

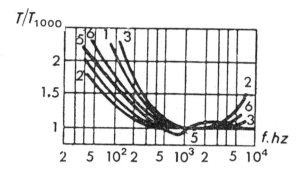

Fig. 2.30. Frequency characteristics of optimum reverberation time according to the data of Lifshits (1) Knudsen (2 for speech, 3 for music) Dreisen (4) Danish Radio (5) Morris & Nickedy (6).

This approach to optimum reverberation time results in the establishment of a relationship represented in Fig. 2.30 by curves 2 and 3, for speech and music, respectively.

The criticisms which we have made earlier about the theoretical conclusions reached by Lifshits and Knudsen about the volume optimum of reverberation, should be fully applied also to the conclusions they have reached in their considerations of the question of the frequency optimum. In both cases, these authors have based their calculations on the same insufficiently accurate premises.

Others who have studied the question of the frequency optimum of reverberation include Watson, Macnair, Richmond, Dreisen, etc. Each of these has begun from a defined assumption differing in some way from the premises of Lifshits or Knudsen.

The basic principle of I. Dreisen's reasoning,[24] for instance, is contained in the statement that fluctuations of the sound decay process at all frequencies should be at a similar level of aural perception. This condition results, in the case of music, in a need for $T_{f=100} = 1{\cdot}4 T_{f=1000}$ (curve 4 in Fig. 2.30).

65

5

Interestingly enough, curve 5 in Fig. 2.30 was obtained experimentally in a Danish radio studio, and corresponds quite closely to curves 4 and 6 in the same diagram. Curve 6 was obtained by Morris and Nickedy in 1936.

All the foregoing indicates the absence of any single point of view on the question of the frequency dependence of optimum reverberation, a fact that is still further stressed by some American experimenters, who point out that an increase in the characteristic $T_{opt} = \psi(f)$ at low frequencies is determined also by the volume of the enclosure. In connection with this, they recommend that optimum reverberation time at 100 Hz should be increased for volumes of 300, 3000 and 30000 cubic metres by 10 per cent, 30 per cent and 50 per cent, compared with this time at 1000 Hz.

Finally, unlike other experimenters, Conturie expresses the view that in view of the danger of masking sound by low frequencies, the shape of the frequency characteristic of T_{opt} should be chosen depending on the character of the sound to be transmitted and the method of performance.

Thus for speech transmissions he recommends a reduction in reverberation time at low frequencies, for light music a reduction in reverberation time at both high and low frequencies, while for classical music, a straight line frequency characteristic may be retained. Only in large studios where symphonic music is being performed does he recommend a slight lift of the low frequency characteristic.

Studies of a large number of enclosures with good acoustic qualities have shown that their reverberation time frequency characteristics correspond to a greater or lesser degree to the optimum reverberation characteristics expressed in curves 4 and 5. Satisfactory results are obtained if the frequency characteristics of reverberation time for studios of various applications are kept within the limits of the shaded areas shown in Fig. 2.31.

It follows from the graph in Fig. 2.23 that for music enclosures of similar volume, a change of 40–50 per cent in reverberation time may not affect a subjective evaluation of the acoustic qualities of these enclosures. This can be explained only by the fact that the acoustic quality of these enclosures cannot be judged merely by the length of reverberation time. The comparatively wide limits within which lie the frequency characteristics of actual radio studios (as shown in Fig. 2.31), and which have satisfactory acoustic properties, confirm this explanation.

As a result of experiments, it has been established that another important factor, apart from optimum reverberation time, which can be used to characterize the acoustics of an enclosure, is the relative value of the direct energy received by a listener and the diffuseness of the sound field in the enclosure. It has been established, for example, that where there is a comparatively high density of sound energy caused by the action of directional sources, even a long reverberation time does not interfere with clear reception of sound. A long reverberation time does not noticeably impair the clarity and good tonal quality of music if the sound field in the enclosure is diffuse.

Later work has shown that optimum conditions of sound reception in an enclosure depend to a great degree on the manner of build-up of the sound

66

field (see Chapter 3.3). It appears that not only are the length and shape of the decay process curve important, but also the time structure of this process, i.e. the change in levels and delay time for each of the series of initial reflections relative to the direct signal.

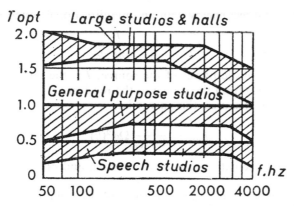

Fig. 2.31. Areas within which lie the frequency characteristics of reverberation of actual radio studios with good acoustic conditions: (1) large studios and halls, (2) general purpose studios, (3) talks studios.

Thus, although the achievement of optimum reverberation time is an essential condition for good sound reception, it is not the only one, and alone it is not sufficient.

2.10 The acoustic ratio and equivalent reverberation

At the conclusion of formula 2.41, which is used to determine reverberation time, assumptions were made to the effect that sound energy is evenly distributed throughout the entire space of an enclosure. But this supposition is not, in fact, true, as every point of the most ideal enclosure receives not only the uniform energy of the reflected waves but also the energy of the direct sound waves. This component diminishes with the distance from the source. As a result, the ratio between reflected and direct energy at various points in the enclosure differs, which has a significant influence on the perception of sound.

In 1936, V. V. Furduev[81] proposed a new criterion for the evaluation of acoustic conditions in an enclosure, which he called the acoustic ratio.

The acoustic ratio is expressed as a ratio between the density of diffuse sound energy to the density of direct energy coming from the source, i.e.

$$R = \frac{E_{\text{reflected}}}{E_{\text{direct}}}$$

The density of reflected or reverberant energy, whatever the direction of the radiator, is equal to the product of density of energy in a steady

67

state and the mean energy coefficient of reflection, and according to formulae 2.29 and 2.28 is defined as:

$$E_{ref} = E_0(1 - \alpha) = \frac{4P_a}{c_0 \alpha S}(1 - \alpha) \qquad 2.82$$

The density of direct energy with an omnidirectional sound source is:

$$E_{dir} = \frac{I}{c_0} = \frac{P_a}{4\pi r^2 c_0} \qquad 2.83$$

where P_a is the acoustic power of the source, I is the intensity of sound, and r is the distance from the source to the receiver.

With a directional source, this energy is:

$$E_{dir} = \frac{P_a}{4\pi r^2 c_0} \Omega \phi^2(\theta) \qquad 2.84$$

where Ω is the directivity factor of the source, and $\phi(\theta)$ is its directional characteristic.

If we divide expression 2.82 by 2.83 or 2.84, we obtain the acoustic ratio for an omnidirectional and a directional sound source respectively.

$$R = \frac{E_{ref}}{E_{dir}} = \frac{16\pi r^2}{S} \cdot \frac{1 - \alpha}{\alpha} \qquad 2.85$$

$$R' = \frac{E_{ref}}{E'_{dir}} = \frac{16\pi r^2}{S\Omega\phi^2(\theta)} \cdot \frac{1 - \alpha}{\alpha} \qquad 2.86$$

If we substitute the value $\alpha S = \dfrac{0 \cdot 164 V}{T}$, from formula 2.43 into the equations

2.85 and 2.86, they can be written thus:

$$R = 306 \frac{r^2 T}{V}(1 - \alpha) \qquad 2.87$$

$$R' = 306 \frac{r^2 T}{V\Omega\phi^2(\theta)}(1 - \alpha) \qquad 2.88$$

The resulting formulae show that the acoustic ratio increases if:

1. The distance between the sound source and the listener is increased.
2. Reverberation time is increased.
3. Less directional sound sources are used (small values of Ω).
4. The volume and the mean coefficient of absorption of the surfaces of the enclosure are reduced.

This last conclusion is confirmed by Fig. 2.32 which shows the curves $R = f(V)$ for optimum reverberation time and an omnidirectional sound source. Curve 1 corresponds to the back rows of an auditorium, curve 2 to the middle rows, and curve 3 to the front rows. As the volume of the enclosure is reduced, the acoustic ratio increases, despite the fact that in

68

all three cases under consideration the distance from the sound source has been reduced. This indicates that as volume is reduced the diffuse energy increases more rapidly than does the direct energy.

This figure also shows that even in the front rows of an auditorium the reflected energy is greater than direct energy. In the rear seats, where the density of reflected energy is many times greater than that of the direct energy, the importance of reflected energy in maintaining the necessary level of loudness in these rows should be stressed. However, attempts to increase the level of loudness received by remote listeners by increasing the acoustic ratio must be kept within limits because time delays between the reception of the direct and the reflected sound lead to a reduction in the clarity and intelligibility of the sound.

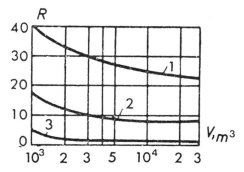

Fig. 2.32. Dependence of acoustic ratio R on the volume of a cinema auditorium for the front, middle and rear rows of seats.

Thus to achieve maximum loudness and optimum clarity simultaneously at all points in the enclosure, we must choose a certain intermediate value of the acoustic ratio.

On the other hand, attempts to create a high sound quality by choosing the optimum value for the acoustic ratio without taking reverberation time into account have proved to be unsuccessful. The acoustic ratio can be used only as a supplementary criterion for the evaluation of the acoustic properties of an enclosure.

If we return to formulae 2.87 and 2.88, we should note that if sound is received unsatisfactorily in an enclosure when using an omnidirectional sound source, the acoustic ratio can be reduced by using a directional source for which $\Omega > 1$, and consequently acoustic conditions improved.

It is clear that a change in the acoustic ratio achieved by moving the listener relative to the sound source is discernible as a change in the subjectively judged reverberation time, which is known as equivalent reverberation time.

Let us consider the relationship between equivalent and actual reverberation time.

Let the level of density of sound energy created by a sound source in an enclosure in a steady state be 60 dB. If the source is switched off, the direct sound energy immediately disappears and the level of sound energy density instantly drops from

$$60 = 10 \log \frac{E_{dir} + E_{ref}}{E_{ps}}$$

(point b in Fig. 2.33) to

$$10 \log \frac{E_{ref}}{E_{ps}}$$

(point c in the same figure).

On the basis of the former of these equations, we get

$$10^6 = \frac{E_{dir} + E_{ref}}{E_{ps}} = \frac{E_{ref}}{E_{ps}}\left(1 + \frac{E_{dir}}{E_{ref}}\right) = \frac{E_{ref}}{E_{ps}}\left(1 + \frac{1}{R}\right)$$

from whence

$$\frac{E_{ref}}{E_{ps}} = 10^6 \frac{R}{1 + R}$$

After logarizing this last expression, we may write that:

$$10 \log \frac{E_{ref}}{E_{ps}} = 60 - 10 \log \frac{1 + R}{R} \qquad 2.89$$

After having dropped instantaneously to this value, the density level of sound energy will, as shown in Fig. 2.33, subsequently decay smoothly

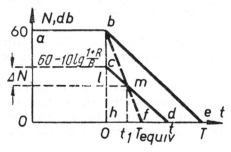

Fig. 2.33. For use in the formula for equivalent reverberation time.

along the straight line cd. The decay time t will be less than the reverberation time T, which corresponds to the decay process of the reverberant sound energy (straight line be).

G. A. Goldberg and S. T. Ter-Osipiancts showed experimentally that the decay process starting with a sharp drop in level, and the process which decays smoothly, are subjectively accepted as equivalent, provided that both decaying signals reach the same level after a definite time t_1. Although the time t_1 varies within certain limits depending on the conditions of transmission, it is always close to 0·2 sec. It follows from this condition that subjectively the decay process moves faster (along the line bf) than a decay

70

with a sharp drop in level at the moment of switch-off. The reverberation time T_{equiv}, which corresponds to this subjective perception, is known as equivalent reverberation time.

The relationship between T_{equiv} and t_1, can be found from the similarity of the triangles hbf and lbm thus:

$$T_{equiv} = \frac{60\,t_1}{\Delta N + 10 \log \dfrac{1+R}{R}} \qquad 2.90$$

The triangles clm and bhe also similar, from whence it follows that:

$$\Delta N = \frac{60}{T}\,t_1 \qquad 2.91$$

If the value of ΔN is inserted into the expression 2.90, and if we say that $t_1 = 0\cdot2$ sec, then we can write that:

$$T_{equiv} = \frac{60\,t_1}{\dfrac{60 t_1}{T} + 10 \log \dfrac{1+R}{R}} = \frac{1\cdot2T}{1\cdot2 + T \log \dfrac{1+R}{R}} \qquad 2.92$$

This expression shows that a reduction in the acoustic ratio, achieved, for instance, by using more directional loudspeakers (with small Ω) will result in a reduction in equivalent reverberation and an improvement in the clarity of the sound. It will thus be clear that the acoustic ratio characterizing acoustic conditions in the enclosure as a whole, partly determines the quality of perception of speech and music signals at various point in the enclosure.

Let us consider the curves of $T_{equiv} = f(T)$ drawn from formula 2.92 for different values of R, and shown on Fig. 2.34 by solid lines. These curves show that equivalent reverberation is little different from actual reverberation time only when reflected energy considerably exceeds direct $(R > 3)$. This happens in large parts of big, normally-treated halls, where listeners are placed fairly distant from the sound source. In these circumstances there is no point in determining equivalent reverberation.

Equivalent reverberation has a special value in determining the influence of acoustic conditions in an enclosure on a sound transmission using microphones. That this is so is not merely a result of the fact that the distance between the sound source and the microphone is usually small, but also because of the directionality of the microphone which increases acoustic ratio due to the fact that it discriminates against the reflected energy.

The electrical power developed in the microphone chain up to the moment of switch-off of the source can be represented as:

$$P_{a_1} \approx \phi_0{}^2 E_{\text{dir}} + \phi^2{}_{\text{ref}} E_{\text{ref}}$$

where ϕ_0 is the sensitivity of the microphone along its axis, and where ϕ_{ref} is its mean sensitivity in the diffuse field.

After the sound source has been switched off the power is:

$$P_{a_2} \approx \phi^2{}_{\text{ref}} E_{\text{ref}}$$

From these two expressions, the change in the power level after the source has been switched off is represented by the equation:

$$10 \log \frac{P_{a_1}}{P_{a_2}} = 10 \log \left[\left(\frac{\phi_0}{\phi_{\text{ref}}}\right)^2 \cdot \frac{E_{\text{div}}}{E_{\text{ref}}} + 1 \right]$$

As the ratio $\left(\dfrac{\phi_0}{\phi_{\text{ref}}}\right)^2 = \Omega$ represents the coefficient of directionality of the microphone, after the corresponding substitution, this equation can be restated in the form:

$$10 \log \frac{P_{a_1}}{P_{a_2}} = 10 \log \left(\Omega \cdot \frac{1}{R} + 1 \right) = 10 \log \frac{\Omega + R}{R} \qquad 2.93$$

The change in the energy level caused by the cessation of the source under direct listening conditions is expressed by the second term of the denominator of the fraction in the right hand side of equation 2.92. If we compare the expressions in 2.93 with this part, we observe that if the sound is heard through a microphone, as opposed to direct listening, the 1 in equation 2.92 must be replaced by Ω. Thus, equivalent reverberation heard through electro-acoustic system is represented by the equation

$$T_{\text{equiv}} = \frac{1 \cdot 2T}{1 \cdot 2 + T \log \dfrac{\Omega + R}{R}} \qquad 2.94$$

If we bear in mind that for cardioid and figure-of-eight microphones $\Omega = 3$, we can find the relationship between T_{equiv} and T for different values of the acoustic ratio. This dependence is shown in Fig. 2.34 by dotted line curves.

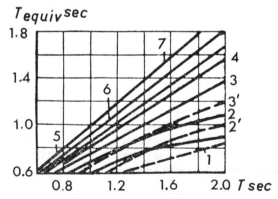

Fig. 2.34. The dependence of T_{equiv} on T for omnidirectional microphones (solid lines) and for directional microphones (dotted lines) for R 0·25, 0·5, 1, 2, 3, 6, ∞ (curves 1, 2, 3, 4, 5, 6, 7).

The curves in Fig. 2.34 show that the use of directional microphones in an electro-acoustic system allows the influence of acoustic conditions in the enclosure on the quality of the sound transmission to be considerably reduced.

2.11 The 'liveliness' coefficient of an enclosure, and factors of clarity and of reverberation interference

Although the acoustic ratio, which allows us to take into account the peculiarities of sound reception at various points in the enclosure, makes it possible to characterize more fully the acoustic properties of the enclosure, it has not become a fully recognized criterion for the evaluation of these properties. This is because it proved to be impossible to establish a correlation between the acoustic ratio and any of the subjective parameters, for instance the articulation index.

1. *The coefficient of liveliness of an enclosure.* Attempts to find a criterion which would allow a sufficiently complete and uniform evaluation of the acoustic properties of an enclosure resulted in the proposal, in 1947, of a new criterion known as the coefficient of 'liveliness' of an enclosure. This coefficient is proportional to the value of articulation, and is represented as the equation

$$G = kA$$

where k is the coefficient of proportion.

The coefficient of liveliness is represented as the ratio between the integral of reverberant sound energy and the energy of the direct sound. Analytically this ratio takes the form:

$$G = \frac{\int_0^\infty E(t)\, dt}{E_{dir}} \qquad 2.95$$

Considering that the reverberation process follows the exponential law, and using equation 2.70, we can write that:

$$E(t) = E_{ref}\, e^{-\delta t} = E_{ref}\, e^{-(13\cdot8/T)t} \qquad 2.96$$

In this case the integral in the numerator of expression 2.95 will be equal to:

$$\int_0^\infty E(t)\, dt = \int_0^\infty E_{ref}\, e^{-(13\cdot8/T)t}\, dt = E_{ref}\, \frac{T}{13\cdot8}$$

By substituting this value into the equation 2.95, we find that the coefficient of liveliness can be expressed by the formula:

$$G = \frac{E_{ref}}{E_{dir}} \cdot \frac{T}{13\cdot8} = 0\cdot027RT \qquad 2.97a$$

73

or, by substituting the value of R from equation 2.87, we see that

$$G = 0.072T \frac{306r^2T}{V} (1 - \alpha) = 22.2 \frac{r^2T^2}{V} (1 - \alpha) \qquad 2.97b$$

From the expressions 2.95, 2.96, and 2.97, we can form some conclusions.

The coefficient of liveliness of an enclosure allows us to consider both optimum reverberation time and the permissible value of articulation. It is expressed in terms of reverberation time and the acoustic ratio.

Like the acoustic ratio, this coefficient depends on the volume of the enclosure, the distance from source to receiver, reverberation time, and the mean absorption coefficient.

Thus the coefficient of liveliness, being connected with the coefficient of articulation, can express rather more fully than can the acoustic ratio the subjective perception of reverberation. However, even this criterion, like the acoustic ratio, is far from giving an accurate evaluation of the acoustic conditions in an enclosure. Its inaccuracy, and this is a very important inaccuracy, lies in the fact that none of the criteria we have referred to takes into consideration the subjective difference between reflected waves which reach the listener with varying delays.

The authors who have put forward this criterion believe that the acoustic conditions will be good if the coefficient of liveliness falls between 0.2 (for speech) and 20 (for music).

2. *The factor of clarity.* To avoid the inaccuracy which is inherent in the determination of the acoustic ratio and the coefficient of liveliness, we must clearly take into account the fact that diffuse energy may impair or improve the way sounds are heard. As was pointed out in Chapter 1.1, the early reflections are heard by the ear as combined with the direct sound, and improve reception by increasing the loudness of the sound if they follow the direct sound with a delay of less than 40–90 msec (which, at average speech speeds is at a time of about one sixteenth of a sec). Interference comes only from later reflections which, as a result of their greater delay, are not heard combined with the direct sound, and which therefore worsen the clarity of the signal.

Similar considerations led to an attempt to evaluate acoustic conditions in an enclosure according to the magnitude of the so-called factor of clarity. According to R. Thiele's definition, the 'factor of clarity' is the ratio of the useful part of the reflected energy to all the energy filling the enclosure. Numerically, this factor is represented by the equation:

$$D = \frac{\int_0^\tau E(t)\,dt}{\int_0^\infty E(t)\,dt} \qquad 2.98$$

The results of measurements of the factor of clarity in a number of music enclosures, which were carried out using specially built apparatus, are

74

shown in Fig. 2.35 which indicates the dependence of this factor on the volume of the enclosure.

It follows from the graph that the mean values of the factor of clarity for various enclosures vary from 31 to 75 per cent, and that these variations are in no way connected with the volume of the enclosures. The absence of any clearly defined dependence of the factor of clarity on the volume of the

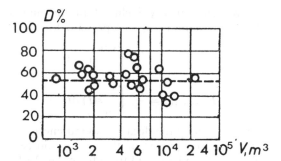

Fig. 2.35. Diagram of the dependence of the coefficient of clarity D on the volume in a number of actual halls.

enclosure allows us to consider that for any music enclosure this factor should be approximately 54 per cent.

3. *The factor of reverberation interference.* A more complete calculation of useful energy is carried out by determining the reverberation interference factor or the acoustic quality factor (Sukharevsky-Strett's factor). This factor is represented as a ratio between useful energy and harmful energy

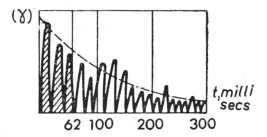

Fig. 2.36. The useful (shaded) and harmful parts of the energy of a decaying signal.

in the sound field. The useful energy is made up of the energy of the direct radiation and that part of the reflected energy which is received by the ear during a time of one sixteenth of a second after the arrival of the direct sound. All the remaining reflected energy is regarded as harmful (see Fig. 2.36). Also harmful is the energy from extraneous sound and noise.

If there are many sound sources and the distances from them to the listener are significantly different one from another, the harmful energy includes not merely parts of the reflected energy, but also parts of the direct radiation reaching the point of reception at a delay of greater than one sixteenth of a second.

From this definition the reverberation interference factor can analytically be represented as an equation:

$$Q = \frac{E_{\text{dir}} + E_{d(t \leqslant 1/16)}}{E_{d(t > 1/16)} + E_{\text{noise}}} \qquad 2.99$$

The densities of reflected energy included in the numerator and the denominator of this last equation can be represented in the form:

$$E_{d(t \leqslant 1/16)} = \frac{P_a}{V} \int_0^{1/16} \varepsilon(t)\, dt \qquad 2.100$$

$$E_{d(t > 1/16)} = \frac{P_a}{V} \int_{1/16}^{\infty} \varepsilon(t)\, dt \qquad 2.101$$

where $\varepsilon(t)$ is a function defining the decay of sound energy in the enclosure

The values of the acoustic power of the sound source P_a and of the volume of the enclosure V can be found from formulae 2.82 and 2.41 in the form:

$$P_a = \frac{c_0 \alpha S}{4(1 - \alpha)} E_d \text{ and } V = -\frac{TS \ln(1 - \alpha)}{0 \cdot 164}$$

while the value $\varepsilon(t)$ in view of the exponential character of the sound decay process, and by calculation from formula 2.70 can be put in the form:

$$\varepsilon(t) = e^{-\delta t} = e^{-(1 \cdot 38/T)t}$$

If we introduce these values into equation 2.100, we get:

$$E_{d(t \leqslant 1/16)} = -\frac{0 \cdot 164 c_0 \alpha}{4(1 - \alpha) \ln(1 - \alpha)} \cdot \frac{E_d}{T} \int_0^{1/16} e^{-(13 \cdot 8/T)t}\, dt$$

As

$$\int_0^{1/16} e^{-(13 \cdot 8/T)t}\, dt = -\frac{T}{13 \cdot 8} e^{-(13 \cdot 8/T)t} \Big|_0^{1/16} = \frac{T}{13 \cdot 8} (1 - e^{-0 \cdot 86/T}) \qquad 2.102$$

then after integration and removal of similar factors, we have:

$$E_{d(t \leqslant 1/16)} = -\frac{\alpha E_d}{(1 - \alpha) \ln(1 - \alpha)} (1 - e^{-0 \cdot 86/T}) \qquad 2.103$$

By a similar method, we find that:

$$E_{d(t \leqslant 1/16)} = -\frac{\alpha E_d}{(1 - \alpha) \ln(1 - \alpha)} \cdot e^{-0 \cdot 86/T} \qquad 2.104$$

The substitution of formulae 2.103 and 2.104 in the expression, after some rearrangements, and assuming no extraneous noise ($E_N = O$), gives:

$$Q = e^{0 \cdot 86/T} \left[1 - \frac{E_{\text{dir}}}{E_d} \cdot \frac{(1 - \alpha) \ln(1 - \alpha)}{\alpha} \right] - 1$$

or, finally, expressing $\dfrac{E_{dir}}{E_d}$ as an acoustic ratio,

$$Q = e^{0 \cdot 86/T} \left[1 - \frac{(1 - \alpha) \ln (1 - \alpha)}{R\alpha} \right] - 1 \qquad 2.105$$

In this case, when the coefficient of sound absorption is comparatively small, i.e. $\alpha \ll 1$

$$Q \simeq e^{0 \cdot 86/T} \left(1 + \frac{1}{R} \right) - 1 \qquad 2.106$$

Note that when reverberation time is large, i.e. $T \gg 0.86$, this last expression simplifies still further, and becomes:

$$Q \simeq \frac{1}{R} \qquad 2.107$$

This last expression shows that the conditions set out above, viz., $\alpha \ll 1$, and $T \gg 0.86$, correspond to an almost complete absence of perceived reflected energy combined with the direct sound. If in the expression 2.99, $E_{d(t \leqslant 1/16)}$ is very small, then in the absence of noise it takes a form inverse to the acoustic ratio. In other words, the reverberation interference factor becomes transformed into a value the inverse of the acoustic ratio when all reflected energy can be considered as harmful.

In order to clarify to what extent the reverberation interference factor, which is the fullest and most accurate criterion of the articulative quality of an enclosure, can be regarded as a uniform criterion, calculations of numerical values of this factor for various values of syllable articulation A were carried out. As a basis for these calculations we used experimental curves of the change in articulation with the distance r between the source and the listener at different reverberation times.

In Fig. 2.37 curves 1, 2, and 3 respectively are drawn for reverberation times of 1.2, 2.2, and 4 sec. In Fig. 2.38 curves 1, 2, and 3 show the dependence, calculated for the same values of reverberation time. The fact that there are three curves for A differing from one another shows the absence of an unique connection between the reverberation interference factor and articulation.

Analysis has shown that this connection becomes more clearly defined if an additional multiplier is introduced into the expression for the factor of reverberation interference. This multiplier takes into account the duration of the action of sound energy on the ear, and represents the reverberation time of the enclosure. The connection becomes better defined if this factor is represented as

$$Q' = TQ \qquad 2.108$$

But this definition is only valid for constant signal level. If the signal level decreases, so does the articulation index for speech (see Fig. 2.16). Consequently for different signal levels, different values are obtained for the more accurate criterion shown at 2.108, and its uniformity can be maintained only by introducing yet another multiplier defining the change in articulation caused by changes in signal level. In this case the factor of reverberation interference is presented in another form, as:

$$Q'' = k_L Q' = k_L T Q \qquad\qquad 2.109a$$

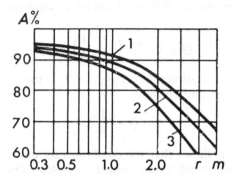

Fig. 2.37. The dependence of articulation A on the distance r between the source and the listener where T=1·2 sec (1), T=2·2 sec (2), T=4 sec (3)

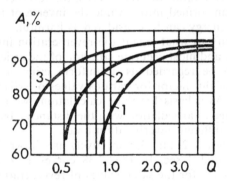

Fig. 2.38. The dependence of articulation A on the factor of reverberation interference Q where T=1·2 sec (1), T=2·2 sec (2), and T=4 sec (3).

It has been shown that the curve in Fig. 2.16 is accurately enough represented by the equation

$$k_L = 0{\cdot}039 N_r \, e^{-N_r/70}$$

where $N_r = 10 \log \dfrac{E_{\mathrm{dir}} + E_{d(t \leqslant 1/16)}}{E_0}$ is the level of the useful signal at the point of reception.

If we introduce the above corrections into equation 2.105, the improved reverberation interference factor may be represented as

$$Q'' = 0{\cdot}039 \left\{ e^{0{\cdot}86/T} \left[1 - \frac{(1-\alpha)\ln(1-\alpha)}{R\alpha} \right] - 1 \right\} T N_r e^{-N_r/70} \quad 2.109b$$

78

Calculations based on formula 2.109b and on experimental data of articulation measurements carried out by various experimenters, resulted in a curve which directly expresses a connection between articulation A and the reverberation interference factor Q''. This curve is shown in Fig. 2.39.

From a consideration of formulae 2.105, 2.108 and 2.109b and the graph in Fig. 2.39, we see that:

(a) the reverberation interference factor, represented in formula 2.105, taking into account the useful part of the sound energy, and also reverberation and noise interference, characterizes the acoustic properties of an enclosure more fully than do the acoustic ratio or the coefficient of liveness.

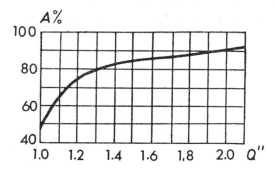

Fig. 2.39. The dependence of articulation A on the coefficient Q''.

(b) this factor, presented in its new form in formula 2.109b, takes in, in addition to the above, the ability of the ear to integrate sound energy received during a limited interval of time, and to receive sounds with greater clarity at a certain level.

(c) the reverberation interference factor, with its correcting coefficients, is connected with articulation, and, in order to ensure good articulation, its numerical value should be not less than 2 (see Fig. 2.39).

Despite the various corrections which we have discussed above, a real error has been allowed in determining the reverberation interference factor, the acoustic ratio and the coefficients of clarity and 'liveliness' of the enclosure. This inaccuracy lies in the fact that in calculating the value of the reflected energy which in some form or another comes into the formulae for these criteria, it has been assumed that the sound field is diffuse and that the decay of sound energy follows the exponential law.

The use of the coefficients we have considered for the evaluation of the acoustic properties of the large number of enclosures, the sound field in which cannot be considered diffuse, inevitably results in error. This fact, and also the difficulties involved in the measurement of these coefficients has resulted in their finding comparatively little use in practice.

2.12 The correlation method of evaluation of the acoustic properties of an enclosure

The statistical approach to the analysis of the sound field in an enclosure is based on the fact that the sound field is diffuse, and that the energy contained in it is a complex function made up of a large number of reflected signals; these premises are permissible where the signals are non-coherent (where the coefficient of coherence $\rho(t) \approx 0$). In real conditions, the diffuseness of the sound field and the degree of non-coherence of the signals depend to a significant degree on the sound absorption coefficients of the separate surfaces of the enclosure, on the form and dimensions of the enclosure and on the positions of the source and receiver of sound within the enclosure.

The acoustic quality of an enclosure which does not satisfy the condition that the field should be diffuse can be evaluated with the help of the

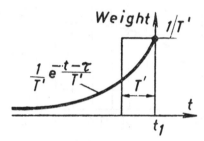

Fig. 2.40. Weighting function for determining the mean audibility of a signal.

statistical characteristics of the sounds produced within the enclosure, for example, of speech or musical sounds which are random processes.

The method used to analyse the statistical links between the values which characterize random processes is known as the correlation method. It allows us to determine the existence and strength of the connections between phenomena.

In its application to room acoustics, as V. V. Furduev[83] has shown, the degree of correlation can be determined with the help of a running auto-correlation function of the signals $f(t)$ and $f(t - \tau)$.

The form of the auto-correlation function, represented by expression 2.5, in which the functions $f(t)$ and $f(t - \tau)$ are multiplied and normalized as a fraction of unity shows that in the interval from $(t_1 - T')$ to t_1 the weights of the instantaneous values of this product remain constant (Fig. 2.40). This weight cannot be constant, as any sound receiver (and in particular the ear which has a limited memory) integrates and averages the sound in such a manner that the relative magnitude of instantaneous values depends on their duration.

Aural perception is determined not only by the value of the signal at a given instant of time, but also by preceding sound which has arrived during the minimum delay time. It is usual to assume that for hearing, the nearer the preceding signal is to the given moment of time the greater is its relative magnitude or weight, and that this time dependence is expressed as an exponential function in the form:

$$H(t) = \frac{1}{T'} e^{-(t-\tau) T'}$$

which is known as the weighting function.

The weighted auto-correlation function is thus shown by the equation

$$r_t(\tau) = \frac{1}{T'} \int_{-\infty}^{t} e^{-(t-\tau)/T'} . f(t) . f(t-\tau) \, dt \qquad 2.110$$

This is called a running auto-correlation function, while the constant value of time T' which is included in the formula is known as the 'memory' of the sound receiver.

Incorporating this weighting, expression 2.6 for aural perception can be restated as

$$P_\Sigma(t) = 2[P' + r_t(\tau)] \qquad 2.111a$$

where

$$P' = \frac{1}{T'} \int_{-\infty}^{t} e^{-(t-\tau)/T'} f^2(t) \, dt \qquad 2.111b$$

The signal $f(t)$ received by the listener in an enclosure can be represented as the sum of the direct signal $\phi(t)$ and of a number of reflected signals which diminish progressively as a result of losses by a factor of β, i.e.

$$f(t) = \sum_i \beta_i \phi(t - \tau_i) \qquad 2.112$$

where τ_i is the delay time of each of the reflected signals. If we substitute the value of $f(t)$ derived from 2.112 into the expression 2.110 we get:

$$P_\Sigma(t) = \sum_i \beta_i^2 P' + 2 \sum_i \beta_i \sum_k \beta_k r_t(\tau) \qquad 2.113$$

In this equation the first part represents the sum of the mean powers of the direct and reflected signals and defines that part of the power vested in by non-coherent signals. The second part indicates a certain coherence of the reflected signals which gives rise to a definite interference pattern.

To evaluate the influence of the enclosure on the character of the radiated signal $\phi(t)$ we can take the ratio of the total power defined in expression 2.113 and the running value of the mean power of the basic signal represented by the equation 2.111b. The mean signal level is defined by this equation as:

$$N(t) = 10 \log \frac{P_\Sigma(t)}{P'}$$

The influence of the enclosure can also be evaluated by calculating the coefficient of clarity, which shows that the role of the second part of expression 2.113 is very small compared with that of the first part.

Thus if we take the value of the mean power P_0 of the original signal $\phi(t)$ outside the limits of the sum of the first element, we can restate expression 2.113 in this form:

$$P_\Sigma = \overline{P_0} \left(1 + \sum_{i=1}^{\infty} \beta_i^2\right) \qquad 2.114$$

The useful part of the signal is composed only of those reflected signals which reach the listener together with the direct signal at a delay no greater than the minimum. If we reckon that n reflections take place in that time, we can write the following equation for the coefficient of clarity:

$$D = \frac{1 + \sum_{i=1}^{n} \beta_i^2}{1 + \sum_{i=n}^{\infty} \beta_i^2} \qquad 2.115$$

Expressions 2.115 and 2.110 show that the useful part of the reverberation signal consists of those reflections which have a close correlation with the basic signal and which create together with it a certain interference pattern. The time during which the useful energy is received is defined by a value of τ at which the running auto-correlation function expressed in equation 2.110 is still far from zero.

The expression 2.110 taken alone has an application in the evaluation of the acoustic conditions in an enclosure. As it is a function which changes only slowly with time, it shows how a signal produced in the enclosure is strengthened or weakened by the interference of reflected signals.

Another way of using the correlation function to evaluate an acoustic enclosure was put forward by S. G. Gershman.[19] She considered it possible to evaluate the diffuseness of the sound field in an enclosure from the cross-correlation function of two signals received in positions separated from one another by a short distance.

The cross-correlation function of the two signals $f_1(t)$ and $f_2(t - \tau)$ is represented by the equation:

$$r_{1\cdot 2}(t) = \frac{1}{T'} \int_0^{T'} f_1(t) \cdot f_2(t - \tau)\, dt \qquad 2.116$$

and the distance between two microphones receiving these signals is expressed by time τ.

The cross-correlation function depends not only on the distance between the receiving positions but also on their disposition within the enclosure and on positions relative to the sound source. However, for any given enclosure and for a fixed position of the radiator and the observation points, given a previously chosen spectral composition of the measuring signal, the cross-correlation function remains constant.

This property of the cross-correlation function is based on the fact that given known statistical properties of the measuring signal (band of noise) this function will take two extreme values. In a standing wave field,

82

i.e. where there is a complete absence of diffuseness, the cross-correlation coefficient $\rho_{1\cdot2}(t) = \dfrac{r_{1\cdot2}(t)}{P}$ is equal to unity, and in an ideal diffuse field at a certain minimum distance between reception points it is equal to zero. Measurement of the magnitude of the cross-correlation coefficient in given enclosures allows us to determine—by the measure to which this value approximates to these limiting values—the degree of diffuseness of the sound field.

Practical measurements confirm these propositions. Thus measurements in an absorbing room at a distance of 9·5 metres from the radiator and with the measuring signal in the form of bands of white noise, gave a cross-correlation coefficient of 0·996. In an echo chamber of the same volume

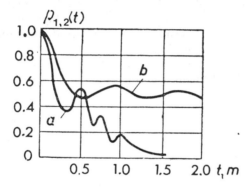

Fig. 2.41. The dependence of the cross correlation coefficient $\rho_{1\cdot2}$ (t) on the distance between receiving position l: (a) with an asymmetrical and (b) with a symmetrical arrangement of these points relative to the axis of the enclosure.

(225 m³), with measurements made under the same conditions and using a directional radiator, $\rho_{1\cdot2}(t) = 0\cdot1 \div 0\cdot16$, and with a non-directional radiator $\rho_{1\cdot2}(t)$ was close to zero.

Fig. 2.41 shows the relationship between the cross-correlation coefficient $\rho_{1\cdot2}(t)$ and the distance between receiving position l. This is derived from measurements carried out in the concert studio of Riga Radio. Curve a is taken from an asymmetrical arrangement, and curve b from a symmetrical arrangement of microphone positions relative to the longitudinal axis of the studio. The first shows that if the enclosure is sufficiently diffuse, the cross-correlation coefficient was close to zero for values of l greater than 1·5 m.

The second curve, in which it will be seen that $\rho_{1\cdot2}(t)$ is almost equal to 0·5 for a considerable distance, demonstrates only a limited symmetry of the sound field. For absolute symmetry $\rho_{1\cdot2}(t)$ would be equal to unity.

Analysis of the correlation methods for evaluating the acoustic properties of an enclosure shows that although these methods have not yet been fully developed, they are of great interest and allow a number of important conclusions to be reached.

1. Correlation methods for evaluating listening conditions in enclosures allow the particular statistical properties of various natural sounds to be taken into account.

2. They create the possibility of using the value of the auto-correlation function to judge the degree of coherence of reflected signals and to study the validity adding the energies contained in signals.

3. The results of the calculation or measurement of the cross-correlation coefficient allow a judgment to be made of the degree of diffuseness of the sound field in an enclosure.

While recognizing the practical value of the correlation methods, we should also point out their various shortcomings.

1. The measurements of the coefficients of auto-correlation and cross-correlation demand complex and carefully controlled apparatus.

2. In measuring these coefficients in the high frequency ranges, inaccuracies appear as the result of the increased directional quality of microphones at these frequencies.

3. The symmetry of the structure of the sound field in the enclosure influences the results of the measurements.

3 The Geometrical Theories of Room Acoustics

3.1 Some general statements of the geometrical theory

The statistical theory is valid only for enclosures characterized by an even distribution of sound energy. But there is a large group of enclosures which do not meet this condition.

If absorbent materials are unevenly distributed in an enclosure, a few good reflecting surfaces can create a strong reflected wave for certain areas of the enclosure. When this strong reflected wave is added to other diffuse reflections, it noticeably disturbs the character of the process of reverberation, and gives the effect of a second sound source within the enclosure.

To overcome this effect of directional reverberation, the surface responsible for the irregularity in the sound field must be located and its action neutralized by changing the form or position of the surface or by treating it with good absorbent materials.

The need to examine the inner surfaces of the enclosure in detail applies also to another irregularity of the sound field. This is the echo which may be heard in an enclosure if the difference in time between the arrival at the listener of the reflected and the direct wave is greater than the minimum delay time.

A significant irregularity in the sound field can be created by the presence in the enclosure of curved concave surfaces, as a result of which there may be local concentrations of sound energy within the enclosure.

The locations of surfaces which are detrimental to acoustic conditions in the enclosure, and also of regions of concentrated sound energy can be discovered by the so-called geometrical method, i.e. by the method of constructing diagrams to show the paths of reflected sound rays. All the rules of geometric optics hold good for these sound rays, which is the term generally accepted to describe the directions of propagation of sound waves occupying an infinitely small solid angle. In particular it is taken that a sound ray falling on a flat surface is reflected at an angle equal to the angle of incidence, and that each reflected ray may be considered as a ray emanating from an image source S' which is a mirror image of the actual sound source in the plane of the wall (Fig. 3.1).

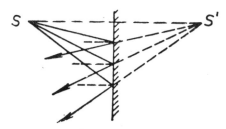

Fig. 3.1. Plan of the construction of reflected rays and of the image sound source S'

When the geometrical method is used in acoustics, just as when it is used in optics, the smaller the wavelength in comparison with the dimensions of the reflecting surfaces the more accurate are the results. From this we conclude that the geometric method of analysis should be used only when considering comparatively large reflecting surfaces. Also, the accuracy of the method increases, and more confidence can be placed in the results, when sounds of middle and high frequencies are the subject of the analysis.

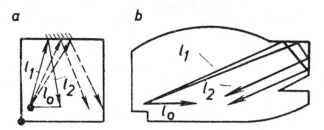

Fig. 3.2. Plan of the path of reflected rays in (a) a high and (b) a long room.

For large enclosures, where there is a danger of the appearance of an echo, for enclosures with large good reflecting surfaces, or for enclosures which have curved surfaces or an unusual shape, the character of the sound field may be determined by drawing sketches of the reflected sound rays. The sketches are drawn in the plane of a horizontal or a vertical section of the enclosure, depending on the plane of those areas capable of disturbing the sound field.

Fig. 3.2 shows sketches of longitudinal sections of two enclosures. Fig. 3.2a shows the path of rays reflected from the ceiling of a high enclosure, and Fig. 3.2b the path of rays reflected from the ceiling and the rear wall of

86

a long enclosure. If the ratio between the lengths of path of rays l_0, l_1 and l_2 for these rooms is such that

$$\frac{l_1 + l_2}{c_0} - \frac{l_0}{c_0} \geqslant \tau$$

where τ is the minimum delay time, an echo arises in the enclosures. Consequently, the ray sketch executed to scale, allowing us to determine the length of any direct and reflected ray, can be useful in finding those surfaces in an existing or projected enclosure which will give rise to an echo.

Drawing sketches of reflected rays for enclosures which have concave surfaces enables an answer to be given to the question as to whether and where the possible and probably undesirable focusing of sound energy will take place. As can be seen from Fig. 3.3a, if the radius of the curve in the

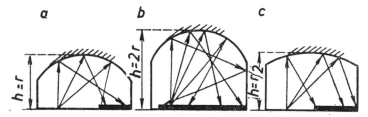

Fig. 3.3. Ray path drawings in enclosures with domed ceilings: (a) with the radius of the curve r equal to the height of the enclosure h, (b) where r is equal to h/2, and (c) where r is equal to 2h.

ceiling is little different from the height of the enclosure, it leads to a concentration of sound rays near the area where listeners are seated, and adversely affects listening conditions mainly because of the disturbance of the correct balance between direct and reflected energy.

Moreover, in this case, energy in the enclosure is redistributed in such a way that there are zones of increased and greatly reduced loudness of sound in the enclosure. Finally there is one more possibility; appearance of an echo caused by the concentration at a point in the enclosure of rays from second-, third-order, etc. reflections, which have been delayed by a time exceeding the minimum delay time, and which have acquired, as a result of focusing, an intensity comparable to the intensity of the direct rays.

All this points towards the need to avoid the creation of concave surfaces in enclosures, or at least, if they are necessary for architectural reasons, to make them so that the radius of the curve is considerably greater or smaller than any of the dimensions of the enclosure. If this is done the points at which the rays focus will be sufficiently remote from the listeners (Fig. 3.3b, c).

If there are focusing surfaces in the enclosure, the sound field may be changed in another way, because the medium and high frequency components of the sound are focused but the low frequency components are not, if the radius of the curve of the surface is less than their wavelength.

87

The geometric method allows us to reveal the conditions under which 'flutter', or repeated echo, arises. Flutter echo appears as a rule when two large parallel surfaces of the enclosure (e.g. the floor and the ceiling) are only weakly absorbing in comparison to the other surfaces, or as the result of the curvature of one of the reflecting surfaces (Fig. 3.4). As can be seen from the diagram a multiple echo is formed in the enclosure as a result of the return to point A of a number of reflections—in fact, of a sextuple reflection.

The geometrical method may be used not only to detect acoustic shortcomings of an enclosure, but also to determine the correct placing of reflecting surfaces to allow the sound energy to be evenly dispersed through the

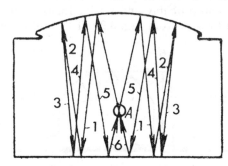

Fig. 3.4. Ray path in an enclosure with a repeated echo.

enclosure. This can be done by sending the rays of the first-order of reflections in a parallel cluster and thus raising the signal level in the area where the listeners are situated, by concentrating them at the end of the hall in the area where the level of the direct signal is at its lowest, or by dispersing them with the help of convex surfaces in order to achieve the maximum diffusion of the sound field.

3.2 The method of constructing pictures of reflections

The geometric method is also used to solve complex problems connected with the complete analysis of the acoustic properties of an enclosure. Such an analysis can be carried out by drawing a diagram of the reflections, a method proposed by G. A. Chigrinski.[92,93] This method is based on straightening a sound ray which has undergone a number of reflections and projecting it on to a grid with cell dimensions proportional to the dimensions of the surfaces which make up the enclosure.

If a sound source S is placed in an enclosure related to a system of coordinates x, y, and z, (Fig. 3.5) the sound rays leaving the source are distributed parallel to the coordinate axes (axial type rays), parallel to the coordinate surfaces (tangential rays) or parallel to neither (oblique rays).

Let us construct a reflection diagram for the second group of rays.

Let the rays be distributed in a plane parallel to the plane xy, and reflected from the side surfaces of the enclosure. In a corner rectangle of the grid representing the horizontal section of the enclosure, one of the rays is drawn and is extended towards the boundaries of the grid. As can be seen from Fig. 3.6 the sections $1-2'$, $2'-3'$, etc. correspond to sections $1-2$, $2-3$, etc. Thus the line $S-5$ represents the straightened form of the

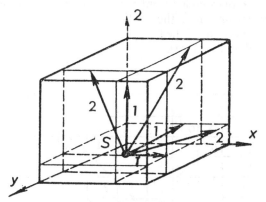

Fig. 3.5. Diagram of the distribution of axial (1) and tangential (2) rays from sources S.

sound ray which has undergone five reflections from the side limits, and each section of this line which lies between the lines of the grid represents the length of path of the sound ray between two adjacent reflections.

Fig. 3.6 shows the straightened forms of a number of rays distributed in the xy plane, and as a result being reflected from two opposite boundaries

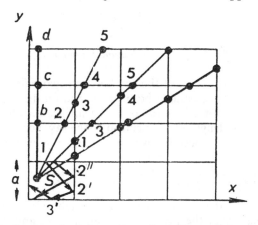

Fig. 3.6. Plan of reflection of axial rays (b, c, d) and tangential waves (1, 2, 3, 4,).

(rays crossing only the vertical or only the horizontal lines of the grid) or being reflected from all four side boundaries of the enclosure (rays crossing both horizontal and vertical lines of the grid).

In order to draw a diagram of reflections for rays which undergo reflection from all six inner surfaces of the enclosure, three grids are constructed, each of which has cells with dimensions corresponding to the dimensions of the enclosure boundaries lying in each of the planes xy, xz, yz (Fig. 3.7). In corner rectangles of the grids we mark the projection points of the sound source (S, S', S''). The projections of each ray are drawn on these surfaces and then the lines are extended right across the grids. It is clear that in this method of ray construction, the sections between the points shown on Fig. 3.7 will be projections of the free path of the ray between two reflections on all three coordinate surfaces.

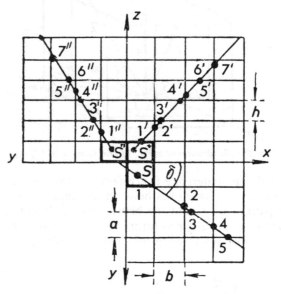

Fig. 3.7. Plan of reflection of oblique rays.

The length of each section of the ray between two reflections depends on the distance between the parallel boundaries and on the angles defining the azimuth and declination of the ray over the surface on which the parallelepiped is based. The length of the section of the ray, according to the boundaries of the enclosure between which it is moving, will be represented as

$$l_1 = \frac{a}{|\cos \gamma \cdot \cos \delta|}, \quad l_2 = \frac{b}{|\cos \gamma \cdot \sin \delta|}, \quad l_3 = \frac{h}{|\sin \gamma|}$$

where a, b and h are the length, breadth and height of the enclosure, γ is the angle of declination of the ray over the surface ab, and δ is the angle of azimuth between the axis x and the projection of the ray along the base surface.

From these equations, the number of reflections n_1, n_2, n_3 from every pair of parallel surfaces of the enclosure for a total ray length L, will be

$$n_1 = \frac{L}{l_1} = \frac{L|\cos \gamma \cdot \cos \delta|}{a}, \quad n_2 = \frac{L}{l_2} = \frac{L|\cos \gamma \cdot \sin \delta|}{b}, \quad n_3 = \frac{L}{l_3} = \frac{L|\sin \gamma|}{h}$$

90

From this, the mean path between reflections for a single ray is defined as

$$l_{\text{mean } 1} = \frac{L}{n_1 + n_2 + n_3}$$

$$= \frac{V}{S_E |\cos \gamma . \cos \delta| + S_S |\cos \gamma + \sin \delta| + S_F |\sin \gamma|} \qquad 3.1$$

where $V = a . b . h =$ volume of the enclosure, and S_F, S_S, S_E are the areas of the floor, the side and end walls.

The mean free path for all rays can be calculated by the formula

$$l_{\text{mean}} = \frac{l_{\text{mean } 1} + l_{\text{mean } 2} + l_{\text{mean } 3} \cdot \cdot \cdot + l_{\text{mean } m}}{m} = \frac{1}{m} \sum_m l_{\text{mean } m}$$

where m is the number of rays.

Once we have determined the mean free path, it is easy to find the reverberation time, for which purpose, following formulae 2.31 and 2.16, we represent the density of sound energy after the switch off of the sound source in the form

$$E = E_0 \, e^{[c_0 \ln (1 - \alpha)/l_{\text{mean}}] \, t}$$

From the definition of reverberation time it follows that

$$e^{[c_0 \ln (1 - \alpha)/l_{\text{mean}}] \, T} = 10^{-6}$$

from where

$$T = -\frac{6}{c_0 \log e} \cdot \frac{l_{\text{mean}}}{\ln (1 - \alpha)} = 0.041 \frac{l_{\text{mean}}}{-\ln (1 - \alpha)} \qquad 3.2$$

It is clear that expression 3.2 is valid only when the coefficient of absorption of all surfaces is the same and equal to α. If different areas of these surfaces have different coefficients of absorption, then by finding the number of times k that the ray meets areas of surface with similar coefficients of absorption, which is done by using the reflection diagram, we may write the density of sound energy in the enclosure after the sound source has been switched off as:

$$E = E_0 \frac{\sum_m e^{\sum \ln (1 - \alpha_n) k_n}}{m}$$

where n is the number of materials with different coefficients of absorption.

From this expression we can compute the reverberation time for any distribution of different sound absorbing materials.

By this method we can also solve problems relating to semi-enclosed spaces, such as a town square, a stadium or an open-air cinema.

Fig. 3.8 shows a diagram of reflections for a volume with an open top. The upper part of the figure shows the grid of rays projected in the vertical plane, and the lower part the projection of rays in the horizontal plane. In

91

this case the density of sound energy after the sound source has been switched off is given by the equation:

$$E = (1 - B)E_0\, e^{(c_0/l_{\text{mean}})\,\ln(1-\alpha)t} \qquad\qquad 3.4$$

where B is the proportion of sound energy leaking through the open top.

The factor $(1 - B)$ is the proportion of the energy remaining in the non-enclosed space. It is clear that if an omnidirectional sound source is being

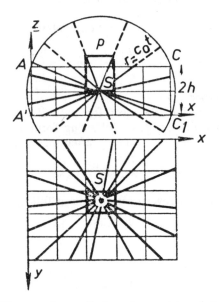

Fig. 3.8. Diagram of ray reflections for a semi-enclosed space.

used this will be equal to the ratio of the surface of the spherical belt AA_1CC_1, which has a radius of $r = c_0t$ and height $2h$, to the surface of a sphere of the same radius, i.e.

$$1 - B = \frac{2\pi r 2h}{4\pi r^2} = \frac{h}{r} = \frac{h}{c_0 t}$$

If we substitute this value of $(1 - B)$ in the expression 3.4 we find that

$$\frac{E}{E_0} = \frac{h}{c_0 t}\, e^{(c_0/l_{\text{mean}})\,\ln(1-\alpha)t}$$

Equating this last ratio to 10^{-6} and taking logarithms, we get an expression by which the reverberation time for a volume with one open boundary may be determined:

$$\log c_0 + \log \frac{T}{h} - \frac{c_0}{l_{\text{mean}}}\ln(1-\alpha)\,T \log e = 6$$

92

whence

$$\log \frac{T}{h} - 145 \frac{\ln (1 - \alpha)}{l_{\text{mean}}} T = 4\cdot47$$

or

$$T = \frac{(4\cdot47 - \log T/h)l_{\text{mean}}}{-145 \ln (1 - \alpha)} \qquad\qquad 3.5$$

where c_0 is given in milliseconds, T in seconds, and l_{mean} in metres.

This expression shows that for an unenclosed space the reverberation time decreases with a decrease in the height of the point of reception h or an increase in the mean free path of a ray l_{mean}.

By considering this method of constructing a reflection diagram, we thus see:

1. The method allows us to describe the acoustic conditions in spaces both of the enclosed and semi-enclosed types.

2. For spaces of the former type, it is advantageous to use this method if the sound field is considerably uneven as a result of poor choice of dimensions or by an uneven distribution of sound absorbing materials.

3. For semi-enclosed spaces the reverberation time depends on the position of the sound source and the receiver relative to one another and to the boundaries of the space.

4. An increase in the area of the open side of such a space results in an increase in the leakage coefficient, resulting in turn in a reduction in reverberation time.

5. The process of sound decay in a semi-enclosed space follows a broken line which differs significantly from the exponential law.

3.3 The method of calculating sound fields formed by a distributed system of radiators

If a non-directional sound source is radiating in any enclosed space, any point in that space receives both a direct ray and a whole series of reflected rays. Every such ray can be considered as coming from a certain image source which is situated on the continuation of a line linking the given point A with the point of the last reflection (Fig. 3.9).

Thus the sound field in the enclosure can be represented as a field created by image sources, grouped in the form of some spatial network. It is clear that the power of each of k image sources in the network is

$$P_k = P_a(1 - \alpha)^k$$

while the total energy created by all these sources at any point in the enclosure is

$$E = \sum_k \frac{P_k \, e^{-\delta r_k}}{r_k^2} \qquad\qquad 3.6$$

where P_a is the power of the *actual* sound source, and r_k is the distance from the point A to each of the image sources.

As it is extremely difficult to perform the summation in equation 3.6 for an arbitrary number of reflected rays, Rozenberg, who worked out this method, proposed that the energy from each discrete source within the limits of a certain line, area or volume should be concentrated in such a way that all the image sources could be combined in one continuous source. This allowed the sum in 3.6 to be replaced by an integral which was easily solved. Calculations have shown that such a substitution results in an insignificant error.

The following factors are used to characterize a field created by a system of image sources: the sound energy in the enclosure in the steady state, the factor of reverberation interference, and the law of decay.

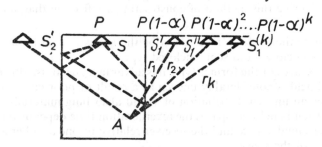

Fig. 3.9. A plan of the distribution of image sources caused by the first-order (S'_1) second-order (S''_1) etc. reflections.

Let us consider how to find these values for the simplest case where a chain of discrete radiators is replaced by one infinite linear radiator.

We first determine the density of sound energy at a point A, whose distance from the sound source is h. As can be seen from Fig. 3.10, if the power of unit length of the source is uniform throughout the whole length of the source, and equal to P_0, then the density of energy dE reaching point A from element dx of the linear radiator is:

$$dE = \frac{P_0 dx}{4\pi r^2 c_0} = \frac{P_0}{4\pi c_0} \cdot \frac{dx}{x^2 + h^2} \qquad 3.7$$

The density of energy reaching point A from the whole infinite radiator is

$$E = 2 \int_0^\infty dE = \frac{P_0}{2\pi c_0} \int_0^\infty \frac{dx}{x^2 + h^2}$$

Let us suppose that instead of the intermittent absorption of energy which actually occurs, every time the ray meets absorbent surfaces there is a continuous absorption of energy in the medium during the passage of the ray along the same path $r = \sqrt{x^2 + h^2}$. If the coefficient of attenuation in the medium is δ, energy losses are given by the expression $e^{-\delta(x^2 + h^2)}$.

On this assumption, the actual energy density in the steady state is expressed by the equation

$$E = \frac{P_0}{2\pi c_0} \int_0^\infty \frac{e^{-\delta(x^2 + h^2)}}{x^2 + h^2} dx \qquad 3.7a$$

94

If x_0 is a section of the radiator from which sound energy reaches the point of reception in the period of time $t = \frac{1}{16}$ sec, then the density of useful energy, in accordance with expression 2.12, should be shown in the form:

$$E_{(t \leqslant 1/16)} = \frac{P_0}{2\pi c_0} \int_0^{x_0} \frac{e^{-\delta(x^2 + h^2)}}{x^2 + h^2}\, dx,$$

and the density of harmful energy as

$$E_{(t > 1/16)} = \frac{P_0}{2\pi c_0} \int_{x_0}^{\infty} \frac{e^{-\delta(x^2 + h^2)}}{x^2 + h^2}\, dx$$

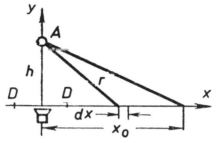

Fig. 3.10. Diagram for determining the density of energy reaching point A from a constant radiator coinciding with axis x.

From formulae 2.99 and 2.83, the reverberation interference factor can easily be found in the form:

$$Q = \frac{E_{\mathrm{dir}} + E_{(t \leqslant 1/16)}}{E_{(t > 1/16)}} + \frac{\dfrac{P_a}{2h^2} + P_0 \displaystyle\int_0^{x_0} \dfrac{e^{-\delta(x^2 + h^2)}}{x^2 + h^2}\, dx}{P_0 \displaystyle\int_{x_0}^{\infty} \dfrac{e^{-\delta(x^2 + h^2)}}{x^2 + h^2}\, dx} \qquad 3.8$$

where x_0 is the coordinate determined from Fig. 3.10 as

$$x_0 = \sqrt{c_0^2 t^2 - h^2} = \sqrt{450 - h^2}$$

The function representing the decay of sound after the source has been switched off, is defined as the ratio of the density of energy reaching a point at any moment after switch off, and the density of energy at a steady state, and is written as

$$H(\tau) = \frac{E_\tau}{E_0}$$

E_τ may be found in the following way. In the replacement of the chain of discrete radiators by one continuous radiator, the process of decay of sound energy may be considered as the arrival at the receiver of increasingly delayed signals from increasingly distant elements of this linear radiator.

This representation is equivalent to the alternative view in which it is supposed that sound energy from all the elements arrives simultaneously at

95

the point of reception, but that the linear source is switched off gradually. The first to be switched off is the central element of this source (between points D on Fig. 3.10), after which the point of switch-off travels at the speed of sound in air to all the other elements on both sides of the central element. In this representation, which is mathematically equivalent to the first view, the density of sound energy reaching the receiver at moment τ, will be

$$E_\tau = 2P_a \int_x^\infty e^{-\delta(x^2+h^2)} \cdot \frac{dx}{x^2 + h^2}$$

and

$$x = \sqrt{c_0^2 \tau^2 - h^2}$$

If we consider the density of energy in a steady state as the density due to the whole infinite radiator at the receiver, we can say, on the basis of equation 3.7a, that the decay function is:

$$H(\tau) = \frac{\displaystyle\int_x^\infty \frac{e^{-\delta(x^2+h^2)}}{x^2+h^2}\,dx}{\displaystyle\int_0^\infty \frac{e^{-\delta(x^2+h^2)}}{x^2+h^2}\,dx} \qquad\qquad 3.9$$

Now let us consider the field of the image sources formed by an actual sound source placed between four surfaces forming a rectangular enclosure (see Fig. 3.9). These image sources form a plane grid which can be replaced by a solid plane radiator placed in a medium with constant attenuation.

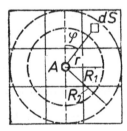

Fig. 3.11. Diagram to determine the density of sound energy created by an elementary area dS.

The density of sound energy due to an elementary area dS at the position of the receiver A which is the origin of the field coordinates (Fig. 3.11), may be represented by expressions similar to 3.7 and 3.7a, viz.,

$$dE = \frac{P_0 dS}{4\pi r^2 c_0} e^{-\delta r}$$

where

$$dS = r \cdot dr \cdot d\phi$$

and

$$\delta = \delta_1 \cos\phi + \delta_2 \sin\phi$$

96

δ_1 and δ_2 being the coefficients of decay of a ray reflected normally between each of two pairs of parallel surfaces bounding the enclosure:

$$dE = \frac{P_0}{4\pi r^2 c_0} e^{-r(\delta_1 \cos \phi + \delta_2 \sin \phi)} dr \cdot d\phi \qquad 3.10$$

The density of sound energy created by images within a ring bounded by circles of radii R_1 and R_2 is derived from 3.10 and 3.7a, giving the equation:

$$E = \frac{P_0}{4\pi c_0} \int_0^{2\pi} \int_{R_1}^{R_2} \frac{1}{r^2} e^{-r(\delta_1 \cos \phi + \delta_2 \sin \phi)} dr \cdot d\phi \qquad 3.11$$

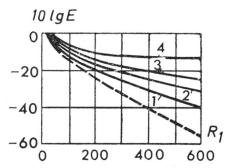

Fig. 3.12. Curves of the decay-function. Curves 1, 2, 3, and 4 are respectively for the cases, $\delta_1/\delta_2 = 1, 5, 10$ and ∞.

The same method can be used to determine the reverberation interference factor and the decay function for a plane grid of image sources.

Fig. 3.12 shows the curves relating the level of sound energy on the radius R_1 which is the lower limit of integration in formula 3.11. These curves define the form of the decay function and show that the rate of the decay of sound energy increases as the ratio δ_1/δ_2 decreases.

In the three-dimensional form of the formula for the determination of the density of sound energy in a steady state, the factors of reverberation interference and the decay functions are analogous to, though more complicated than, the formulae for the one- or the two-dimensional problems.

The theory developed by L. D. Rosenberg and schematically set out above, allows a number of conclusions to be reached:

1.   For a regular grid of linear sources, the decay function differs to a greater or lesser degree from the exponential form depending on the values of the coefficients δ_1 and δ_2.

2.   Where there is great difference between decay in one direction compared with the other two directions, the field of image sources ceases to be a three dimensional field and becomes a plane field, and thus highly uneven.

3.   To achieve an evenly distributed field and the more effective use of absorbent materials, they should be disposed in such a way that the sum of the absorption of each pair of parallel surfaces of a rectangular enclosure are the same.

97

3.4 Method of construction of areas for an equal number of reflections

Individual parts of the inner surfaces of an enclosure play various roles in the setting up of a sound field in the area where listeners are placed. From certain parts of these surfaces the listeners receive the sound waves after only one reflection, from others after two or more reflections. In the first case the reflected waves arrive with a comparatively small delay after the direct wave, and carry energy weakened by $1/\beta$ times, and in the second case the delay increases and the energy decreases as the number of reflections undergone by the wave increases.

This particular quality in individual parts of the surfaces was the basis for a new method of studying the sound field in an enclosure which was worked out in detail by A. N. Kacherovich.[41,42]

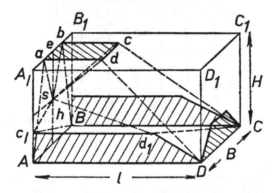

Fig. 3.13. The construction of areas of first-order reflections on the ceiling and walls

This method proposes in the first place the construction of areas for reflections of the first-, second-order etc., i.e. areas on which the sound wave falls and then goes on to reach the listener after the corresponding number of reflections.

For an enclosure in the form of a rectangular parallelopiped, and with a sound source S placed in the plane of one of the walls, the area of first-order reflections (Fig. 3.13) is bounded on the ceiling by straight lines intersecting at the points $abcd$ from which the reflected rays move off to the corner points of the floor $ABCD$. The width of this area can be found from the triangles aeS and aA_1A, as

$$b_a = \frac{H - h}{2H - h} B \qquad\qquad 3.12$$

Similarly the triangles aSd and ASD are used to determine the length of the area in the form:

$$l_n = \frac{H - h}{2H - h} L \qquad\qquad 3.13$$

In these expressions, L, B, and H are the length, breadth, and height of the enclosure.

It follows from expressions 3.12 and 3.13 that the dimensions of the area of primary reflections are greater the lower the mounting point of the sound source h.

The areas of first-order reflections on the side walls are bounded at their upper limits by points from which the rays of the first-order reflections, as in the cases indicated above, move on to the corner points of the floor. Such points on the side surfaces AA_1D_1D of the enclosure illustrated in Fig. 3.13 are points b_1 and d_1, which define the trapezoidal form of the side wall areas of primary reflections.

In this case, lowering the height of the sound source results not in an increase but in a reduction of the areas of first-order reflections. This is clearly shown also in Fig. 3.14 which shows the relationship between the

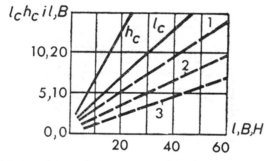

Fig. 3.14. The relationship between the dimensions of the areas of 1st order reflections and the dimensions of the enclosure. The solid lines are for side walls for the determination respectively of height h_c and of length of the upper limit of the area l_c; broken lines – for the ceiling (1) where h equals 0·2H, (2) where h equals 0·5H and (3) where h equals 0·7H.

dimensions of the areas of first-order reflections on the ceiling and on the walls (b, l) and the dimensions of the enclosure for varying heights of the sound source h.

Using the same method of construction, we can show that the rear wall has only one area of first-order reflections which is an equal sided trapezium with a height equal to half the height of the sound source $h/2$ and whose upper side is half the width of the enclosure $B/2$.

The method under consideration allows us to construct areas of second- and third-order, etc. reflections. In particular, as can be found by tracing the rays which reach the corner points of the floor after two reflections (Fig. 3.15), the areas of second-order reflections on the ceiling are demarcated by straight lines which intersect at points $abcdd_1b_1$. If we bear in mind that the area of first-order reflections lies within the limits of the polygon thus obtained, we can say that the ceiling has three areas of second-order reflections.

Rays falling on area 1 (see Fig. 3.15) are reflected first to the rear wall and only then to the area of the floor. From areas 2 and 3 rays are reflected from the side walls before falling on the floor.

The areas of first-order, second-order, etc. reflections on the ceiling have a rectangular shape and identical dimensions. Areas on the side and rear walls, however, have the form either of a trapezium or of an imperfect rectangle, and decrease rapidly in size as the number of reflections increases. They are all situated below a horizontal line at the height of the sound source.

In practical conditions, where the sound source is placed low in the room, there are large areas on the ceiling, and comparatively small areas on the walls.

If, from the point of view of explanation of the role of individual surfaces in the reverberation process it is important to define the form and dimensions of the areas of reflection, it is equally important, from the point of

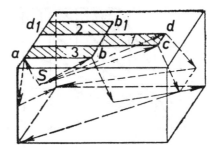

Fig. 3.15. Construction of areas of second-order reflection on the ceiling.

view of making a quantitative evaluation of this process, to determine the length of any ray reaching any point in the enclosure after n reflections. By determining the length of rays successively reaching the given point, it is possible to calculate the delay time of each of them relative to one another and to find the time during which any given number of such rays reach the point.

As an analytical calculation requires the use of a large number of formulae, a nomogram is used to determine the length of the rays and their delay time. One of these (for rays reflected from the ceiling) is shown in Fig. 3.16. These nomograms are used to determine the coefficients M, S and Q, which allow us to determine the length of any ray reaching any point after n reflections, according to the formula

$$l_n = \sqrt{M^2 + S^2 + Q^2}$$

3.14

The indices of the points of the nomogram refer to the corresponding reflection areas.

As can be seen from Fig. 3.16 the length of a ray falling on the ceiling within the limits of one of the areas of third-order reflections for which

$M = 2H - h$ will be:

$$l_3 = \sqrt{(2H - h)^2 + (2l - x)^2 + (2B - y)^2}$$

where x and y are the coordinates of the point of reception.

The delay time of the ray of any order of reflection relative to the direct ray can be found from the formula

$$t_n = \frac{l_n - l_0}{c_0} \qquad\qquad 3.15$$

where l_0 is the length of the direct ray falling on a given point of the floor. This length, as can be seen from Fig. 3.17 is

$$l_0 = \sqrt{h^2 + R^2} = \sqrt{h^2 + x^2 + y^2} \qquad\qquad 3.16$$

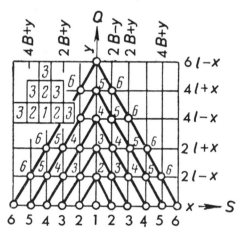

Fig. 3.16. Nomogram to determine coefficients M, S, and Q for calculations on formula 3.14.

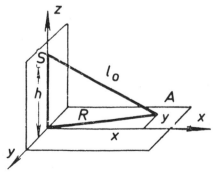

Fig. 3.17. Diagram for the determination of the length of direct ray l_0.

101

By using formula 3.15, the delay time for rays with any number of reflections can be easily calculated. Fig. 3.18 shows a ray delay picture calculated from this formula. The picture is of the delay of rays of the first- and second-orders of reflection in an enclosure of normal dimensions $(L:B:H = 3:2:1)$ with a sound source placed low $(h = 0\cdot2H)$ and symmetrically relative to the side walls. The reflection delay diagrams in Fig. 3.18 correspond to receiving points where $x = 0\cdot5H$, $1\cdot5H$, $2\cdot5H$ respectively, and $y = 0$.

A comparison of the delay pictures in Fig. 3.18 reveals that as the distance between the receiving point and the sound source increases, the delay time

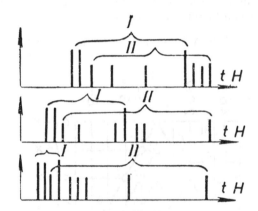

Fig. 3.18. Time sequence of arrival or rays of the first and second orde rs of reflection

of rays of both first and second orders of reflection (ordinates of lines I and II) significantly decreases, and the density of the time spectrum of reflections increases. Calculations by formula 3.15, moreover, show that an increase in the height of the sound source results in a reduction of the delay time of the rays of all reflections and particularly of those which have been reflected from the ceiling.

An alteration in the proportions of the dimensions of the enclosure leads to a change in the order in which rays of different orders of reflection arrive. If, for example, the length of the enclosure is significantly longer than the breadth, rays of first order reflection from the side walls reach the point of reception earlier than rays of first-order reflection from the ceiling.

Fig. 3.18 shows that the number of reflected rays reaching a single point increases as the order of the reflection increases. If the number of rays of first order reflection is four, then there will be eight rays of the second order, twelve of the third order and so on. It is easy to see that the number of rays, and consequently the areas of the nth order will be $4n$ (ignoring rays reflected from the floor).

If we were to continue to construct delay pictures for rays with yet more reflections, we would see that the lines of this picture would be concentrated in the right hand part of the graph, and that their density constantly

102

increases while the energy corresponding to the length of these lines decreases. It follows from this that the beginning of the reverberation process (the left hand part of the graph in Fig. 3.19) is caused by a small number of first- and second-order reflections which exclude the possibility of the establishment of an evenly distributed sound field. The final part of the process in indicated by the large number of multiple-order reflections which create a dense time spectrum. This meets the conditions necessary for the appearance of an evenly distributed field.

Thus the need arises to divide the process of the decay of sound energy in an enclosure into two qualitatively different parts. One of them can be

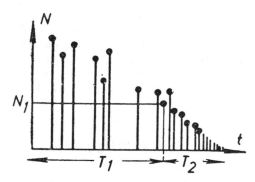

Fig. 3.19. Graph showing the increasing density of reflected rays with time.

defined from the geometric theory, and the second is easily defined by the formulae of the statistical theory.

Thus

$$T_{\text{eff}} = T_1 + T_2 \qquad 3.17$$

where T_{eff}, in contrast with normally defined reverberation time, is known as effective reverberation, T_1 is the time before the reflection sequence starts to become dense (normally the last of the secondary reflections), and T_2 is the time interval starting from the moment coinciding with the beginning of the densification of the sequence to the moment when the density of the decaying energy has fallen by 60 dB compared with the density of the direct energy.

Time T_1 can be defined by a method of graphical calculation. The first step is to find the difference between the level of direct energy and that of the energy of the primary and secondary reflections for a certain mean point in the enclosure.

Bearing in mind that rays with whatever number of reflections meet along their path materials with varying coefficients of absorption (α_1, α_2, α_3, etc.) the density of sound energy reaching this point after reflection from k surfaces is

$$E_k = \frac{P_a}{4\pi l_k^2 c_0} (1 - \alpha_1)(1 - \alpha_2) \ . \ . \ . \ (1 - \alpha_k)$$

103

If we divide this last expression by expression 2.83 which defines the density of direct energy, we get:

$$\frac{E_k}{E_{\text{dir}}} = \frac{l_0{}^2}{l_k{}^2}(1 - \alpha_1)(1 - \alpha_2) \ldots (1 - \alpha_k)$$

The difference in level between the direct energy and the energy which has been reflected from k surfaces is thus:

$$N = 10 \log \frac{E_k}{E_{\text{dir}}} = -10[2 \log l_0 - 2 \log l_k + \sum_k \log(1 - \alpha_k)] \qquad 3.18$$

Equations 3.14, 3.15 and 3.18 also allow the levels of energy of all first and second-order reflections to be established if the coefficients of absorption are known. By using formula 3.15 to find the delay time for these reflections we can construct a graph such as that shown in Fig. 3.19, showing the decrease in levels. This graph allows us to determine not only the time T_1 but also the difference in levels at this time, which is necessary for the calculation of time T_2.

This difference is

$$-N_1 = 10 \log \frac{E_1}{E_0}$$

where E_1 is the density of sound energy at the moment T_1, which may be written:

$$E_0 = E_1 10^{-0 \cdot 1N_1} \qquad 3.19$$

Time T_2 can be calculated from formulae of the statistical theory. If in expression 2.31 we replace E_0 with its value in 3.19, we get:

$$\frac{E_1}{E_1 . 10^{-0 \cdot 1N_1}} = 10^{-6} = e^{(c_0 S \ln (1 - \bar{\alpha})/4 V)T_2}$$

or

$$e^{(c_0 S \ln (1 - \bar{\alpha})/4 V)T_2} = 10^{-(6 - 0.1N_1)}$$

whence, after logarizing and calculating the numerical coefficients,

$$T_2 = 0 \cdot 027(6 - 0 \cdot 1N_1) \cdot \frac{V}{-S . \ln (1 - \bar{\alpha})} \qquad 3.20$$

In formula 3.20 the value $\bar{\alpha}$ does not always correspond to the value of the mean coefficient of absorption.

Let us consider the following two cases. Let the material covering the areas of the first and second-order reflections be much more efficient than the material covering all the other surfaces. In this case the process of decay of energy up to the moment T_1 is rapid. From moment T_1 the process slows down considerably, because the remainder of the decay is caused by reflections coming from poorly absorbent surfaces. This case is illustrated by curve 1 in Fig. 3.20.

If effective absorbent material of the same area is placed instead on the areas of the higher-order reflections, the progress of the decay differs-from the first case, the beginning being slower and the end quicker. Acceleration of the decay process in the second stage results from the fact that rays which have undergone a large number of reflections are reflected mainly from areas with a higher coefficient of sound absorption than rays of the first- and second-orders (curve II on Fig. 3.20).

Curve III on Fig. 3.20 illustrates the progress of the same reverberation process calculated according to statistical theory.

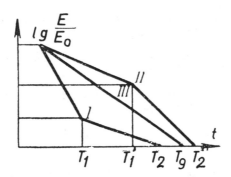

Fig. 3.20. Graph of decay of sound energy in an enclosure with effective deadening of the areas of first-order reflections (I) areas of final reflections (II) and where deadening is evenly distributed (III).

A comparison of the right hand sections of curves I and II shows that the mean effective coefficient of absorption $\bar{\alpha}$ is not the same for both of them. This coefficient can be connected with the diffuse coefficient of absorption by the equation

$$\bar{\alpha} = A\alpha$$

where A is a coefficient depending on the siting of the sound absorbent material. For the first of the cases considered above this coefficient is smaller, and in the second greater, than 1.

The cases we have considered show that the distribution of sound absorbent materials on the surfaces of the enclosure has an influence both on the decay process and on reverberation time.

To determine the coefficient of absorption used in expression 3.20 it must be borne in mind that the probability of sound rays meeting surfaces making up the enclosure is not the same for every position of the sound source in the plane of the front wall. This will become clear if the decay of sound energy in the enclosure is represented in the form of two independent processes. The first is the decay of energy which falls first on the ceiling and walls and subsequently reaches the listener (in the plane of the floor); the second is connected with the decay of energy which reaches the floor first and then after many reflections from other surfaces returns to the listener again.

If we follow the progress of rays of the first type we shall see that for enclosures of any volume, of rectangular or even trapezoidal shape, and whatever the angle of impact of the sound rays, the number of times they meet the upper part of the walls (i.e. that part above the line on which the sound source is sited) is on average 2·2 times greater than the number of times they meet the lower part.

An analysis of the progress of the second type of ray shows that from the point of view of the reflection of rays from the various surfaces, the process takes place as if the area of the walls was approximately 3·3 times larger than the area of the floor.

Thus the coefficient of absorption $\bar{\alpha}$, as it were, consists of two parts, the first of which, presented (as in formula 2.8) as

$$\bar{\alpha}_1 = \frac{2 \cdot 2\alpha_U S_U + \alpha_L S_L + \alpha_{\text{ceiling}} S_{\text{ceiling}}}{2 \cdot 2 S_U + S_L + S_{\text{ceiling}}} \qquad 3.21$$

expresses the way the reflection of rays of the first type conforms to established laws, and the second, in the form

$$\bar{\alpha}_2 = \frac{3 \cdot 3\alpha_{\text{wall}} S_{\text{floor}} + \alpha_{\text{floor}} S_{\text{floor}} + \alpha_{\text{ceiling}} S_{\text{ceiling}}}{4 \cdot 3 S_{\text{floor}} + S_{\text{ceiling}}} \qquad 3.22$$

expresses the interrelation indicated above of rays meeting with the corresponding surfaces if the rays from the source fall first on the plane of the floor. Consequently:

$$\bar{\alpha} = \bar{\alpha}_1 + \bar{\alpha}_2 =$$
$$\frac{2 \cdot 2\alpha_U S_U + \alpha_L S_L + 2\alpha_{\text{ceiling}} S_{\text{ceiling}} + (3 \cdot 3\alpha_{\text{wall}} + \alpha_{\text{floor}}) S_{\text{floor}}}{2 \cdot 2 S_U + S_L + 2 S_{\text{floor}} + 4 \cdot 3 S_{\text{floor}}}$$

where α_U, α_L, α_{wall}, α_{floor}, α_{ceiling} are mean coefficients respectively of the upper part, the lower part of the walls, the walls as a whole, the floor and the ceiling; and where S_U, S_L, S_{wall}, S_{floor}, S_{ceiling} are the areas of these surfaces.

If we consider that $S_{\text{floor}} = S_{\text{ceiling}}$ this last formula can be rewritten in the form:

$$\bar{\alpha} = \frac{2 \cdot 2\alpha_U S_U + \alpha_L S_L + (3 \cdot 3\alpha_{\text{wall}} + 2\alpha_{\text{ceiling}} + \alpha_{\text{floor}})}{2 \cdot 2 S_U + S_L + 6 \cdot 3 S_{\text{floor}}} S_{\text{floor}} \qquad 3.23$$

As can be seen from a comparison of this expression with expression 2.8, which enables us to determine the mean coefficient of sound absorption given an even distribution of sound absorbers, the concentration of absorbers on the ceiling and the lower part of the walls leads to a distinct change in the total absorption and to a reduction of effective reverberation compared with reverberation time as normally defined. Moreover, as calculations show, such a concentration of sound absorbers results in the effective reverberation remaining largely unaffected by the presence of audience in the hall (where a hall is full, T_{eff} is from 10–15 per cent less than in an empty hall). This is characteristic for an effective treatment of the ceiling; an

106

even distribution of sound absorbers results in a difference in reverberation between a full and empty hall of something of the order of 40–50 per cent.

The relationship between T_{eff} and T is apparent from Fig. 3.21. Each of the curves in this diagram is drawn for a specific value of fall of the level of sound energy ΔL during a time corresponding to the time of decay in the first stage. The uppermost curve is for small halls, as it is difficult in such halls to reach a value of ΔL greater than 6 dB. The second and third curves are more characteristic of large halls.

The difference in value of standard and effective reverberation is a result not merely of a different disposition of absorbent materials, but also of the fact that the former value presupposes that a steady-state condition has been

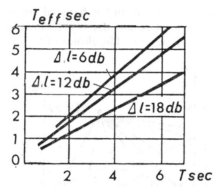

Fig. 3.21. Connection between the reverberation time and effective reverberation.

reached in the enclosure before cut-off. The approach involved in the determination of effective reverberation time corresponds better to practical conditions because in the transmission of speech or music comparatively short signals are emitted into the enclosure.

Thus the method of constructing areas for an equal number of reflections shows that:

1. The reverberation process, by the character of its formation, is divided into two stages: (a) an initial stage caused by the basically small number of reflections from comparatively large areas disposed on the ceiling and on the lower part of the walls; (b) a final stage resulting from the large number of reflections of a high order (i.e. many times reflected) proceeding from small areas disposed on the middle and upper parts of the walls.

2. The intervals between the first and second order reflections are comparatively large, and thus it is necessary, in determining the characteristics of the initial stage of the process of decay, to use the method of consecutive calculation of the energy of each reflection. For the second stage, characterized by the increase in density of the time sequence of the rays from multiple reflections, it is possible to use statistical averages.

3. As the different sectors of the surfaces of the enclosure play different roles in the formation of the reverberation process, a change in the placing

of the sound absorbent material leads to a change in the effective reverberation time in the enclosure.

4. The effective reverberation time, defined as the sum of the time sections corresponding to the first and second stages of the decay of sound energy, differs from standard reverberation time also, because the reverberation process itself is considered as the result of the decay of impulses of brief duration which arrive sequentially at the point of reception. Standard reverberation envisages a sound field initially in a steady state.

3.5 The analysis by the geometrical method of the field which arises in stereophonic transmissions

The use of geometric acoustics is not exhausted by the original theories quoted above. An example of recent use is the analysis of the sound field produced by a stereophonic system. The choice of the geometrical method is

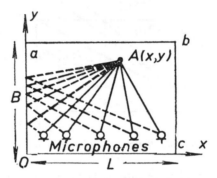

Fig. 3.22. The disposition of a stereophonic system in an enclosure before sound source A. The solid lines are direct rays, the broken lines rays of first-order reflections from the wall Oa.

preferred because it allows us to see clearly how the direct and reflected sound energy, received by each microphone in a stereophonic system and by the system as a whole in any disposition of sound sources and sound receivers, varies in accordance with the properties of the system itself and of the enclosure. This method also allows us to determine time shifts in the arrival of signals from primary sound sources variously sited both in depth and transversely to the source.

As these are the basic factors which allow an accurate perception of the spatial movement of the apparent source of sound in stereophonic transmissions, the determination of their values allows us to establish a connection between the parameters of the system and the enclosure on the one hand and these factors on the other.

To determine the magnitude of sound energy received by each microphone of a stereophonic system of an *AB* type (see Fig. 3.22) we have to determine the length of path of the sound waves by any reflection between the sound source and any microphone of the system.

108

Let microphones of a multi-channel stereophonic system be disposed in a rectangular enclosure $Oabc$ at points $M_1(x_1, y_0)$, $M_2(x_2, y_0)$. If the performer moves parallel to the line of microphones, stopping at any point in the enclosure A (XY), then we can construct rays coming from the performer to any of the microphones after k reflections. To do this we must first consider the cruciform grid, the cell dimensions of which correspond to the dimensions of the enclosure L and B. If each ray which reaches the microphone, for example microphone M_1, is straightened out after multiple reflections from the parallel walls we can find the position of image sound sources corresponding to $1, 2, 3 \ldots k$ reflections (points $A', A'' \ldots A^{(k)} \ldots$).

The image sound sources corresponding to multiple reflections from the walls will be placed on a line passing through point A and parallel with the X axis. The image sources created as a result of the reflections from the end wall will also be on a line passing through point A, but this line will be parallel to the Y axis.

From the direct construction of straightened rays we can establish that the image sources of first- and second-order etc. reflection are disposed in the corresponding cells constructed to the right, left, above and below the cell depicting the enclosure. Moreover each consecutive image sound source is, as it were, a mirror reflection of the previous one and the line of the grid which divides them is the axis of symmetry.

Referring to Fig. 3.23, we can determine the length of the ray for any reflection reaching microphone M_1, after having undergone reflection from the side walls alone, from the right-angled triangles $M_1x_1A_z'$, $M_1y_0A_y'$.

The length of the rays for reflections arriving at the microphone from the right will be:

$$d'_{1_z} = \{(y_0 - y)^2 + [2L - (x_1 + x)]^2\}^{\frac{1}{2}};$$
$$d''_{1_z} = \{(y_0 - y)^2 + [2L - (x_1 - x)]^2\}^{\frac{1}{2}}; \qquad 3.24$$

$$\cdots \cdots \cdots \cdots \cdots \cdots \cdots \cdots$$

$$d^{(k)}_{1_z} = \{(y_0 - y)^2 + [aL - (x_1 + (-1)^{k-1}x)]^2\}^{\frac{1}{2}}$$

The length of the rays reaching the microphone from the left are given by the equations:

$$d'_{1_{-z}} = [(y_0 - y)^2 + (x_1 + x)^2]^{\frac{1}{2}}$$
$$d''_{1_{-z}} = \{(y_0 - y)^2 + [2L + (x_1 - x)]^2\}^{\frac{1}{2}} \qquad 3.25$$
$$d^{(k)}_{1_{-z}} = \{(y_0 - y)^2 + [aL + (x_1 + (-1)^{k-1}x)]^2\}^{\frac{1}{2}}$$

Here $a = k$ for even values of k, for odd values in expressions 3.24. $a = k + 1$ and in expressions 3.25 $a = k - 1$.

It will be seen from the equations 3.24 and 3.25 that the general expression for the length of a ray which has undergone any number of reflections from the side walls of the enclosure will have the form:

$$d^{(k)}_{r \pm z} = \{(y_0 - y)^2 + [aL \mp (x_r + (-1)^{k-1}x)]^2\}^{\frac{1}{2}} \qquad 3.26$$

where r is the number of the microphone along the array.

Expression 3.26 allows us also to determine the lengths of the direct rays from the performer to each of the microphones. For this case we should consider $k = 0$.

By the use of the plan in Fig. 3.23 we can similarly find the lengths of the rays which have been reflected only from the front and rear walls of the enclosure. The expression for this case will have a form similar to that of 3.26.

Fig. 3.23 allows us to determine the lengths of rays only for sound waves reflected from mutually parallel vertical surfaces of the enclosure (for waves of the axial type) on condition that the sound source and the microphones of the stereophonic system are placed in a plane parallel to the surface of

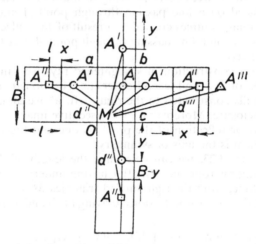

Fig. 3.23. A plan of the disposition of image sound sources A', A" etc. caused by reflections from each pair of the parallel walls of the enclosure.

the floor. However there will also be in this plane tangential waves caused by reflection from all four vertical surfaces of the enclosure.

Simple geometrical constructions show that to determine the position of image sources of this type the grid shown in Fig. 3.23 has to be extended. It is clear that image sources for second, third, etc. order reflections of the tangential type will be placed in cells situated next to cells of image sources of the axial type forming a rectangular grid shown in Fig. 3.24.

These new image sources will be placed at the points of intersection of lines parallel to the lines of the image sources of the axial type and passing through the points at which the latter are situated. Thus a second grid is formed, shown in Fig. 3.24 by dotted lines, at the intersection of the lines of which the image sources are placed. Their order increases in all directions with distance from the position of the real source. In this figure, as in Fig. 3.23, image sources of any given order are indicated by specific signs. Moreover at the limits of the grid showing the situation of the image sources the co-ordinates of the lines which compose the grid are indicated, which

110

allows us to write out the expression for the length of the ray of any reflection for rays of the axial and tangential types distributed in the plane xy.

If we consider the grid of image sources in Fig. 3.24 we can establish a certain conformity in their placing which allows us to write an expression common for all these sources and allows us to define the length of path of a sound wave which completes any number of reflections between the performer and the microphone. This expression, bearing in mind the fact that

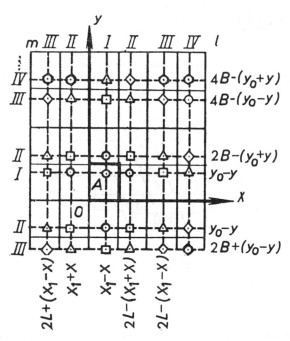

Fig. 3.24. Sketch of arrangement of virtual sources of a sound, created by reflections from the walls of a room.

A, Real source. O, Origin of co-ordinates. ⊙ Virtual source of first order, ⊡ second order, △ third, and ◇ fourth orders.

the height of the position of the microphones and the primary source may be different, appears from Fig. 3.24 and equation 3.26 in the form:

$$d^2{}_{l,m} = (z_0 - z)^2 + \{cB \mp [y_0 + (-1)^m y]\}^2 + \{aL \mp [x_2 + (-1)^l x]\}^2$$

$$3.27$$

where l and m are the horizontal and vertical co-ordinate numbers of the image in the grid; z and z_0 are the heights of the primary source and the microphones.

Expression 3.27 shows that the image grid is in effect a spatial grid, each cell of which corresponds to one image. There are many such cells in directions x and y (l and m), while in direction z there is only one cell. Thus

111

expression 3.27 corresponds to a single layer spatial grid as shown in Fig. 3.25. In this figure, similar numbers indicate cells the image sources of which have been created as a result of a similar number of reflections.

As in Fig. 3.24 we can construct a grid of image sources obtained from reflections of axial and tangential sound waves from the floor, ceiling and side walls, i.e. for image sources lying in the plane yz. A similar grid can be constructed for sources lying in the plane xz.

From these grids we can construct formulae to determine the length of a ray linking every image source in the plane yz or the plane xz with one of the microphones of a stereophonic system. These formulae will be similar to formulae 3.27.

In order to obtain a more general expression which will take in the reflection of all types of waves, including waves of the oblique type which

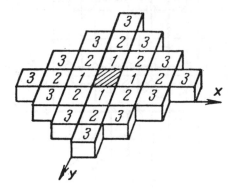

Fig. 3.25. A spatial single layer grid of images.

are reflected from any of the six surfaces, of the enclosure we must construct spatial grids of image sources created by primary, secondary, etc. reflections for waves of any type.

Fig. 3.26 shows the cells for image sources created by a single reflection. The inner parallelepiped represents a scale reduction of the enclosure and the point A within it is the position of the real source of sound. Reflections from the six surfaces create six image sources ($A'_x, A'_y, A'_z, \ldots$) situated in rectangular parallelepipeds adjacent to these surfaces. All these relate to axial reflections. Thus the number of image sources of the first order is expressed by the equation:

$$N(k) = (1)\,2 + 4k = 6$$

where k is the number of reflections, here equal to 1.

The number of double reflections or the corresponding number of image sources of the second order ($k = 2$) can be obtained from Fig. 3.27. Each of these image sources is situated in the corresponding external parallelepiped, the number of which can be calculated as:

$$N(k) = (1 + 4)\,2 + 4k = 18$$

112

The number of image sources of the third order ($k = 3$) will be equal to the number of external cells constructed on the cells in Fig. 3.27 and will comprise, as can be seen from Fig. 3.28.

$$N(k) = (1 + 4 + 8) 2 + 4k = 38$$

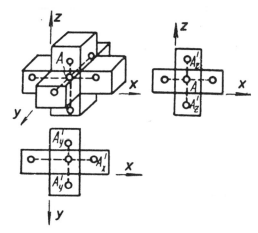

Fig. 3.26. Spatia lgrid of image sound sources of the first order

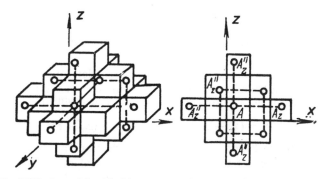

Fig. 3.27. A spatial grid of image sound sources of the second order.

If we consider the three last equations we can derive a general expression for the determination of the number of image sound sources of any order. It will be in the form:

$$N(k) = [1 + 4 + 8 + \ldots + 4(k - 1)] 2 + 4k \qquad 3.28$$
$$= (2k^2 + 1)$$

As follows from Fig. 3.28 in which cells relating to oblique reflections are shaded, image sources formed by this kind of reflection do not lie in any of the layers of spatial cells situated within the three co-ordinate planes. Consequently in order to determine the length of path of a sound wave

113

after any number of reflections of any type (axial, tangential, oblique) we must move outside the layer with any of the co-ordinate planes, which means that the first, round-bracketed, part of expression 3.27 must be replaced by something more general.

A general expression for the length of a sound ray which has completed any number of reflections k of any type between the sound source and any microphone M_r will be:

$$d^2{}_{l,m,n} = \{aL \mp [x_2 + (-1)^l x]\}^2 + \{cB \mp [y_0 + (-1)^m y]\}^2 +$$
$$+ \{bH \mp [z_0 + (-1)^n z]\}^2 \quad 3.29$$

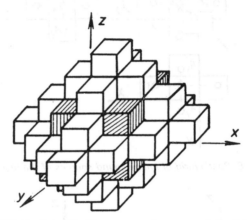

Fig. 3.28. Spatial grid of image sound sources of the third order.

Expression 3.29, being general, allows us to determine the lengths of direct rays linking sound sources directly with each of the microphones in the system. In this case, taking $l = m = n = 1$, we will find that

$$d_r{}^2 = (x_r - x)^2 + (y_0 - y)^2 + (z_0 - z)^2 \quad 3.29a$$

When we know the length of the rays for reflections of any order we can determine the energy of the first-order, second-order, etc. reflections and also the total energy received by any microphone in the system. This may be considered proportional to the sum of the squares of the sound pressures acting on it from real and image sources situated in a layer coinciding with plane xy and in layers lying above and below this basic layer. Thus for any of the microphones in the system

$$p_r{}^2 = p_{1l}{}^2 + p_{2l}{}^2 + p_{3l}{}^2 + \ldots + p_n{}^2 = \sum_{n=1}^{\infty} p_n{}^2 \quad 3.30$$

where $1l$, $2l$, $3l$, etc. are the numbers of the layers in the direction of axis z.

In their turn, the energy of the sources situated in the layer lying in the plane xy will be:

$$p_{Ir}{}^2 = p_I{}^2 + p_{II}{}^2 + p_{III}{}^2 + \ldots + p_m{}^2 = \sum_{m=1}^{\infty} p_m{}^2 \quad 3.31$$

114

where $p_I, p_{II}, p_{III}, \ldots p_m$ are pressures created by groups of image sources situated at I, II, III etc. horizontals of the grid (see Fig. 3.24).

We can find the energy of the group of sources situation on horizontal I by means of Fig. 3.18 as

$$p_I{}^2 = p_0{}^2\phi^2(\theta_r)\cdot\left(\frac{y_0 - y}{d_1}\right)^2 + p_0{}^2\phi^2(\theta_r{}^1)\left(\frac{y_0 - y}{d_1'}\right)^2(1-\alpha) + \ldots$$

$$+ p_0{}^2\phi^2(\theta_r{}^{k-1})\left(\frac{y_0 - y}{d_1{}^k}\right)^2(1-\alpha)^{k-1}$$

or, in order to simplify the calculations, assuming that the microphones are omnidirectional, i.e. $\phi(\theta) = 1$:

$$p_I{}^2 = p_0{}^2(y_0 - y)^2 \sum_{k=1}^{\infty} \frac{(1-\alpha)^k}{(d_r{}^{(k)})^2} \qquad 3.32$$

where α is the mean coefficient of sound absorption and $d_r{}^{(k)}$ is the length of the ray determined from expression 3.26.

The energy of image sources situated along the second, third or mth horizontal as is seen from Fig. 3.24 is represented by the equation

$$p_m{}^2 = p_0{}^2(y_0 - y)^2 \sum_{l=1}^{\infty} \frac{(1-\alpha)^{l+m-2}}{d^2{}_{l,m}} \qquad 3.33$$

where d_{lm} is defined by expression 3.27.

Correspondingly, starting from expression 3.31, we can derive the energy of all image sources situated in the layer coinciding with the plane xy. This energy is:

$$p_{1l} = \sum_{m=1}^{\infty} p_m{}^2 = p_0{}^2(y_0 - y)^2 \sum_{m=1}^{\infty} \sum_{l=1}^{\infty} \frac{(1-\alpha)^{l+m-2}}{d^2{}_{l,m}} \qquad 3.34$$

The energy of the image sources situated in layers above and below this layer is found by means of equation 3.28 to be given by the equation:

$$p_n{}^2 = p_0{}^2(y_0 - y)^2 \sum_{m=1}^{\infty} \sum_{l=1}^{\infty} \frac{(1-\alpha)^{l+m+n-3}}{d^2{}_{l,m,n}} \qquad 3.35$$

And finally, the total energy received by any of r microphones of the stereophonic system is represented by the equation:

$$p_r{}^2 = p_0{}^2(y_0 - y)^2 \sum_{n=1}^{\infty} \sum_{m=1}^{\infty} \sum_{l=1}^{\infty} \frac{(1-\alpha)^{l+m+n-3}}{d^2{}_{l,m,n}} \qquad 3.36$$

Expression 3.36 in combination with the other expressions we have quoted above allows us to find values for all the quantities which define the perception of a spatial picture in stereophonic transmissions. In fact, the total energy received by all the microphones of the system and which determines the perspective of the apparent sound source can be found from

$$p_s{}^2 = \sum_{r=1}^{\infty} p_r{}^2 \qquad 3.37$$

115

where $p_r{}^2$ is the energy received by one of the microphones of the system and defined by the formula 3.36.

By using formula 3.36 and the grids in Figs. 3.24, 3.26 and 3.27 we can define the energy of the first-order, second-order, etc. reflections or find the energy of all the reflections received by one or by all the microphones of the system.

If from formula 3.37 we calculate the energy of all the reflections and relate this to the energy in the direct ray as expressed by the first part of the sum in 3.32, we can find the acoustic ratio which characterizes the influence of the acoustic enclosure on spatial perception in a stereophonic transmission. This acoustic ratio, for any of the microphones, is expressed by the equation:

$$R_r = \frac{p_r{}^2}{p^2}$$

Finally, we can calculate the time delays in the arrival of signals created by primary sources of sound placed in a variety of positions both in depth and laterally relative to the stereophonic system.

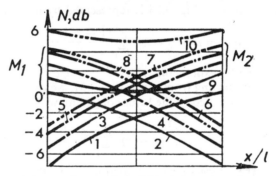

Fig. 3.29. Graphs of change of energy in individual channels as the sound source moves.

The formulae we have worked out have been used for the calculations, and it has been supposed for the sake of simplicity that the microphones in the system are placed in the same horizontal plane as the performer, i.e. that $z_0 = z = 0$.

The following dimensions were selected for the room: $L = 5l$, $B = 4l$ and $H = 2{\cdot}5l$ where $2l$ is the distance between the microphones at the ends of the array.

If we consider that the performer moves parallel to the lines of microphones of a 2-channel system at a distance y from the line given by $y/l = 1$, and that the mean coefficient of sound absorption for the selected enclosure is $\alpha = 0{\cdot}1$, then calculations carried out using these values allows us to construct the curves shown in Fig. 3.29. These curves show how the direct energy received by each of the microphones (curves 1 and 2) changes as the performer moves between the acoustic axes of the end microphones (from $x/l = 0$ to $x/l = 2$).

116

The curves which lie above depict the change of energies received by one microphone (curves 3, 5 and 7) and the other (curves 4, 6 and 8) of the system, taking into account only the first-order, first- and second-order and, finally, the first-, second- and third-order reflections respectively. Curve 9 shows the change in the total direct, and curve 10 the change in the total direct and reflected energy.

A consideration of Fig. 3.29 will show that part of the reflected energy received by any of the microphones in the system, even with such a small coefficient of absorption, is not very great and allows an increase of total energy of four to five dB. Moreover, as is shown by a comparison of curves 1 and 7 and 2 and 8, movement of the primary sound source along the line of the microphones has comparatively small influence on the change of reflected energy received by each microphone. In the given case a movement of the source from point $x/l = 0$ to point $x/l = 2$ resulted in the share of reflected energy received by the first microphone increasing by 0·4–0·5 dB and the share of the energy received by the second microphone decreasing by the same amount.

It should also be noted that the growth of reflected energy decreases as the number of the reflections increases. As follows from curves 1, 3, 5 and 7, if the first order reflections increase the energy received by the microphone 1·8–2 dB, and second order reflections by 1–1·5 dB, then third order reflections increase the total level of energy only by 0·5–0·6 dB.

Calculation shows that where an enclosure is heavily absorbing the growth of total energy level as a result of reflections significantly decreases and where $\alpha = 0·5$ amounts to a value of 1·2 dB for the first-order reflections, to 0·5 dB for the second-order and only 0·2 dB for the third-order. An increase in the coefficient of absorption is accompanied by minor variations in the reflected energy received by each microphone as the performer moves within the limits from $x/l = 0$ to $x/l = 2$. These variations are of the order of only 0·2 dB.

If by formula 3.36 we determine first the direct energy received by each microphone (taking $l = m = n = 1$) and then the reflected energy, we can find the acoustic ratio for cases corresponding to various positions of the performer relative to the microphones. The dependence of the acoustic ratio R on the position x/l of the primary sound source as it moves at a distance given by $y/l = 1$ from the line of microphones is shown in Fig. 3.30. The value of the coefficient of sound absorption is a parameter of each curve. Curves, 1, 2, 3, 4 and 5 are for one microphone and for values of this coefficient respectively of 0·1, 0·3, 0·5, 0·7 and 0·9. The symmetrically positioned curves with the same numbers refer to the second microphone.

Judging by the curves, the movement of the performer along the line of the microphones causes noticeable changes in the acoustic ratio. The quantity of reflected compared with the direct energy received by one of the microphones, grows particularly quickly as the performer moves in the zone of action of the second microphone, and the steepness of the growth becomes greater as the coefficient of sound absorption decreases.

117

In Fig. 3.30 dotted lines show the curves of the total acoustic ratio. As can be seen from the curves, the total acoustic ratio which is almost constant for changes in x/l, gradually grows as the coefficient of absorption increases.

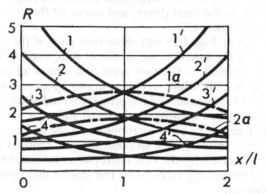

Fig. 3.30. The dependence of the acoustic ratio R on the situation of the sound source x/l at a distance from the line of the microphones of $y/l = 1$.

In order the more fully to understand the influence of the acoustic conditions of the enclosure on the amount of reflected energy received by a stereophonic system, Fig. 3.31 gives graphs of the change in the level of this energy as a function of the coefficient of sound absorption where the

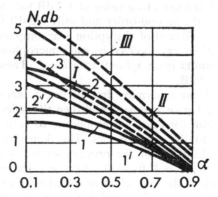

Fig. 3.31. The dependence of the level of reflected energy on the coefficient of sound absorption.

performer is placed in front of the first microphone ($x/l = 0$). The solid lines show the change of energy of the first-order reflections, broken lines the change of energy of first- and second-order, and dotted lines the change in energy of first-, second- and third-order reflections. Moreover, figures 1, 2 and 3 indicate curves relating to the first microphone, 1', 2' and 3' are curves relating to the second microphone, and I, II, III are curves expressing the sum of the energies received by the two microphones.

118

It follows from Fig. 3.31 that a reduction in the coefficient of absorption noticeably changes the amount of reflected energy received by the microphones. Where α is reduced by 0·2 the value of reflected energy increases on average by 1 dB.

Formula 3.36 enables us to calculate some values which characterize the working of a three-channel system. In particular Fig. 3.32 shows curves for $R = f(x/l)$; 1, 1′, 1″ indicate curves of the change of acoustic ratio for each of the microphones of the system where $\alpha = 0·1$; the group of curves 2 and 2″ correspond to the case when $\alpha = 0·5$. Curves 1a and 2a express the change in the total acoustic ratio for, respectively, the first and the second of the above cases.

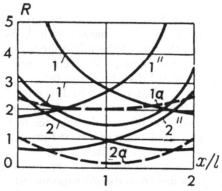

Fig. 3.32. The dependence of the acoustic ratio R on the movement of the performer x/l for a three-channel system.

If we compare curve 1a in Fig. 3.30 with the same curve in Fig. 3.32, which has similar parameters, we can observe that the total acoustic ratio for two- and three-channel systems, which are almost identical for $x/l = 0$ and for $x/l = 2$, are slightly different where $x/l = 1$ (approximately by 0·8–1). The same thing, although to a lesser degree, is seen in a comparison of curve 3a in Fig. 3.30 with curve 2a in Fig. 3.32 which correspond to the same value of α, i.e. equal to 0·5.

The conclusions which arise from the above analysis can be expressed as follows:

1. The method of geometrical construction of a reflection diagram enables us to find the positions of image sound sources whose energy, together with the energy of the real sources, is received by the microphones of a stereophonic system set up in an enclosure.

2. Image sound sources formed by different reflections are situated in the cells of a spatial grid, each of which represents the enclosure under analysis reduced in scale. The position of each image source in its respective cell is determined by the position of the primary sound source and of the microphone. Moreover the image sound sources of two adjacent orders are situated in cells and symmetrically placed relative to the common surfaces of the cells.

3. If there are several real sources distributed in the enclosure, a similar number of image sources acting on one of the microphones in the system appears in each cell. The other microphones of the stereophonic system have their own systems of image sources situated in each of the spatial cells.

4. The laws governing the positions of image sources in the cells of the spatial grid allow us to derive a general expression by which we can calculate the direct and reflected energy, the acoustic ratio, and the time shifts between signal for seach microphone and for the whole stereophonic system for various positions of the sources and receivers relative to one another and relative to the reflecting surfaces of the enclosure, and also for various values of α.

5. The share of reflected energy received by any microphone of a two- or three-channel system changes comparatively little as the primary sound source moves relative to the line of the microphones. With a movement of from $x/l = 0$ to $x/l = 2$ the changes in reflected energy are about 0·4 to 0·5 dB where $\alpha = 0·1$, and 0·2 dB where $\alpha = 0·5$.

6. The growth of the total energy received by any of the microphones of the system increases as the coefficient of sound absorption decreases. For a distance $y/l = 1$ from the line of the microphones to the line along which the performer is moving, a reduction of α by 0·2 produces an increase in the share of reflected energy of approximately 1 dB.

7. The growth of energy decreases as the number of the reflections increases. The energy coming from the microphone after the third reflection is so small that the increase in level resulting from it is 0·5 to 0·6 dB where $\alpha = 0·1$, and 0·2 dB where $\alpha = 0·5$.

8. The acoustic ratio for each microphone changes within wide limits as the performer moves parallel to the microphones and increases considerably if the coefficients of absorption of the bounding surfaces of the enclosure are small. This fact plays an important role in ensuring the naturalness of stereophonic transmissions, as an increase in the acoustic ratio is connected with the perception of a retreating sound source.

9. The acoustic ratio caused by the working of all microphones of two- and three-channel systems also depends on the coefficient of sound absorption, but changes comparatively little as the performer moves laterally. For all values of α this change is no greater than 0·5–1.

120

4 The Wave Theory of the Acoustics of an Enclosure

4.1 Normal modes of vibration of the air space of an enclosure

The analysis of sound processes in an enclosure can be carried out not only by using a particular idealization of the sound field, as in the statistical and geometrical theories of the acoustics of an enclosure. It can be carried out also on the basis of an actual physical picture of the wave phenomena in the enclosure. The importance of carrying out this kind of analysis in a systematic way lies in the fact that unlike the statistical or geometric analyses, this method reveals the reality of sound phenomena in the enclosure, and their basic characteristics.

The volume of air in a closed enclosure is a complex vibratory system which can be regarded as a combination of a number of simple systems. When a sound source is switched on in the enclosure it sets up a vibratory process which has both forced vibrations and normal modes of vibration. When the latter die away, a steady state situation is reached in the enclosure. The vibrations of this steady state situation, unlike the effects when they are distributed in open space, can be expressed as the sum of a large number of standing waves. When the sound source is switched off, the system, being thrown out of balance, will carry out only its normal modes of vibration, which also have the form of standing waves, and which gradually decay according to the exponential law which is common to all normal modes of vibration.

Experience shows that in normal enclosures the phenomenon of reverberation is observed at any frequency of the exciting sound, which is clearly only possible if these enclosures have a great many characteristic frequencies close to one another. It is moreover well known that an infinitely large number of characteristic frequencies will be associated with an equal number of vibratory systems with evenly distributed constants.

Thus from the point of view of the wave theory the volume of air in an enclosure is seen as a complex vibratory system with distributed parameters, a system which when excited by a sound impulse carries out its own, gradually decaying, normal vibration modes.

As the phenomenon of reverberation may be represented as the decay of characteristic frequencies of the air space of an enclosure, it is necessary to study the nature of these vibrations very closely.

Let there be a rectangular enclosure in which the sound field is non-uniform and in which the absorption of energy is negligible.

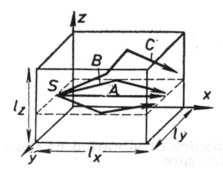

Fig. 4.1. The distribution of axial, tangential and oblique waves in an enclosure.

When a sound source situated in one of the walls is radiating, waves are formed in the enclosure and distribute themselves in a variety of directions. Some are distributed parallel to the edges l_x, l_y, l_z which coincide with the co-ordinate axes x, y, z (Fig. 4.1) and are formed as a result of reflection from two walls. Such waves are called axial. Others are distributed parallel to one pair of the surfaces of the enclosure and are formed as a result of reflection from four walls. These waves are called tangential. Finally there are waves which derive from reflection from all the surfaces of the enclosure. These are called oblique.

For waves which are distributed in direction x we can write a differential equation in the following form (equation for a plane wave):

$$\frac{\partial^2 p}{\partial x^2} = \frac{1}{c_0^2} \frac{\partial^2 p}{\partial t^2} \qquad\qquad 4.1$$

the solution of which is represented by the equation: $p = A\, p(x)\, e^{j(\omega t - kx)}$.

The product $Ap(x)$ represents the amplitude of the sound pressure, which has a constant value when a wave is propagated in unlimited space. If in its path the wave meets a barrier, as in the enclosure under consideration, the wave is reflected and, combining with the direct wave, forms a standing wave. As a result of this, the amplitude of sound pressure for vibrations distributed along one of the axes of the enclosure, and bearing in mind the

122

possibility of the creation of a large number of standing waves, should be written as:

$$A_n p_n(x) = \sum_n A_n \cos k_x x$$

and the sound pressure itself, bearing in mind the decaying character of the vibrations which is caused by absorption at the boundaries of the enclosure, should be written in the form:

$$p(x, t) = \sum_{n=1}^{\infty} A_n \cos k_x x \cdot e^{j\omega_{nx}t - \delta_{nx}t} \qquad 4.2$$

where A_n is an arbitrary constant defining the amplitude of each of n modes of vibration; $\cos k_x x$ is the x-direction cosine; ω_{nx} is the frequency of the nth natural vibration in the direction of the x-axis; and δ_{nx} is the index of decay caused by absorption at the boundaries of the enclosure.

Equation 4.2 shows that along the direction of the x-axis normal modes of vibration are created having the form of standing waves and a frequency of

$$f_n = \frac{\omega_{nx}}{2\pi} = \frac{1}{2\pi} \sqrt{\omega_{nx}^2} \qquad 4.3$$

The nodal points of these waves can be found from the condition that the pressure at the points of reflection of a sound wave should be maximal, i.e. that where $x = l_x$

$$\cos k_x x = 1$$

But $\cos k_x x = 1$, when $k_x l_x = n_x \pi$, where n_x is any whole number indicating the number of standing waves arising in the direction of axis x.

As k_x is the wave number, if we insert its value into this last equation:

$$k_x l_x = \frac{\omega_{nx}}{c_0} l_x$$

whence

$$\omega_{nx} = n_x \frac{c_0}{l_x} \pi \qquad 4.4$$

After replacing k_x and ω_{nx} in expressions 4.2 and 4.3 by their values derived from Equation 4.4, expressions 4.2 and 4.3 can be rewritten as:

$$p_n(x,t) = \sum_{n=1}^{\infty} A_n \cos \frac{n_x \pi x}{l_x} e^{j\omega_{nx}t - \delta_{nx}t}; \qquad 4.5$$

$$f_n = \frac{1}{2\pi} \sqrt{\omega_{nx}^2} = \frac{1}{2\pi} \sqrt{\left(\frac{n_x c_0 \pi}{l_x}\right)^2} = \frac{c_0}{2} \sqrt{\left(\frac{n_x}{l_x}\right)^2} \qquad 4.6$$

It is clear that in expressions 4.5 and 4.6, if index x is replaced by index y or index z, these expressions will define axial waves distributed parallel to axes y and z.

123

If we remember that $n_z = 0, 1, 2, 3 \ldots$, normal vibration modes for axial waves can be presented in accordance with Equation 4.6 as graphs (Fig. 4.2).

For waves which are distributed parallel to one of the surfaces of the enclosure, e.g. parallel to the xy plane, the wave equation can be written thus:

$$\frac{\partial^2 p}{\partial x^2} + \frac{\partial^2 p}{\partial y^2} = \frac{1}{c_0^2} \frac{\partial^2 p}{\partial t^2}$$

Its solution under the above mentioned conditions, by analogy with equation 4.2, can be presented in the form:

$$p_n(x, y, t) = \sum_{n=1}^{\infty} A_n \cos \frac{n_x \pi x}{l_x} \cos \frac{n_y \pi y}{l_y} e^{j\omega_n t - \delta_n t} \qquad 4.7$$

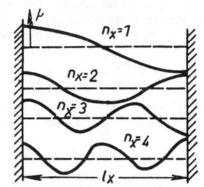

Fig. 4.2. The form of axial waves for different values of n_z.

Frequencies of the normal modes of vibration which take place in the xy plane, by analogy with Equation 4.6 can be expressed by equation

$$f_n = \frac{1}{2\pi} \sqrt{\omega_x^2 + \omega_y^2} = \frac{c_0}{2} \sqrt{\left(\frac{n_x}{l_x}\right)^2 + \left(\frac{n_y}{l_y}\right)^2} \qquad 4.8$$

These last two equations express the relationship for axial and tangential waves which are distributed parallel to only one of the surfaces of the enclosure. To find similar relationships for waves which are distributed parallel to other surfaces it is necessary, as it was for axial waves, to change the indices and variables.

For all forms of waves distributed in an enclosure the wave equation takes the form:

$$\frac{\partial^2 p}{\partial x^2} + \frac{\partial^2 p}{\partial y^2} + \frac{\partial^2 p}{\partial z^2} = \frac{1}{c_0^2} \frac{\partial^2 p}{\partial t^2} \qquad 4.9$$

By analogy with expression 4.5 its solution is:

$$p(x, y, z, t) = \sum_{n=1}^{\infty} A_n \cos \frac{n_x \pi x}{l_x} \cos \frac{n_y \pi y}{l_y} \cos \frac{n_2 \pi z}{l_z} e^{j\omega_n t - \delta_n t} \qquad 4.9a$$

124

and the characteristic frequencies are defined by an equation similar to Equation 4.6.

$$f_n = \frac{c_0}{2} \sqrt{\left(\frac{n_x}{l_x}\right)^2 + \left(\frac{n_y}{l_y}\right)^2 + \left(\frac{n_z}{l_z}\right)^2}$$ 4.10

If we compare expressions 4.6 and 4.8 with the equation 4.10, it is easy to see that the former is derived from the last where $n_y = 0$ and $n_z = 0$, while the second is derived by inserting into the last $n_z = 0$. Thus expression 4.10 is the most general and defines the frequencies of natural vibrations of the oblique as well as the tangential and axial types.

It follows from equation 4.9a that normal modes of vibration of the air space of an enclosure are characterized by a complex interweaving of standing waves and the decay of the various waves which comprise these vibrations can proceed with varying speeds defined by the index of decay δ. Oblique waves, as they have the shortest mean free path and undergo the largest number of reflections in each second will decay faster than the others. The next to decay are the tangential, and finally the axial waves. The sound field, which is fairly uniform and diffuse at the first instant after the source has been switched off, becomes progressively less uniform as the oblique and tangential waves decay. This is because after the decay of oblique waves, the randomly distributed waves are increasingly dominated by the more orderly tangential or axial waves distributed parallel to surfaces and to the edges of the surfaces respectively.

It follows from expression 4.10 that the natural frequencies of the enclosure are defined by its parameters l_x, l_y, l_z and by the number of standing waves which arise in directions x, y, z. It will be seen from equation 4.10 that if two or three dimensions of the enclosure are equal or if their lengths are multiples of one another, coincidence of characteristic frequencies may result. Thus for example, if the dimensions of the enclosure are identical ($l_x = l_y = l_z$), identical values of characteristic frequency are obtained where $n_x = 1$, $n_y = 1$, $n_z = 1$ while if these dimensions are all different there are three of these frequencies.

From what has been said we can make the following conclusions:

1. The volume of air in an enclosure, being a distributed vibratory system has a large number of characteristic frequencies (equation 4.10).
2. Normal modes of vibration are represented as a complex pattern of standing waves (equation 4.9).
3. Each characteristic frequency decays according to the exponential law.
4. Waves of different types may decay at different speeds which depend on the value of the index of decay.
5. The process of decay of sound energy in the enclosure (the decay of all waves) approaches more closely to the exponential, as the differences between the times of decay of the various types of normal vibration modes diminish.
6. The frequencies of normal modes of vibration of an enclosure are

125

arranged in a more dense spectrum, i.e. the sound field in the enclosure is more even, if its length, breadth and height are not equal and not multiples of one another.

4.2 The density of the spectrum of characteristic frequencies of an enclosure

If we consider axial vibrations distributed in the x direction ($n_y = 0$, $n_z = 0$), we observe that in accordance with equation 4.6 each of the frequencies of these vibrations can be represented as a vector coinciding in direction with the x-axis. The length of the first vector is $c_0/2l_x$ (where $n_x = 1$), the second $2\,c_0/2l_x$ (where $n = 2$), etc.

This interpretation allows us to represent each characteristic frequency by a certain point in frequency space whose co-ordinates are defined as $c_0/2l_x$, $c_0/2l_y$, $c_0/2l_z$ and to consider the total volume of the space as filled by a lattice of frequency points.

The cells of this frequency space define the frequencies of normal vibrations of the axial (vectors 1, 2, 3), of the tangential (vectors 1', 2', 3') and of the oblique types (vector 1") as shown in Fig. 4.3.

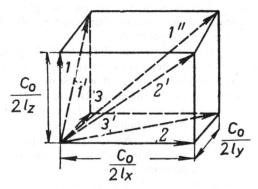

Fig. 4.3. A cell of the spatial grid of frequencies with frequencies of normal modes of the axial (1, 2, 3), tangential (1', 2', 3') and of the oblique (1") types.

It may happen that among the characteristic frequencies calculated by the above method there are some whose values lie beyond the limits of the range which concerns us. In this case, to determine the number of natural frequencies lying below a given frequency f_n, a section of a sphere is constructed within the system of co-ordinates xyz. The sphere has its centre at the junction of the coordinates and has a radius equal to f_n (Fig. 4.4).

To find the number of characteristic frequencies of the axial type for such a spherical frequency space we must first find their number for a certain rectangular space shown in Fig. 4.4 as a dotted line and having length, breadth and height equal to f_n.

126

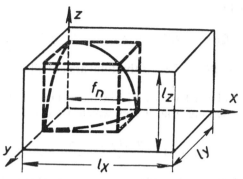

Fig 4.4. The construction of a frequency space for limiting value of frequency value f_n.

The number of frequencies for this rectangular space along axis x is found by dividing the maximum permitted frequency by $c_0/2l_x$, i.e.

$$n_x(f_n) = \frac{f_n}{c_0} \cdot 2l_x = \frac{2l_x}{c_0} \cdot f_n$$

along axis y

$$n_y(f_n) = \frac{f_n}{c_0} \cdot 2l_y = \frac{2l_y}{c_0} f_n \qquad 4.11$$

along axis z

$$n_z(f_n) = \frac{f_n}{c_0} \cdot 2l_z = \frac{2l_z}{c_0} f_n$$

The total number of characteristic frequencies of vibrations of the axial type is thus

$$n_{tc}(f_n) = \frac{2l_x}{c_0} f_n + \frac{2l_y}{c_0} f_n + \frac{2l_z}{c_0} f_n = \frac{2}{c_0}(l_x + l_y + l_z) f_n \qquad 4.12$$

or, as the total length of all the edges of the enclosure $L = 4(l_x + l_y + l_z)$ then:

$$n_{tc}(f_n) = \frac{L}{2c_0} f_n \qquad 4.13$$

Within the volume of the octant of a sphere inscribed in the cube considered above, are included firstly the frequencies of these axial modes represented by points along three of the twelve edges of the cube; from 4.13 their number therefore is:

$$n'_{tc}(f_n) = \frac{n_{tc}(f_n)}{4} = \frac{L}{8c_0} f_n \qquad 4.14$$

The number of characteristic frequencies for tangential waves distributed parallel to plane xy is:

$$n_{xy}(f_n) = \frac{S_{1/2\,\text{circ}}}{S_{\text{rect}}} = \frac{\pi f_n^2}{4} \div \frac{c_0^2}{4l_x l_y} = \frac{\pi}{c_0^2} S_{xy} f_n^2$$

127

For all tangential waves:

$$n_{\text{tan}}(f_n) = \frac{\pi}{c_0^2}(2S_{xy} + 2S_{xz} + 2S_{yz})f_n^2 = \frac{\pi}{c_0^2}Sf_n^2$$

where $S = 2S_{xy} + 2S_{xz} + 2S_{yz}$ is the total area of all the bounding surfaces of the enclosure.

As in the case of axial waves within the octant of the sphere we shall calculate only those tangential waves which are distributed along three of the six planes, so

$$n'_{\text{tan}}(f_n) = \frac{n_{\text{tan}}(f_n)}{2} = \frac{\pi}{2c_0^2}Sf_n^2 \qquad 4.15$$

Finally to find the number of frequencies for all the oblique waves, the volume of the octant of the sphere must be divided by the volume of the elementary parallelepiped i.e.

$$n'_{\text{obl}}(f_n) = \frac{\pi f_n^3}{6} \div \frac{c_0^3}{8V} = \frac{4\pi}{3c_0^3}Vf_n^3 \qquad 4.16$$

The total number of all characteristic frequencies is found as the sum of expressions 4.14, 4.15 and 4.16:

$$n(f_n) = \frac{4\pi}{3c_0^3}Vf_n^3 + \frac{\pi}{4c_0^2}Sf_n^2 + \frac{L}{8c_0}f_n \qquad 4.17$$

To determine the density of the frequency spectrum, i.e. the number of natural frequencies in the interval from f_n to $f + df_n$ it is necessary to differentiate the last equation, as a result of which we get:

$$\frac{dn}{df_n} = 4\frac{\pi V}{c_0^3}f_n^2 + \frac{\pi S}{2c_0^2}f_n + \frac{L}{8c_0}$$

If the interval with which we are concerned is in the high frequency range, then the last two elements of the equation we have obtained will be small compared with the first and this equation can be rewritten as:

$$\frac{dn}{df_n} \approx 4\frac{\pi V}{c_0^3}f_n^2$$

from which the density of the spectrum (the number of characteristic frequencies) in a given interval Δf_n will be:

$$\Delta n = 4\frac{\pi V}{c_0^3}f_n^2\Delta f_n \qquad 4.18$$

By calculating from this last formula we can construct graphs of the variation of Δn with frequency for enclosures of different volumes (Fig. 4.5). It will be seen from them that:

1. Only for very small enclosures and particularly at low frequencies is the spectrum density of characteristic frequencies low. This causes the unevenness of the sound field in them.

128

2. Larger enclosures have a greater spectrum density even at low frequencies. At high frequencies this spectrum becomes almost continuous.

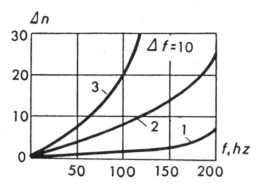

Fig. 4.5. The dependence of the number of characteristic frequencies △n on the frequency of the signal, for enclosure volumes V equal to 26, 135 and 400 m³ for curves 1, 2 and 3 respectively.

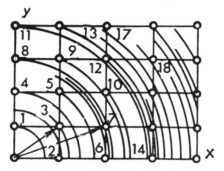

Fig. 4.6. The junction points and lines of characteristic frequencies of the frequency space of an enclosure for plane xy.

Fig. 4.6 shows a network of characteristic frequencies for tangential waves with junction points numbered in order of increasing frequency. As radius f increases, corresponding to an increase in frequency, the arcs of circles drawn through the junction points some closer together, indicating an increase in the density of the spectrum.

4.3 Decay of standing waves. Reverberation

Normal modes of vibration excited in the form of standing waves gradually decay, as can be seen from expression 4.7. This happens as a result of the loss of energy at the boundaries of the enclosure. In this case, in accordance with expression 2.22, the index of decay δ equals $-\dfrac{dE}{Edt}$, i.e. it is equal

9

to the ratio of density of energy lost in a unit of time to the total density of energy in the enclosure.

As the density of sound energy in the presence of absorption decreases in one second by a value equal to $E\alpha n$, then $\delta = -\dfrac{E\alpha n}{E} = -n\alpha.$

If we use the known relationship between the intensity of sound and sound pressure for a plane sound wave, we can write that

$$\alpha = \frac{I_{abs}}{I_{inc}} = \frac{p_n^2(x, y)}{2\rho_1 c_1} \cdot \frac{2\rho_0 c_0}{p_n^2(x,y,z)}$$

where $p_n(x, y, z)$ and $p_n(x, y)$ are mean values of pressure for sound waves falling on and passing through the bounding surfaces of the enclosure. These values are defined by equations 4.9 and 4.7; $\rho_1 c_1$ is the mean value of the specific acoustic impedance for the materials surfacing the walls of the enclosure.

If, in the equation written above for the coefficient of decay we insert the values of α and of n, derived by means of equation 2.15, the equation may be rewritten:

$$\delta = -n\alpha = -\frac{c_0 S}{4V} \cdot \frac{p_n^2(x, y)}{p_n^2(x, y, z)} \cdot \frac{\rho_0 c_0}{\rho_1 c_1} = \frac{c_0^2 \rho_0 S}{4 c_1 \rho_1 V} \cdot \left(\frac{p_n(x, y)}{p_n(x, y, z)}\right)^2 \quad 4.19$$

In accordance with equation 2.39 reverberation time can be expressed in terms of the coefficient of decay in the following way:

$$T = -\frac{6}{\delta \log_{10} e} \quad 4.20$$

If we insert in this expression the value of δ from equation 4.19 we get:

$$T = \frac{6}{\log_{10} e} \cdot \frac{4V}{c_0^2 \rho_0 S} \cdot \frac{c_1 \rho_1}{\left(\dfrac{p_n(x, y)}{p_n(x, y, z)}\right)^2}$$

or after inserting the numerical values of the constants e, c_0 and ρ_0

$$T = \frac{0\cdot04 V c_1 \rho_1}{S \left(\dfrac{p_n(x, y)}{p_n(x, y, z)}\right)^2} \quad 4.21$$

Thus reverberation time can be defined after formulae 4.9 and 4.7 have been used to find the amplitudes of sound pressure in waves falling on and passing through the surfaces of the enclosure.

Let us look more closely at the special case of axial waves distributed in the enclosure in direction x. This case can be met by equation 4.1 whose solution is represented in equation 4.2. However this solution can be found as:

$$p(x, t) = p_m \cosh\left[(\delta_x - j\omega)\frac{x}{c_0}\right] e^{(j\omega - \delta)t} \quad 4.22$$

and this meets the case in which the origin of coordinates is situated within the enclosure (at the bounding surfaces of the enclosure $x = l_x/2$).

The wave velocity of the particles corresponds to the sound pressure to be found by equation 4.22 this velocity is:

$$v(x, t) = \frac{p_m}{c_0 \rho_0} \sinh \left[(\delta_x - j\omega) \frac{x}{c_0} \right] e^{(j\omega - \delta)t} \qquad 4.23$$

From this the wave impedance of the medium is defined as:

$$R(x) = \frac{p(x, t)}{v(x, t)} = c_0 \rho_0 \coth \left[(\delta_x - j\omega) \frac{x}{c_0} \right] \qquad 4.24$$

Bearing in mind that at the boundaries of the enclosure (where $x = l_x/2$) the sound wave meets another medium with a specific acoustic impedance $c_1 \rho_1$, equation 4.24 for this boundary case can be written in the form:

$$R' = \frac{c_1 \rho_1}{c_0 \rho_0} = \coth \left[(\delta_x - j\omega) \frac{l_x}{2c_0} \right]$$

If the sound absorption of the material of the wall is large enough we can consider that $\delta \gg \omega$, and then approximately

$$R' = \coth \delta_x \frac{l_x}{2c_0} \qquad 4.25$$

For normal absorbent materials $R' \gg 1$, which corresponds to a small value of the argument of the hyperbolic cotangent, then if we set out coth x in a series, it is clear that we can limit ourselves to the first term, i.e. we can write that:

$$\coth x = \frac{1 + \dfrac{x^2}{2!} + \dfrac{x^4}{4!} + \cdots}{x + \dfrac{x^3}{3!} + \dfrac{x^5}{5!} + \cdots} \approx \frac{1}{x}$$

Taking this into account, expression 4.25 can be rewritten as

$$R' = \frac{2c_0}{\delta_x l_x}$$

from whence

$$\delta_x = \frac{2c_0}{R' l_x} \qquad 4.26$$

As can be seen from expression 5.16, where we have a high value of R' the coefficient of sound absorption of the material can be selected as approximately equal to $\alpha_\perp = 4/R'$. Knowing this and substituting the value of δ_x

from equation 4.26 into equation 4.20 this last can be put into the following form:

$$T_z = \frac{6}{\log e} \cdot \frac{4l_x}{c_0 \alpha} \cdot \frac{2l_y l_z}{2l_y l_z} = \frac{0.164V}{\alpha 2S_{yz}} \qquad 4.27$$

This formula is interesting primarily because it underlines the common ground between the wave and statistical theories which when considering the reverberation process result in identical final expressions. Moreover formulae 4.27 and 4.21 indicate real differences in a number of conclusions which arise from similar formulae in these two theories.

Unlike the conclusions of the statistical theory on the basis of formulae 4.26 and 4.27 we can make the following conclusions:

1. The group of axial modes formed between each pair of opposite surfaces of the enclosure has its own index of decay and its own group reverberation time.

2. Where axial vibrations are simultaneously present in all three directions the decay process is made up of separate processes proceeding at different speeds which does not result in conditions permitting a diffuse field in the enclosure.

3. As tangential and oblique vibrations have decay indices still more different from the decay indices of the axial vibrations, this too has an effect on the evenness of the field.

4.4 The group coefficient of sound absorption for waves of various types

In finding the solution of the wave equation describing the sound field in an enclosure in the form 4.5 or 4.7 or 4.9a, we supposed that loss of energy occurs at the bounding surfaces of the enclosure. This supposition was indicated in these solutions by the presence of a factor in the form $e^{-\delta_n t}$.

If we consider, as we did in solving equation 4.1, that the amplitude of sound pressure is represented as $A_n p(x, y, z)$, all these solutions can be written in a general form:

$$p(x, y, z, t) = A_n p(x, y, z) \, e^{(j\omega_n - \delta_n)t} \qquad 4.28$$

This supposition, particularly in the derivation of expression 4.4, made some approximations, as the condition that the maximum value of sound pressure occurs at the boundaries of the enclosure is accurate only when the walls are completely rigid, i.e. when they are totally non-absorbent. If any absorption is possible, then the sound pressure acting on the walls brings them, or the air in their pores, into a state of vibration, as a result of which, at the boundaries of the enclosure, (where $x = l_x$, $y = l_y$ or $z = l_z$), the value of $\cos k_x x$, $\cos k_y y$ or $\cos k_z z$ becomes progressively less than unity the further the ratio $\dfrac{p}{v_n} = \dfrac{1}{\sigma}$ differs from zero. In this last expression, v_n is the

132

normal component of particle velocity, and σ is the inverse of the specific mechanical impedance of the surface of the wall, known as the admittance.

Let us consider a case in which absorption is small, which is shown by the low value of admittance σ and of the index of decay δ, i.e. a case corresponding to expression 2.23, which was obtained by a statistical consideration of the decay process for low values of the absorption coefficient α.

This case is fitted by the equation:

$$E = E_0\, e^{-2\delta_n t} \qquad\qquad 4.29$$

which is derived from expression 4.28, if we consider that energy is proportional to the square of the sound pressure and that the amplitude of this pressure for each natural vibration having a frequency ω_n, changes exponentially.

If we equate the indices of the order of reflections 2.23 and 4.29, we can write that

$$2\delta_n = \frac{c_0 \alpha S}{4V}$$

from whence

$$\alpha = 2\delta_n \cdot \frac{4V}{c_0 S} \qquad\qquad 4.30$$

The value δ_n can be calculated if in the initial equation 4.9 we substitute the solution 4.28 and take into account conditions at the boundaries. It will be defined by the equation:

$$2\delta_n = \rho_0 c_0^2 \frac{\displaystyle\int_S \sigma A_n^2 p^2(x, y, z)\, \mathrm{d}S}{\displaystyle\int_V A_n^2 p^2(x, y, z)\, \mathrm{d}V} \qquad\qquad 4.31$$

In order to find the connection between the mean coefficient of absorption defined by the statistical theory, and the admittance which characterizes the properties of the surfaces of the enclosure according to the wave theory, let us consider the rectangular enclosure (Fig. 4.7), each pair of

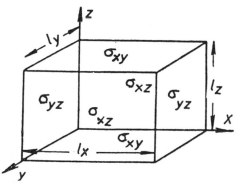

Fig. 4.7. Plan of the disposition of the surfaces of an enclosure having varying degrees of admittance σ.

133

parallel surfaces of which we will assume, for the sake of simplicity, to have the same material and the same admittance. The upper and lower surfaces of this enclosure have an admittance equal to σ_{xy}, the side surfaces σ_{xz} and the end surfaces σ_{yz}.

Let us first find the values of the surface and volume integrals which come into expression 4.31 for axial waves distributed in the x direction. As for this case

$$A_n p(x, y, z) = A_n \cos \frac{n_x \pi x}{l_x}$$

the first of these will be calculated as

$$\int_S \sigma A_n^2 p^2(x, y, z) \, dx \, dy = A_n^2 \left(\sigma_{xy} \int_0^{l_x} \int_0^{l_y} \cos^2 \frac{n_x \pi x}{l_x} \, dx \, dy + \right.$$

$$\left. + \sigma_{xz} \int_0^{l_x} \int_0^{l_z} \cos \frac{n_x \pi x}{l_x} \, dx \, dz + \sigma_{yz} \int_0^{l_y} \int_0^{l_z} dy \, dz \right)$$

$$= A_n^2 \left(\sigma_{xy} \frac{l_x l_y}{2} + \sigma_{xz} \frac{l_x l_z}{2} + \sigma_{yz} l_y l_z \right)$$

$$= \tfrac{1}{2} A_n^2 (\sigma_{xy} S_{xy} + \sigma_{xz} S_{xz} + 2\sigma_{yz} S_{yz}) \qquad 4.32$$

and the second as

$$\int_V A_n^2 p^2(x, y, z) \, dx \, dy \, dz = \int_0^{l_x} \int_0^{l_y} \int_0^{l_z} A_n^2 \cos^2 \frac{n_x \pi x}{l_x} \, dx \, dy \, dz$$

$$= A_n^2 \frac{l_x l_y l_z}{2} = \tfrac{1}{2} A_n^2 V \qquad 4.33$$

If we substitute the values of the integrals into expression 4.31, we obtain:

$$2\delta_n = \rho_0 c_0^2 \frac{\sigma_{xy} S_{xy} + \sigma_{xz} S_{xz} + 2\sigma_{yz} S_{yz}}{V} \qquad 4.34$$

After the substitution in expression 4.30 of the value of $2\delta_n$ from expression 4.34, we have:

$$\alpha_x = 4\rho_0 c_0 \frac{\sigma_{xy} S_{xy} + \sigma_{xz} S_{xz} + 2\sigma_{yz} S_{yz}}{S} \qquad 4.35$$

or for the case in which all surfaces of the enclosure have equal admittance, ($\sigma_{xy} = \sigma_{xz} = \sigma_{yz} = \sigma$):

$$\alpha_x = 4\rho_0 c_0 \sigma \frac{S_{xy} + S_{xz} + 2S_{yz}}{S} = 4\rho_0 c_0 \sigma \left(1 - \frac{S_{xy} + S_{xz}}{S} \right) \qquad 4.36$$

The coefficient of absorption of axial waves distributed in directions y or z is expressed in an analogous way, the only difference being that the suffixes of S in the fraction in the right hand side of expression 4.36 are replaced by the two which include the direction of wave propagation, (y or z).

134

For waves of the tangential type distributed in the plane xy:

$$A_n p(x, y, z) = A_n \cos \frac{n_x \pi x}{l_x} \cdot \cos \frac{n_y \pi y}{l_y}$$

then, as can be calculated, the surface integral for this type of wave is represented by the equation:

$$\int_S \sigma A_n^2 p^2(x, y, z) \, dx \, dy = \tfrac{1}{4} A_n^2 (\sigma_{xy} S_{xy} + 2\sigma_{xz} S_{xz} + 2\sigma_{yz} S_{yz}) \qquad 4.37$$

and the volume integral by:

$$\int_V A_n^2 p^2(x, y, z) \, dx \, dy \, dz = \tfrac{1}{4} A_n^2 V \qquad 4.38$$

The group coefficient of absorption for waves of this type, from equations 4.31 and 4.30, is:

$$\alpha_{xy} = 4p_0 c_0 \frac{\sigma_{xy} S_{xy} + 2\sigma_{xz} S_{xz} + 2\sigma_{yz} S_{yz}}{S} \qquad 4.39$$

If the walls of the enclosure are of equal admittance,

$$\sigma_{xy} = 4p_0 c_0 \sigma \left(1 - \frac{S_{xy}}{S}\right) \qquad 4.40$$

It is clear that for tangential waves propagated in plane xz or yz, the group coefficient of absorption is expressed by an equation similar to 4.39 or 4.40, in which the indices are altered. Particularly, in expression 4.40 the index of S, (xy), must be replaced by an index xz or yz.

By analogy with expressions 4.35 and 4.39 the group coefficient of absorption for oblique waves can be written without more ado as

$$\alpha_{xyz} = 4p_0 c_0 \frac{2\sigma_{xy} S_{xy} + 2\sigma_{xz} S_{xz} + 2\sigma_{yz} S_{yz}}{S} = 8p_0 c_0 \frac{\sum \sigma_k S_k}{S} \qquad 4.41$$

or by analogy with equations 4.36 and 4.40 for surfaces with the same admittance as:

$$\alpha_{xyz} = 8p_0 c_0 \sigma \qquad 4.42$$

If we compare expressions 4.35, 4.39 and 4.41, we note that only for oblique waves are the roles of all surfaces of the enclosure of equal importance in the formation of sound absorption within the enclosure. For axial and tangential waves only those surfaces of the enclosure on which the waves impinge have a real importance. Those surfaces which the waves graze, e.g. S_{xy} and S_{xz} for axial waves distributed in direction x, and S_{xy} for tangential waves distributed in plane xy, are only half as effective as other surfaces in their contribution to the total absorption, as can be seen from expressions 4.35 and 4.39.

The principal case in which all surfaces of the enclosure play similar roles in the decay processes is that of a diffuse field, as considered in the statistical theory.

135

As a result of the fact that progressively fewer surfaces of the enclosure take part in the effective decay of axial and tangential waves, the group coefficient of absorption for these waves decreases. The result is that the process of energy decay in the enclosure falls into three stages if the areas of pairs of parallel surfaces and their corresponding admittances are of the same order.

The first stage shows a sharp fall in the decay curve as a result of the rapid absorption of the oblique waves; the second stage is less steep, as the group coefficient of absorption for tangential waves is less than for oblique waves; and finally we have the third stage, corresponding to the still slower decay of axial waves.

If the areas of the pairs of parallel surfaces and their admittances differ significantly from one another, the number of stages of decay increases, because under these conditions each of the three groups of tangential and axial waves has its own speed of decay. In this case the actual decay curve can be replaced by an exponential curve as predicted by the statistical theory.

As differences in the rates of decay of separate groups of tangential or axial waves is connected with differences in areas and values of admittance for the pairs of opposite surfaces in the enclosure, the effect that the shape of the enclosure and the distribution of sound absorbent materials in it have on the progress of the process of decay of sound may be predicted.

If the surfaces of the enclosure are treated with materials having similar admittance, as can be seen from equations 4.36, 4.40 and 4.42, the difference between the group coefficients of absorption for waves of various types decreases, which results in a smoothing out of the decay curve of characteristic frequencies of the enclosure and brings it closer to the exponential curve of the statistical theory. This approximation becomes more noticeable at high frequencies, where the main part of the decaying energy is caused by oblique waves which have a comparatively high and even density of spectrum.

4.5 Practical value of the wave theory

The basic propositions of the wave theory of the acoustics of an enclosure developed by P. Morse and R. Bolt reveal quite fully the physical reality of sound processes in an enclosure. However, these propositions are so general that they do not permit a simple solution of concrete problems in acoustic design by the direct application of the wave theory.

The wave theory yields better practical results in the consideration of problems connected with the description (*a*) of the influence of absorption at the boundaries of the enclosure on different types of waves arising in the enclosure, (*b*) of the character of the sound field in a steady state condition, and (*c*) of the character of the decay of normal modes of vibration.

The wave theory shows how the sound field changes in a rectangular enclosure with rigid bounding surfaces of similar materials, and facilitates the calculation of some of the values which define this field. In particular

136

it can predict the sound pressure and the air particle velocity in various parts of the enclosure at any given moment. However as a result of the fact that in actual enclosures the surfaces are normally covered in dissimilar materials with fairly high absorption coefficients, the solution of specific problems by means of the wave theory is also extremely complex.

The sound field formed in an enclosure can be defined by the spectrum of natural frequencies of the enclosure.

If, on the basis of formula 4.10, we construct such a spectrum for an enclosure in which l_x, l_y, l_z are respectively equal to 6, 4 and 3 metres, it will be as shown in Fig. 4.8. It follows from this figure that in the low frequency

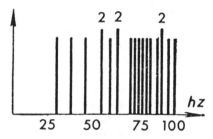

Fig. 4.8. Frequency spectrum of a room with dimensions 6 × 4 × 3 m. The figures 2 indicate duplicated frequencies.

area the characteristic frequencies are well separated. Such gaps between characteristic frequencies are undesirable because individual components of a complex signal which fall in a gap are suppressed by others, which, as they correspond with the characteristic frequencies, are amplified by resonance.

As the frequency increases, the density of the spectrum also increases, becoming continuous at frequencies which become lower as the volume of the enclosure increases. As the spectrum becomes more dense, the sound field in the enclosure becomes more uniform.

The mean interval between neighbouring characteristic frequencies within the limits of a band Δf_n can be determined from expression 4.18 in the form:

$$f_\sigma = \frac{\Delta f_n}{\Delta n} = \frac{c_0{}^3}{4\pi V f_n{}^2} = \frac{3.10^6}{V f_n{}^2} \qquad 4.43$$

This expression shows that in large enclosures, even for comparatively low frequencies, the spectrum density of characteristic frequencies is sufficient to prevent the conditions indicated above which would create irregularity of response.

One can also judge the quality of an enclosure by the nature of the decay curve after the sound source has been shut off. If the various normal modes of vibration decay in the enclosure at the same speed, the process of their common decay is exponential, corresponding to a random sound field.

The general characterization of the sound field in an enclosure from these three points of view may be said to exhaust the practical possibilities of the wave theory. However, in analysing its value, we should not overlook the fact that, as P. Morse and R. Bolt[51] put it paradoxically: 'the practical role of wave acoustics is that it can indicate how to design an enclosure for which geometrical acoustics and statistical acoustics are valid, and in which there is no need of wave acoustics.'

The wave theory shows that the spectrum density of characteristic frequencies is small only for regularly shaped enclosures of small volume, which preordains that the sound field within them will be uneven. In practice, enclosures used for sound transmissions are often made with their dimensions not in simple ratio to each other and are so large that an even sound field can be ensured over the necessary frequency range. The evenness of this field is an essential condition for the accuracy of the statistical theory of acoustics of an enclosure. Consequently the wave theory confirms the possibility of using the statistical theory to evaluate acoustic conditions in large enclosures, particularly where they are of irregular form.

The wave theory, when it considers enclosures with parallel, materially uniform and rigid walls, indicates that although each sound wave in such an enclosure decays exponentially, the general process of decay, as a result of the varying decay speeds of a number of components, does not follow the exponential law. In practice, the majority of enclosures do not have a regular shape, and have a comparatively high absorption coefficient with an uneven distribution of sound absorptive materials. In such enclosures the movement of sound waves is more disorderly and the decay rates of individual characteristic frequencies even out; this also confirms the possibility of using the statistical theory of acoustics of an enclosure for practical calculations.

In considering the value of the wave theory, we must note the fact that its authors, as was shown by L. Brekhovskikh,[9] come to the incorrect conclusion that only by use of the wave theory is it possible to discover the reason for variations in the steepness of the decay curve and for variations in the absorptive properties of materials when their position in the enclosure is altered. It appears that these phenomena are explained just as fully by using the geometric theory, which shows that field analysis carried out by the construction of image sources gives accurate results for totally reflecting surfaces. When the coefficient of reflection β is less than unity, but does not depend on the angle of incidence of the wave, correct results are obtained if the energy radiated by the image source is considered proportional to β. Finally, if the coefficient of reflection depends on the angle of incidence of the sound wave, then it must be reckoned that the image source has a defined directional characteristic, the calculation of which leads to true results.

5 The Sound Absorbing Capacity of Materials and Constructions

5.1 The absorption of sound energy by materials. Classification of sound absorbing materials

The acoustic quality of an enclosure is to a large degree determined by the sound absorbing capacity of the materials within the enclosure, and in the first place of the materials at the boundaries. The sound absorbing capacity, caused by losses in sound energy over the surfaces or within the thickness of the material, in turn depends on its structure, density, elasticity, and other physical properties.

The connection between the sound absorbing capacity and the physical properties of the material is easily seen if we examine the mechanism of sound energy loss.

Fig. 5.1 shows that only part of the energy falling on a wall surface is reflected from it back into the enclosure (ε_{ref}). To the reflected energy must be added the energy radiated by the wall as a result of its elastic vibrations (ε_{el}).

The remaining energy is partly transmitted into the adjoining enclosure through pores in the material and by elastic or bending vibrations (ε_{tr}), and partly propagated by longitudinal vibrations in the wall and parallel to its surface, passing out of the enclosure through other walls or through the foundation (ε_b). The remainder is lost in the material itself. Energy may be lost in the material basically as a result of,

1. Friction at the walls of the pores (ε_{fr}).
2. Heat conductivity of the material (ε_h).

3. Dissimilar deformation of its elements (ε_{def}).
4. Residual deformation (ε_{res}).

As the air vibrates within the pores of the material, the friction of particles at the walls of the pores is of great importance. Considerable friction results in the conversion to heat of part of the sound energy which has penetrated into the thickness of the material. Losses from friction can be expressed in terms of the acoustic impedance (impedance to the passage of

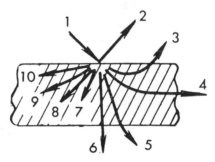

Fig. 5.1. Diagram of the propagation of sound energy into a material.

air) which is the ratio of the pressure difference between the opposite sides of the material (Δp) to the velocity of air passing through the layer (v) i.e.

$$R_a = \frac{\Delta p}{v}$$

The air within the pores of the material is periodically compressed under the action of the sound wave. This causes heating of the air. The heat is transferred to the walls of the pores, and the pressure of the cooled air decreases. As a result of this reduction in pressure part of the sound energy is lost.

Variations in the elastic properties of the material and in the thickness of the walls of its pores results in thinner sections of the walls heating up more strongly as a result of greater deformation under pressure. If, following compression of the air in the pores, its temperature falls before the following rarefaction a reduction in pressure results and consequently a loss of energy.

When a periodic force is acting on the material, the rise in pressure, which causes an interaction between the particles, lags behind the deformation. The pressure depends not only on the degree of deformation at the given moment, but also on its previous value, which is a result of irreversible deformation.

This process is shown in the hysteresis loop in Fig. 5.2. The value of deformation q retained by the material after a reduction in pressure F to zero, is called the residual deformation. It can be seen from Fig. 5.2 that the area of the loop of elastic hysteresis expresses the work irreversibly transformed into heat during the period of one vibration. This work,

140

referred to as losses as a result of residual deformation, can be expressed by an approximate equation, similar to the corresponding equation for magnetic hysteresis:

$$A \approx kq^2_{max}$$

Thus residual deformation is the internal property of materials which absorbs part of the energy falling on them. Losses resulting from residual deformation can be presented in a general form as the difference between

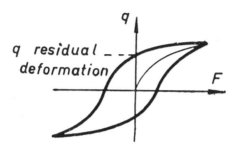

Fig. 5.2. Curve of residual deformation.

the energy exerted by the wave in deforming an insufficiently elastic material and the energy returned by the material when its former shape is restored by removal of the sound pressure.

Depending on the physical properties of the material, and on the predominance in the material of one or other form of loss, absorbing materials may be divided into a number of groups:

1. Porous materials, the sound absorption of which is caused by viscous friction in pores with rigid, non-elastic walls. Losses in sound energy are a result not only of friction but also of the heat conductivity of the material. In this group are included pumice-concrete, consisting of porous fillers such as pumice or slag bound by cement, or acoustolite, which is a thin fabric cemented to a clay tile, etc.

2. Porous elastic materials, the walls of whose pores form an elastic skeleton. The deformation of the individual elements of such a skeleton varies because of the variations in their thickness. Absorption of sound energy in these materials results from heat exchange of energy caused by friction with the walls of the pores, and from their deformation. In this group we include ZP–5, which is a solidified foam of a urea-formadehyde resin; insulit and arborite, made from waste products of the paper industry; fibro-acoustite, pressed from sawdust; and mineral wool, fibreglass, felt, etc.

3. Vibratory systems, the sound absorption of which results from deformation when the whole system vibrates as the result of the action of a sound wave. Losses of energy are caused by internal friction associated with the elastic vibrations of the whole material. Examples of such systems are plywood panels, so-called hemi-cylindrical diffusers, Bekesha screens, etc.

141

4. Porous-vibratory (resonance) systems, which absorb sound energy as a result of friction in the pores of the air-permeable parts of the system, and of internal friction in the parts impermeable to air, which vibrate under the action of the sound wave. Examples of such systems are porous fabrics, oilcloth with apertures, and rigid perforated sheets of steel or of plywood.

Losses of sound energy in materials may be characterized by the sound absorption coefficient α. However, some of the losses, particularly those resulting from the viscosity and heat conductivity of the medium, have values which vary with the frequency of the sound vibrations. Moreover, it is clear that losses are determined to a significant degree by the angle at which the sound wave impinges on the surface of the absorbent material.

Thus the absorption capacity of a material may be defined by:

1. The absorption coefficient at a mean frequency α.
2. The dependence of the absorption coefficient on frequency.
3. Its dependence on the angle of incidence of the sound wave.

5.2 The dependence of the absorption coefficient on the angle of incidence of the sound wave

Let us suppose that a plane sound wave falls at an angle θ on the surface of a material of specific acoustic impedance $c_1\rho_1$. (Fig. 5.3).

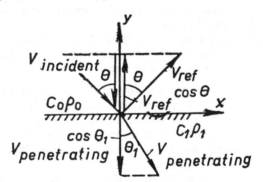

Fig. 5.3. Graph to determine the normal components of particle velocity for a wave impinging at angle θ.

The total sound pressure of the incident and reflected wave at the boundary between two media should be balanced by the pressure of the wave which passes through:

$$p_{\text{inc}} + p_{\text{reb}} = p_{\text{tr}} \qquad 5.1$$

The vertical components of the particle velocity of the incident, reflected and transmitted waves will be equal respectively to $v_{\text{inc}} \cos \theta$, $v_{\text{ref}} \cos \theta$, and $v_{\text{tr}} \cos \theta_1$. As the first and third components have the same direction, and the second is opposite to them, the condition of balance is expressed in the form:

$$v_{\text{inc}} \cos \theta - v_{\text{ref}} \cos \theta = v_{\text{tr}} \cos \theta_1$$

142

whence

$$v_{ref} = \frac{v_{inc} \cos \theta - v_{tr} \cos \theta_1}{\cos \theta}$$ 5.2

For a plane wave sound pressure is represented as the product of the particle velocity and the wave impedance; therefore, equation 5.1 can be rewritten in the form:

$$v_{inc} c_0 \rho_0 + v_{ref} c_0 \rho_0 = v_{tr} c_1 \rho_1$$

If we replace v_{ref} by its corresponding value from expression 5.2, we get

$$v_{inc} c_0 \rho_0 + \frac{v_{inc} \cos \theta - v_{tr} \cos \theta_1}{\cos \theta} c_0 \rho_0 = v_{tr} c_1 \rho_1$$

After rearrangement, we can write:

$$2 v_{inc} c_0 \rho_0 \cos \theta = v_{tr}(c_1 \rho_1 \cos \theta + c_0 \rho_0 \cos \theta_1)$$

If we divide both parts of the equation by v_{inc}, we find that the ratio of particle velocities in the transmitted and the incident waves is:

$$\frac{v_{tr}}{v_{inc}} = \frac{2 c_0 \rho_0 \cos \theta}{c_1 \rho_1 \cos \theta + c_0 \rho_0 \cos \theta_1}$$

For a plane wave $I = v^2 c \rho$, and so according to equation 2.1 we can write the coefficient of absorption for a wave with an angle of incidence θ as:

$$\alpha_\theta = \frac{I_{tr}}{I_{inc}} = \frac{v_{tr}^2 c_1 \rho_1}{v_{inc}^2 c_0 \rho_0} = \frac{4 c_0 \rho_0 c_1 \rho_1 \cos^2 \theta}{(c_1 \rho_1 \cos \theta + c_0 \rho_0 \cos \theta_1)^2}$$ 5.3

If we suppose that the sound wave is distributed in the absorbing material only along its thickness (e.g. in its longitudinal pores), we put $\theta_1 = 0$, and;

$$\alpha_\theta = \frac{4 c_0 \rho_0 c_1 \rho_1 \cos^2 \theta}{(c_1 \rho_1 \cos \theta + c_0 \rho_0)^2}$$ 5.4

or, adding to the numerator and subtracting from it $(c_0 \rho_0 \cos \theta)^2$ and $(c_1 \rho_1 \cos \theta)^2$,

$$\alpha_\theta = 1 - \frac{(c_1 \rho_1 - c_0 \rho_0)^2 \cos^2 \theta}{(c_1 \rho_1 \cos \theta + c_0 \rho_0)^2}$$ 5.5

and finally, when $\theta = 0$, i.e. when the direction of the sound ray coincides with the vertical to the surface of the material:

$$\alpha_\perp = \frac{4 c_0 \rho_0 c_1 \rho_1}{(c_1 \rho_1 + c_0 \rho_0)^2}$$ 5.6

or

$$\alpha_\perp = 1 - \left(\frac{c_1 \rho_1 - c_0 \rho_0}{c_1 \rho_1 + c_0 \rho_0}\right)^2$$ 5.7

If graphs $\alpha_\theta = f_\theta$ (Fig. 5.4) are now constructed according to formula 5.4, we reach the following conclusions from them.

1. The coefficient of sound absorption is greatest for the material for which the acoustic impedance is closest to the acoustic impedance of the air.

2. For every material, an increase in the angle of incidence of the sound wave results in a reduction in the coefficient of absorption, and this reduction becomes progressively more rapid as the angle of incidence approaches 90°.

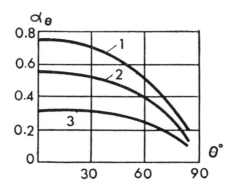

Fig. 5.4. The dependence of the coefficient of sound absorption α on the angle of incidence of the sound wave for materials with $c_1\rho_1 = 2\,c_0\rho_0$ (1), $c_1\rho_1 = 5c_0\rho_0$ (2) $c_1\rho_1 = 10\,c_0\rho_0$ (3).

3. The speed at which the coefficient of absorption decreases as the angle of incidence increases is greatest for materials of low acoustic impedance.

Thus the coefficient of sound absorption is not a constant but depends upon the angle of incidence of the sound wave.

However, in the treatment by the statistical theory, the coefficient of sound absorption was taken to mean the coefficient which was obtained when all angles of incidence of the wave were equally probable. Consequently, for practical purposes, the so-called coefficient of diffuse sound absorption is important, as it defines the reverberation time in the enclosure.

As can be seen from formula 2.12, the probability that a sound wave is incident at angle $d\theta$ on a unit of area dS in unit time ($dt = 1$) is:

$$\omega_3 = \frac{c_0}{4v} \sin 2\theta d\theta$$

If the reserve of energy in the enclosure is ε_0, then on to this area will fall energy:

$$\varepsilon_{\text{inc}} = \varepsilon_0 \frac{c_0}{4v} \sin 2\theta d\theta$$

144

Absorbed energy in this case is

$$\varepsilon_{abs} = \alpha_\theta \varepsilon_{inc} = \frac{\alpha_\theta c_0 \varepsilon_0}{4v} \sin 2\theta d\theta$$

and for all possible angles of incidence of the sound wave:

$$\varepsilon_{abs} = \frac{c_0 \varepsilon_0}{4v} \int_0^{\pi/2} \alpha_\theta \sin 2\theta d\theta \qquad 5.8$$

On the other hand, it is plain that the absorbed energy is $n\alpha$ times less than the incident energy, where n is the mean number of reflections in 1 sec., as defined by formula 2.15. From this formula, all absorbed energy:

$$\varepsilon'_{abs} = n\alpha\varepsilon_0 = \frac{c_0 S}{4V} \alpha\varepsilon_0$$

while the energy absorbed by a unit of area is

$$\varepsilon_{abs} = \frac{\varepsilon'_{abs}}{S} = \frac{c_0 \varepsilon_0}{4V} \alpha \qquad 5.9$$

If we equate the right hand sides of expressions 5.8 and 5.9, we get

$$\frac{c_0 \varepsilon_0}{4V} \alpha = \frac{c_0 \varepsilon_0}{4V} \int_0^{\pi/2} \alpha_\theta \sin 2\theta d\theta$$

or finally

$$\alpha = \int_0^{\pi/2} \alpha_\theta \sin 2\theta d\theta \qquad 5.10$$

Thus, to define the diffuse coefficient of sound absorption α it is necessary to know the coefficient of absorption α_θ for a directional beam of sound rays at an angle of incidence θ.

5.3 The acoustic impedance of a material and its connection with the coefficient of absorption

The coefficient of absorption for a diffuse field can be found by formula 5.10 from measurements on plane waves incident on the given material at a series of angles, or by direct measurement in a diffuse field. In this case the coefficient of absorption is defined by the physical properties of the material, and can serve as the acoustic characteristic of the material.

But if the sound field in the enclosure is not diffuse, then, as Morse and Bolt[51] show, referring to the work of a number of experimenters, the acoustic properties of the bounding surfaces of the enclosure cannot be described by using the random-incidence coefficient. This happens because where the field is non-diffuse, the dependence of the coefficient of absorption on the physical properties of the material attains different values depending on its size and the nature of the action of the sound wave on the materials placed in various parts of the enclosure.

145

In fact, as was shown in paragraph 4.4, where the sound field is non-diffuse, the decay time of oblique, tangential and axial modes is always different, as also are the group coefficients of sound absorption for each of these types of mode. And as first oblique and then tangential modes decay away, and the field becomes more uneven, the influence of absorbent material on the decay process changes in accordance with its position and with the form and dimensions of the enclosure.

To evaluate the absorbing properties of materials placed at the boundaries of an enclosure, it is more convenient to use what is known as the normal acoustic impedance of the material. This impedance is represented as a ratio between the sound pressure and the normal component of the velocity of air particles, both measured at the surface of the material.

The normal component of velocity is governed by the fact that the air particles either move through the pores or move together with the vibrations of the material caused by the action of the wave. From this we can conclude that the acoustic impedance consists of two parts: the active component R, representing the impedance of friction of particles against the pore walls, and the reactance component X, expressing the inertial impedance of the material brought into a state of vibration by the action of the sound wave.

Thus the acoustic impedance of a material may be written:

$$Z = R + jX \qquad\qquad 5.11$$

The acoustic impedance of a material is a more general criterion than the coefficient of absorption for the evaluation of losses of sound energy at the boundaries of an enclosure, because with its help we can find the index of decay both for enclosures in which the field is even and for which the formulae of the statistical theory hold good, and also for enclosures with uneven sound fields for which these formulae fail.

However, inasmuch as both the coefficient of absorption and the acoustic impedance correctly describe the sound absorbing properties of the bounding surfaces of an enclosure if the sound field within it is diffuse, there is reason to believe that in this particular case there is a constant value relationship between α and Z.

Similarly, remembering that for a plane wave, $Z = c_1\rho_1$, we observe that formulae 5.7 and 5.5 connect the absorption coefficient with the acoustic impedance for a certain particular case. By inserting the value of Z from equation 5.11 these formulae can be re-written in the form:

$$\alpha_\perp = 1 - \left|\frac{(R+jX) - c_0\rho_0}{(R+jX) + c_0\rho_0}\right|^2 = 1 - \left|\frac{(R - c_0\rho_0) + jX}{(R + c_0\rho_0) + jX}\right|^2 \qquad 5.12$$

$$\alpha_\theta = 1 - \left|\frac{[(R+jX) - c_0\rho_0]\cos\theta}{(R+jX)\cos\theta + c_0\rho_0}\right|^2 = 1 - \left|\frac{[R - c_0\rho_0] + jX]\cos\theta}{(R\cos\theta + c_0\rho_0) + jX\cos\theta}\right|^2$$

$$5.13$$

As we are interested in an absolute value for the coefficient of sound absorption, we can find it from the modulus of each of the complex expressions 5.12 and 5.13.

146

For vertical incidence of a sound ray, remembering that the square of the modulus is equal to the product of the conjugate complex expressions, formula 5.12 becomes:

$$\alpha_\perp = 1 - \frac{[(R - c_0\rho_0) + jX][(R - c_0\rho_0) - jX]}{[(R + c_0\rho_0) + jX][(R + c_0\rho_0) - jX]} = 1 - \frac{(R - c_0\rho_0)^2 + X^2}{(R + c_0\rho_0)^2 + X^2}$$

or, reducing to a common denominator:

$$\alpha_\perp = \frac{4Rc_0\rho_0}{(R + c_0\rho_0)^2 + X^2} \qquad 5.14$$

By a similar method we find from 5.13 that the modulus of the coefficient of sound absorption at an angle of incidence θ, is:

$$\alpha_\theta = \frac{4Rc_0\rho_0 \cos\theta}{(R\cos\theta + c_0\rho_0)^2 + X^2\cos^2\theta} \qquad 5.15$$

If we express the sound absorption coefficient in terms of the impedance normalized by dividing by $c_0\rho_0$, i.e. putting;

$$R' = \frac{R}{c_0\rho_0} \quad \text{and} \quad X' = \frac{X}{c_0\rho_0}$$

the above expressions may be rewritten as

$$\alpha_\perp = \frac{4R'}{(R' + 1)^2 + X'^2} \qquad 5.16$$

$$\alpha_\theta = \frac{4R'\cos\theta}{(R'\cos\theta + 1)^2 + X'^2\cos^2\theta} \qquad 5.17$$

Rearranging equation 5.16 gives:

$$R'^2 + 2R' + 1 + X'^2 = \frac{4R'}{\alpha_\perp}$$

whence

$$R'^2 - 2R'\frac{2 - \alpha_\perp}{\alpha_\perp} + 1 + X'^2 = 0$$

Adding and subtracting the fraction $\left(\frac{2 - \alpha_\perp}{\alpha_\perp}\right)^2$ we get:

$$\left\{R'^2 - 2R'\frac{2 - \alpha_\perp}{\alpha_\perp} + \left(\frac{2 - \alpha_\perp}{\alpha_\perp}\right)^2\right\} - \left(\frac{2 - \alpha_\perp}{\alpha_\perp}\right)^2 + 1 + X'^2 = 0$$

Expressing the first three terms as a square and bringing the next two terms to a common denominator, we can write:

$$\left((R' - \frac{2 - \alpha_\perp}{\alpha_\perp}\right)^2 - \frac{4(1 - \alpha_\perp)}{\alpha_\perp^2} + X'^2 = 0$$

or

$$\left(R' - \frac{2 - \alpha_\perp}{\alpha_\perp}\right)^2 + X'^2 = \frac{4(1 - \alpha_\perp)}{\alpha_\perp^2} \qquad 5.18$$

147

This expression is the equation for a circle with a radius $\dfrac{2\sqrt{1-\alpha_\perp}}{\alpha_\perp}$, the centre of which is displaced along axis R' at $\dfrac{2-\alpha_\perp}{\alpha_\perp}$. If we give a series of values to $\alpha_\perp$, we can construct a family of circles (Fig. 5.5) which establish the connection between the coefficient of sound absorption at normal incidence and the components of acoustic impedance of the given material.

The connection between the diffuse coefficient of absorption and the acoustic impedance can be established by inserting in expression 5.10 the

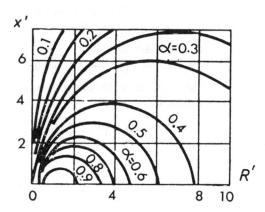

Fig. 5.5. Curves of equal absorption for a vertical impact of sound.

value of α_θ, given by equation 5.17 and taking into account the phase angle ϕ, between the impedance vector and the resistance axis. In this case

$$\alpha = \frac{1}{\pi} \int_0^{2\pi} d\phi \int_0^{\pi/2} \frac{4R'\cos\theta}{(R'\cos\theta + 1)^2 + X'^2 \cos^2\theta} \cdot \sin 2\theta \, d\theta \qquad 5.19$$

Calculations from formula 5.19 result in the following expression for the diffuse coefficient of sound absorption where $Z' \to \infty$:

$$\alpha = \frac{8\cos\phi}{Z'} - \frac{16\cos^2\phi}{Z'^2} \ln Z' + \frac{\phi}{Z} \cdot \frac{\cos 2\phi \cdot \cos\phi}{\sin\phi} \qquad 5.20$$

where $Z' \to 0$:

$$\alpha = \frac{8}{3} Z' \cos\phi - 4 Z'^2 \cos^2\phi \qquad 5.21$$

The curves in Fig. 5.6, drawn from the formulae 5.20 and 5.21, enable the value of the diffuse coefficient of sound absorption to be determined if the acoustic impedance of the material is known. Such a method of determining α is more accurate than methods of direct measurement. This is because the use of non-diffuse fields for measurements, introduces a significant error.

148

The curves in Figs. 5.5 and 5.6 show that:

1. The maximum coefficient of absorption is possessed by those materials for which R' is close to unity, i.e. the characteristic impedance of which is close to the that of air.

2. The coefficient of sound absorption falls as the active or the reactive components of acoustic impedance increase.

3. An increase in the elastic and the inertial impedance (see Fig. 5.5) or an increase in the phase angle (see Fig. 5.6) causes a more rapid reduction of the coefficient of absorption than does an increase in the resistive component of impedance.

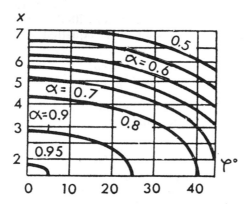

Fig. 5.6. *Curves of equal diffuse absorption for different phase angles of acoustic impedance.*

5.4 Absorption of sound by porous materials

Bearing in mind that the absorption coefficient is most accurately expressed through acoustic impedance, let us determine how the impedance of a rigid porous material depends on the basic parameters of the material.

Let a plane wave be propagated in the air-filled pores of a porous material with rigid walls. The process of absorption of energy within the thickness of the material may be analysed with the help of a wave equation, applied to the vibrations of the air within the material.

To obtain this wave equation we must solve simultaneous equations of continuity, state and motion which describe the changes of particle velocity v, of the density ρ of the medium and of sound pressure.

The equation of continuity for free air has the form:

$$-\frac{\partial v}{\partial x} = \frac{1}{\rho}\frac{\partial \rho}{\partial t} \qquad 5.22$$

The volume of air in a porous material taking part in the vibratory process is $\dfrac{V_{mat}}{V_{pore}}$ times smaller than for free air, and therefore, in order to keep the

149

value $\dfrac{\partial v}{\partial x}$ in 5.22 constant, we must divide the right hand side of 5.22 by $\dfrac{V_{\text{mat}}}{V_{\text{pore}}}$. In this case the equation of continuity takes on the form:

$$-\frac{\partial v}{\partial x} = \frac{P}{\rho_0} \frac{\partial p}{\partial t} \qquad 5.23$$

where $P = \dfrac{V_{\text{pore}}}{V_{\text{mat}}}$ and is known as the porosity of the material, and V_{mat} is the volume of the whole material and V_{pore} is the volume of the air contained in its pores.

We will consider that the volume deformation of the air in the pores takes place too rapidly for any heat transfer with the material of the skeleton, i.e. adiabatically. Under this condition the equation of state, as for free air, may be written in the form:

$$\frac{\partial p}{\partial t} = \frac{K}{\rho_0} \frac{\partial \rho}{\partial t} \qquad 5.24$$

where K is the coefficient of volume elasticity of the air.

It follows from this equation that

$$K = \rho_0 \frac{\partial p}{\partial \rho} \qquad 5.25$$

If deformation of the air in the pores takes place slowly enough to allow complete heat exchange with the skeleton (isothermal process), the coefficient of volume elasticity of air has a different numerical value from that for the adiabatic process. This value is in both cases always substantial.

In a number of practical cases (see Chapter 5.1) heat transfer is only partial. During compression the air heats up, and having given up part of the heat to the skeleton, remains warmer than the skeleton. Under rarefaction, on the other hand, there is a partial heat transfer from the skeleton to the air. As warmer air at the same pressure has a smaller density than cold air, this causes a delay in changes of pressure relative to changes in density, i.e. a phase shift between them.

Thus, in considering the most general case the coefficient of elasticity, expressed by the ratio $\dfrac{\partial p}{\partial \rho}$, is a complex value and can be represented as

$$K = K_1\, e^{j\phi} = K_1\, (\cos\phi - j\sin\rho) = K_0(1 - j\tan\phi)$$

where $K_0 = K_1 \cos\phi$, a substantial part of the coefficient of elasticity, and ϕ is the angle of losses equal to the phase difference between pressure and density.

In the equation of movement for the air in the pores, as distinct from the corresponding equation for free air, the density of the medium has a value $\dfrac{k}{P}$ times greater. This apparent increase in density is connected with the fact that the structural constant of the material k, which depends on the

150

dimensions, form and disposition of the pores (Fig. 5.7) is always greater than unity. Moreover, the apparent density increases with porosity, as will be seen from the consideration of the equation of continuity above.

In deriving the equation of motion, we must take into account the force of friction of air at the walls of the pores, which may be expressed as the product of the resistive component of impedance R and the velocity of the air particles v. Thus the equation for the velocity of air in the pores is:

$$-\frac{\partial p}{\partial x} = \frac{k}{P} \rho_0 \frac{\partial v}{\partial t} + Rv \qquad 5.26$$

In order to obtain the wave equation, we first solve equations 5.23 and 5.25 together. We multiply the numerator and denominator of the right

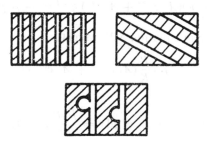

Fig. 5.7. Samples of materials of equal porosity and various structural constants k.

hand side of equation 5.23 by ∂p and substitute in it the value $\frac{\partial p}{\partial \rho}$ derived from 5.25. As a result:

$$-\frac{\partial v}{\partial x} = \frac{P}{\rho_0} \frac{\partial \rho}{\partial p} \frac{\partial p}{\partial t} \quad \text{or} \quad -\frac{\partial v}{\partial x} = \frac{P}{K} \frac{\partial p}{\partial t} \qquad 5.27$$

Substituting:

$$\frac{k\rho_0}{P} + \frac{R}{j\omega} = \rho \qquad 5.28$$

The equation of motion 5.26 can be rewritten in the form:

$$-\frac{\partial p}{\partial x} = \rho \frac{\partial v}{\partial t} \qquad 5.29$$

Differentiating equations 5.27 and 5.29, the former with respect to t and the latter with respect to x, we have

$$-\frac{\partial^2 v}{\partial x \cdot \partial t} = \frac{P}{K} \frac{\partial^2 p}{\partial t^2} \qquad 5.30$$

$$-\frac{\partial^2 p}{\partial x^2} = \rho \frac{\partial^2 v}{\partial t \cdot \partial x} \qquad 5.31$$

151

Substituting the value of $\dfrac{\partial^2 v}{\partial x . \partial t}$ from equation 5.30 into equation 5.31, we obtain an equation of propagation of sound in a porous material with a rigid skeleton:

$$-\frac{\partial^2 p}{\partial x^2} = \frac{P}{K}\rho \frac{\partial^2 p}{\partial t^2} \qquad 5.32$$

The solution of this equation, assuming a harmonic exciting force, has the form:

$$p(x, t) = (A_1 e^{-bz} + A_2 e^{bz}) e^{j\omega t} \qquad 5.33$$

Differentiating with respect to t we find that

$$\frac{\partial p}{\partial t} = -j\omega(A_1 e^{-bz} + A_2 e^{bz}) e^{j\omega t} = -j\omega p(x, t)$$

Inserting this value in equation 5.27 we get:

$$\frac{\partial v}{\partial x} = j\omega \frac{P}{K}(A_1 e^{-bz} + A_2 e^{bz}) e^{j\omega t}$$

whence

$$v = j\omega \frac{P}{K} e^{j\omega t} \int_x (A_1 e^{-bz} + A_2 e^{bz}) \mathrm{d}x = \frac{j\omega P}{Kb} e^{j\omega t}(A_1 e^{-bz} - A_2 e^{bz}) \quad 5.34$$

By means of expressions 5.33 and 5.34 we can define pressure and particle velocity as functions of the distance between the surface of the material and any point in its thickness

$$p(x) = A_1 e^{-bz} + A_2 e^{bz};$$

$$v(x) = \frac{1}{z}(A_1 e^{-bz} - A_2 e^{bz})$$

where

$$z = \frac{Kb}{j\omega P} \qquad 5.35$$

Let us define the value b, which is called the propagation constant, and z which has the dimension of impedance and which is called the wave impedance of the medium.

If we insert the value $p(x)$ from equation 5.33 into equation 5.32 and differentiate, we find that

$$b^2 = \frac{P}{K}\rho\omega^2 \quad \text{or} \quad b = \pm j\omega\sqrt{\frac{P}{K}\rho} \qquad 5.36$$

Replacing b by its value in the expression for wave impedance, we obtain:

$$z = \frac{bK}{j\omega P} = \frac{K}{j\omega P} \cdot j\omega \sqrt{\frac{P}{K}\rho} = \sqrt{\frac{K}{P}\rho} \qquad 5.37$$

152

By using Fig. 5.8 we determine the acoustic impedance for the surface layer of material z_1, which is represented in the form of the ratio of sound pressure to particle velocity, i.e. for our case $z_1 = \dfrac{P(0)}{v(0)}$. Let the impedance z_2 of its rear surface ($x = l$) be known, i.e. boundary conditions given in the form $z_2 = \dfrac{p(l)}{v(l)}$.

From equations 5.35 we find that:

$$p(0) = A_1 + A_2, \qquad v(0) = \frac{1}{z}(A_1 - A_2)$$

and

$$z_1 = \frac{p(0)}{v(0)} = z\frac{A_1 + A_2}{A_1 - A_2} \qquad\qquad 5.38$$

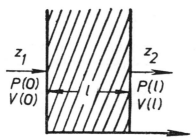

Fig. 5.8. For the calculation of the acoustic impedance of the surface layer of the material z_1.

If we insert into the equations 5.35 the value of x defined by the boundary conditions, then

$$p(l) = A_1\,e^{-bl} + A_2\,e^{bl}$$

$$v(l) = \frac{1}{z}(A_1\,e^{-bl} - A_2\,e^{bl})$$

From these expressions we find the constants of integration of A_1 and A_2.

$$A_1 = \frac{p(l) + v(l)z}{2}\,e^{bl} \quad \text{and} \quad A_2 = \frac{p(l) - zv(l)}{2}\cdot e^{-bl} \qquad 5.39$$

If now the values of A_1 and A_2 are inserted into expression 5.38, the acoustic impedance takes the form:

$$z_1 = z\frac{p(l)(e^{bl} + e^{-bl}) + zv(l)(e^{bl} - e^{-bl})}{p(l)(e^{bl} - e^{-bl}) + zv(l)(e^{bl} + e^{-bl})}$$

or, dividing the numerator and denominator of the right hand side by $v(l)$ and replacing the exponential functions by hyperbolic functions:

$$z_1 = z\frac{z_2 \cosh bl + z \sinh bl}{z_2 \sinh bl + z \cosh bl} \qquad\qquad 5.40$$

153

This expression is identical to the expression for the input impedance of a long electric line or for the input impedance of mechanical vibratory systems with distributed constants.

It follows from formula 5.40 that the acoustic impedance of a material with a rigid skeleton depends on the acoustic impedance of the wall on which the material z_2 is placed, on the thickness of the material l, on the wave impedance z, and on the propagation constant b.

After calculating the acoustic impedance from formula 5.40 we can find the coefficient of absorption for a plane wave and the diffuse coefficient of absorption by inserting the calculated value of z_1 into formulae 5.12, 5.13 or 5.20 and 5.21.

5.5 The connection between the coefficient of absorption in porous material and its parameters

Using the general expression for acoustic impedance (5.40) we can find its value for a number of cases widely used in practice. An analysis of these cases allows us to determine how the acoustic impedance, and consequently the coefficient of absorption, depends on the basic parameters of porous material.

Case 1. $z_2 = z$ which in practice corresponds to a large thickness of absorbent material. From expression 5.40 we have

$$z_1 = z \qquad \qquad 5.41$$

or, taking into account equations 5.37 and 5.28:

$$z_1 = \sqrt{\frac{K}{P}} \rho = \sqrt{\frac{K}{P} \left(\frac{k}{P} \rho_0 + \frac{R}{j\omega} \right)} \qquad \qquad 5.42$$

This case allows us to explain the physical meaning of wave impedance and of propagation. It follows from equation 5.41 that the wave impedance is here the input acoustic impedance for an infinitely thick layer of the absorbent material.

In equation 5.33, if we put $b = \infty$, $A_1 = 0$ the first of the expressions 5.35, takes the form:

$$p(x) = A_2 e^{bx} = A_2 e^{(\delta + j\gamma)x}, \text{ since } b \text{ is complex} \qquad 5.43$$

or, if where $x = 0$, $A_2 = P(0)$, then

$$p(x) = p(0) e^{(\delta + j\gamma)x}$$

It follows from this last expression, that the real part of the propagation constant represents the decay of the amplitude of sound pressure $P(0)$ at a point distant x from the surface of the material, and the imaginary part γ is the relative phase change of sound pressure at the same point.

As the coefficient of absorption always increases as acoustic impedance decreases, starting from expression 5.42 we can say that the absorbing capacity of a porous material increases with an increase in porosity P and

154

of sound frequency ω, and falls with an increase in the resistance R and in the structural coefficient k. This conclusion is confirmed by Fig. 5.9, which shows experimental curves of absorption for samples of the same type of material (felt) with different porosities and resistances. Curve 1 corresponds to the absorption of felt with a thickness of 50 mm, curves 2 and 3 for the absorption of the same layer compressed to thicknesses of 25 and 16 mm.

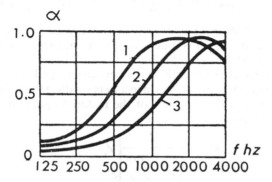

Fig. 5.9. Curves of sound absorption against frequency for various porosities of a material.

Case 2. $z_2 = \infty$ which corresponds in practice to an absorbing material backed by a very hard wall. This case is analogous to that of a long electric line or of a mechanical vibratory system with distributed parameters, with an open-circuit termination and an immobile ending respectively. (Fig. 5.10).

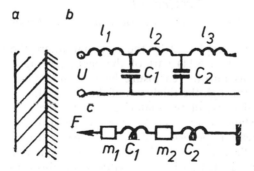

Fig. 5.10. Diagrams of electrical (b) and mechanical (c) vibratory systems equivalent to the sound absorbing system (a) backed by a rigid wall.

Where $z_2 = \infty$, in view of the relatively small values of the second terms of the numerator and denominator, expression 5.40 takes the form:

$$z_1 = z \coth bl. \qquad 5.44$$

If we bear in mind that the values of z and b are defined by Equations 5.37

155

and 5.36, this last equation can be rewritten as

$$z_1 = \sqrt{\frac{K}{P}\rho} \, \coth j\omega l \sqrt{\frac{K}{P}\rho}$$

5.45

or from equation 5.28:

$$z_1 = \frac{K}{P}\sqrt{\left(p_0 + \frac{RP}{j\omega K}\right)} \, \coth j\omega l \sqrt{\left(p_0 + \frac{RP}{j\omega K}\right)}$$

5.46

We can note from expression 5.46 that as the frequency ω comes into the argument, coth, which is a periodic function, then z and consequently α have periodic maxima and minima.

Bearing in mind that the coth varies between ∞ to 1, the acoustic impedance for each given case is always larger (α is smaller) than for the preceding case. A reduction in the thickness of the material results in an increase in z_1 and a reduction in α.

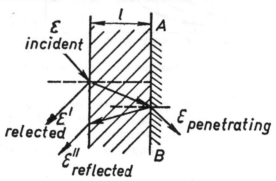

Fig. 5.11. Diagram of the propagation in material of small thickness.

Physically this is explained by the fact that where the thickness of the material is finite (Fig. 5.11) part of the energy, being reflected from the hard wall, returns into the enclosure having undergone losses during its second passage through the thickness of the material.

Where the thickness of the material is finite, the influence of porosity and of the resistance may differ. On the one hand, for a certain thickness of layer an increase in porosity is desirable because it results in a reduction of energy ε'_{ref} reflected from the front surface of the material (see Fig. 5.11). On the other hand such an increase reduces the absorbing capacity of the material as a result of the significant increase in energy ε''_{ref} reflected from the hard wall AB. This underlines the internal contradiction of the process of sound absorption by porous materials.

At low frequencies when the argument of the hyperbolic cotangent is significantly less than unity, the function itself can be expressed by the following two terms of the series into which it is expanded:

$$\coth y = \frac{1}{y} + \frac{y}{3}$$

In accordance with this, expression 5.46 can be rewritten thus:

$$z_1 = \sqrt{\frac{K}{P}\rho} \cdot \left(\frac{1}{j\omega l \sqrt{\frac{P}{K}\rho}} + \frac{j\omega l \sqrt{\frac{P}{K}\rho}}{3} \right)$$

or, removing the brackets,

$$z_1 = \frac{K}{j\omega l P} + \frac{1}{3} j\omega l \rho$$

and, finally, substituting for ρ its value from equation 5.28:

$$z_1 = \frac{K}{j\omega l P} + \frac{1}{3}(j\omega K\rho_0 + lR) \qquad 5.47$$

Formula 5.47 shows that at low frequencies a porous material with a rigid skeleton has an impedance consisting of a significantly high value of stiffness $\left(\frac{K}{lP}\right)$ and of small inertance $\frac{1}{3} lK\delta_0$ and resistance $\frac{1}{3} lR$. Remembering that the resonance frequency is equal to the square root of the ratio of the two first impedances, we can find the frequency of the first maximum for the absorption coefficient:

$$\omega_1 = \frac{1}{l}\sqrt{\frac{3}{P\rho_0}} \qquad 5.48$$

From this it becomes plain that a reduction in the thickness of porous material l or of its porosity P results in a shift of maximum absorption towards the high frequencies.

Finally, from expression 4.57 we can draw one more conclusion, that a reduction in the frequency ω of the vibrations acting on the material causes a significant increase in z or, which amounts to the same thing, a significant reduction in the coefficient of absorption.

The conclusions we have made are in full agreement with the practical data, as can be seen from the curves in Fig. 5.12, for felt of various thicknesses (l, $2l$, etc.).

The presentation of the acoustic impedance as the sum of three components, as in expression 5.47, enables us to consider a porous material as equivalent to an electrical circuit with the elements in series (Fig. 5.13).

Case 3. $z_2 = 0$, which happens when the sound absorbent material is installed at a distance $l = \frac{\lambda}{4}$ from a hard wall.

To confirm this, let us determine the impedance of a layer of air of the indicated thickness from expression 5.44. If we take the decay of sound in air $\delta = 0$ from equation 5.43 we obtain $b = j\gamma$ or

$$z_{\text{air}} = z \coth bl = z \coth j\gamma l = z \coth j\frac{\omega}{c_0}\cdot\frac{\lambda}{4} = 0$$

If we substitute the value of z_2 corresponding to the given case into formula 5.40 we obtain:

$$z_1 = z \tanh bl \qquad\qquad 5.49$$

If the argument is reduced, the hyperbolic tangent tends towards zero, and therefore in this case if the thickness of the material is reduced the acoustic impedance falls and the coefficient of absorption increases.

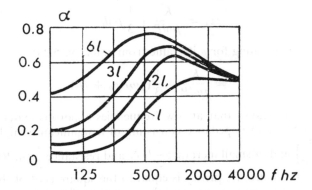

Fig. 5.12. Changes in the frequency dependence of the coefficient of sound absorption for 2 times, 4 times and 6 times increases in the thickness of the material.

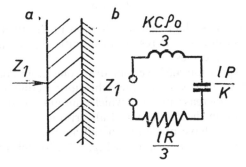

Fig 5.13. Sound absorbing material (a) and its equivalent electrical circuit

As can be seen from formula 5.47, the impedance of porous material at low frequencies is basically defined by the stiffness (first element). Therefore, to obtain an increase in the coefficient of absorption it is necessary to reduce the stiffness of the air by moving the absorbent material away from a hard wall. Such a movement results in a shift of the maximum absorption towards the low frequencies, with a simultaneous increase in the maximum coefficient. This can be seen from Fig. 5.14, which shows curves of the dependence of α on frequency for acoustic plaster with $l = 21$ mm, for various thicknesses of air space.

158

It follows from the above that for a porous material with a rigid skeleton:

1. The coefficient of absorption of thick layers increases with an increase in porosity and a reduction in the resistance R, and of the structural coefficient k.

2. The absorbing capacity decreases with a reduction in the thickness of the layer.

3. The effectiveness of absorption may be increased by choosing a great thickness of material having high porosity and low resistance. This follows from the dual character of the absorption process.

4. A reduction in the thickness or the porosity of the material results in a shift of the maximum absorption towards the high frequencies.

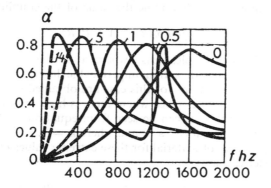

Fig. 5.14. Frequency characteristics of the absorption coefficient for a layer of plaster at different distances (14; 5; 1; 0·5; and 0 mm) from a rigid wall.

5. The frequency characteristic of absorption falls off at low frequencies.

6. The presence of an air space between the material and a rigid wall results in an increase of absorption at low frequencies, and an increase in the air space is accompanied by a shift of the maximum absorption towards the low frequencies and to an increase in the maximum coefficient.

5.6 The absorption of sound by porous-elastic materials

If the absorbing material has a skeleton which is not rigid, but elastic, the solution of the problem of the interrelationship between the absorbing capacity and the parameters of the material becomes more complicated. Not only is the air in the pores subject to vibration, but so, too, is the elastic skeleton. As a result of this the need arises to formulate not one but two equations of continuity and the same number of equations for state and velocity. As well as an increase in the number, the equations themselves

159

are more complicated. The equations of movement have the form:

$$\frac{\partial p_1}{\partial x} = \rho_1 \frac{\partial v_1}{\partial t} + s(v_1 - v_2) \qquad 5.50$$

$$-\frac{\partial p_2}{\partial x} = \rho_2 \frac{\partial v_2}{\partial t} + s(v_2 - v_1) \qquad 5.51$$

where suffix 1 relates to the values characteristic of the skeleton and suffix 2 to values characterizing the air in the pores, s is the coupling coefficient which represents the force acting on the unit of space where $v_1 - v_2 = 1$.

If in equation 5.51 we take $v_2 = 0$, then the material has a rigid skeleton, i.e. it should become identical with equation 5.26. A comparison of these two equations allows us to determine the value of the coupling coefficient:

$$s = j\omega P\rho_0(k - 1) + P^2 R \qquad 5.52$$

Let us consider the simplest case where a porous-elastic material is placed on a hard wall with a thin impermeable layer stretched over its front surface. Simple solutions for this case are obtained only where coupling is complete, which takes place where there is high resistance, or where there is no coupling; this, as can be seen from equation 5.52, is observed when the resistance is small ($R \approx 0$) and where $k = 1$.

Acoustic impedances of material for these extreme values of the coupling coefficient respectively are:

$$z = \sqrt{(K_1 + K_2)(\rho_1 + \rho_2)} \coth j\omega l \sqrt{\frac{\rho_1 + \rho_2}{K_1 + K_2}} \qquad 5.53$$

$$z = \sqrt{K_1\rho_1} \coth j\omega l \sqrt{\rho_1/K_1} + \sqrt{K_2\rho_2} \coth j\omega l \sqrt{\rho_2/K_2} \qquad 5.54$$

The second term in the latter equation is identical to expression 5.45 (where $P = 1$), i.e. it characterizes the impedance of rigid porous material. Hence, the first term of equation 5.54 expresses the difference between the impedance of an elastic-porous material with its pores covered from the front, and that of a rigid porous material.

If the elasticity of the skeleton and the density of its material are large, ($K_1 \gg K_2$, $\rho_1 \gg \rho_2$), which is typical of very rigid materials, then the acoustic impedance resulting from the first term in expression 5.54 is large and the coefficient of absorption very small. This conclusion fully agrees with practice, as it is well known that where a rigid material (e.g. plaster) is covered with an impermeable surface (e.g. oil paint) the plaster becomes almost non-absorbent.

Light materials ($\rho_1 > \rho_2$) and friable materials ($K_1 \approx K_2$) which have a low resistance, behave differently.

As a result of the fact that the density of these materials ρ_1 is great by comparison with the density of air in the pores ρ_2, the first maximum of the coefficient of absorption created by the vibrations of the skeleton, as will

160

be seen from equation 5.48, moves towards the low frequencies. As the resonance frequencies of the skeleton in the given case are lower than the resonance frequencies of the air, the frequency characteristic of absorption take the form shown at curve 2 on Fig. 5.15. This figure shows frequency curves of absorption for a layer of porous resin with low impedance from friction. Curve 1 corresponds to the front surface being uncovered, and curve 2 to its being covered.

As can be seen from Fig. 5.14, movement of rigid porous material away from a hard wall moves the absorption curve to the left. The same effect is obtained for porous-elastic materials, as will be seen from curve 3 in Fig. 5.15, which is obtained for the same material as curves 1 and 2 but with the material placed at 8 cm from the wall.

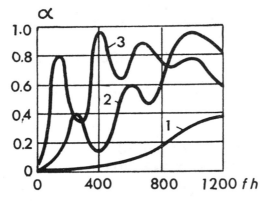

Fig. 5.15. The frequency characteristics of sound absorption of a porous resin firmly fastened to the wall and with its frontal surface uncovered (1), with its frontal surface covered (2), and when it is placed at a distance of 8 cm from the wall (3).

As the analysis of porous-elastic material with open pores is complicated, it seems wise here to reproduce some practically obtained frequency characteristics of absorption for these materials. Fig. 5.16 shows such characteristics for two kinds of porous resin: curves 1 and 2 for a resin with low R but of different thicknesses and curve 3 for a resin for which R is ten times larger than for curve 1.

From Fig. 5.15 and 5.16 it is clear that for porous-elastic materials:

1. The coefficient of absorption increases with an increase in the thickness of the material (curves 1 and 2 on Fig. 5.16).
2. Where there is an absence of coupling (low R) the frequency characteristic is not different in form from that of rigid porous materials.
3. An increase in impedance R results in the appearance of coupled vibrations of the skeleton and the air in the pores and result in irregular absorption at the resonant frequencies of the skeleton (curve 3 Fig. 5.16).

161

4. The presence of a thin covering layer for light and soft materials re-sults in an increase of the absorption as a result of coupled vibrations of the skeleton, with an increase in its unevenness (curve 2 in Fig. 5.15).

5. An air space between the material and the wall raises the absorption at low frequencies (curve 3, Fig. 5.15).

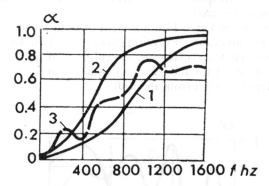

Fig. 5.16. Frequency characteristics of absorption for porous resin with a low (1) and a high (3) impedance from friction and with small thickness (1) and great thickness (2).

5.7 The absorption of sound by rigid vibratory systems

The sound absorbing capacity of vibratory systems, which has been put to practical use recently, has been known for a long time. The physical reality of this phenomenon was clearly formulated by Professor Stoletov in lectures which he delivered in 1893 in Moscow University. He observed that 'absorption of a wave occurs when it meets in its path bodies able to vibrate in its own rhythm. The energy of the initial wave also diminishes, but as a result of this energy, standing waves are set up in the bodies en-countered, which in their turn become sources of such waves. If the body which has been met has sharply expressed natural tones, it absorbs only selected waves, and absorption becomes selective'.

Thus sound absorption is greatest where it induces strong vibrations in the system, i.e. at a resonance where deformation losses in the system have their maximum value.

Let there be a system comprising a number of resilient plates (Fig. 5.17) on which impinges a plane sound wave for which

$$p(x, t) = A_1 e^{(bx + j\omega t)}$$

It would seem that as each of the plates bends under the vibrations, the reflected wave would lose its plane form and would appear as the sum of waves reflected in different directions. However, this is observed for the highest harmonic components and only in the immediate vicinity of the

162

system, so that for comparatively low frequencies at a certain distance from the system, the sound field can be represented by pressure:

$$p(x, t) = (A_1 + A_2)\, e^{(bz + j\omega t)}$$

where A_2 is the amplitude of the reflected wave.

As we are interested in pressure as a function of distance x, and considering decay in air to be $\delta = 0$, 5.43 becomes:

$$p(x, t) = (A_1 + A_2)\, e^{j\gamma x}$$

From expression 5.38 it follows that to define specific acoustic impedance we must find the coefficients A_1 and A_2. As in previous cases they can be found by solving the wave equation for the given vibratory system, which can be done by regarding each of the plates as a resilient rod.

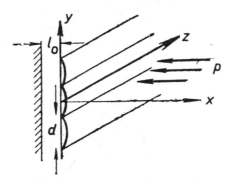

Fig. 5.17. A diagram for analysis of the absorbing capacity of rigid vibratory systems.

The calculations for the solution of the wave equation are complicated: we give here the equation derived from these calculations for $B = \dfrac{A_1}{A_2}$

$$B = \frac{A_1}{A_2} = \frac{\pi^6 s\eta - 8d^2 c_0 \rho_0\, \sqrt{s/m}}{\pi^6 s\eta + 8d^2 c_0 \rho_0\, \sqrt{s/m}} \qquad 5.55$$

where η is the coefficient of hysteresis losses, defined by the formula

$$\eta = \frac{2\Delta\omega}{\omega_p} = \frac{2\Delta\omega d^2}{\pi^2 \sqrt{s/m}} \qquad 5.56$$

s is the shear modulus of the plate, m is the mass of the plate for unit area, d is the breadth of the plate, $2\Delta\omega$ is the breadth of the resonance curve (the frequency difference between the points at which the amplitude of the vibrations is 1·41 times less than the maximum).

The specific acoustic impedance is found from formula 5.38, writing it in the following way:

$$z_1 = z\,\frac{1 + B}{1 - B}$$

163

As in this case $z = c_0\rho_0$, by substituting the value of B from expression 5.55 we find that

$$z_{1p} = \frac{\pi^4 \eta \sqrt{sm}}{8d^2} \qquad 5.57$$

Recalling that

$$\alpha = 1 - \frac{I_{\text{ref}}}{I_{\text{inc}}} = 1 - \left(\frac{p_{\text{ref}}}{p_{\text{inc}}}\right)^2$$

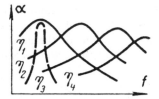

Fig 5.18. The frequency characteristics of sound absorption for different values of deformation losses.

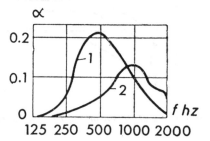

Fig. 5.19. The absorption characteristics of sheets of plywood (1) and hardboard (2) at a distance of 5 cm from the wall.

we can, by substituting for the ratio of pressures the ratio of the coefficients A_1 and A_2, write that

$$\alpha_p = 1 - \left(\frac{A_1}{A_2}\right)^2 = 1 - B^2 = \frac{32 d^2 c_0 \rho_0}{\pi^4 \eta \sqrt{sm} \left(1 + \dfrac{8 d^2 c_0 \rho_0}{\pi^4 \eta \sqrt{sm}}\right)^2} \qquad 5.58$$

In order to find z_1 and α for any other frequencies, instead of frequency ω_p in expression 5.56 we should take the frequency $\omega_p + \Delta\omega$.

If we use the expressions we have obtained, we can construct a family of absorption characteristics which correspond to different values of the co-efficient of hysteresis losses η. From these curves (Fig. 5.18) it can be seen that where η is increased, as is achieved by increases in d and m or by a reduction in s, the maximum of absorption moves towards low frequencies.

Practice fully confirms this conclusion. In fact as for plywood m is greater and s is smaller than for cardboard of the same thickness, then, as follows from Fig. 5.19, curve L constructed for the former of these materials, has

164

a maximum of absorption at a lower frequency than curve 2 which corresponds to the second of these materials.

It has been established in practice that a change in the distance between a vibratory system and a rigid wall, as is true also for porous absorbent materials, has an influence on the value of the absorption coefficient and its frequency characteristic. If this distance is increased, the resonance frequency becomes lower (Fig. 5.20).

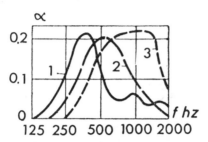

Fig. 5.20. The absorption characteristics of a plywood sheet at 10 cm (1) 5 cm (2) and 2 cm (3) from the wall.

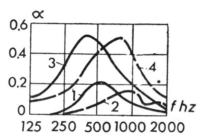

Fig. 5.21. The dependence of the absorption coefficient on frequency for plywood (1, 3) and hardboard (2, 4) with the gap left empty (1, 2) and filled with porous material (3, 4).

A still greater shift to the left can be achieved by filling the space between the wall and the rigid vibratory system with sound absorbing materials—wool or felt. In this case the shift in the maximum of absorption is accompanied by an increase and a sharpening of the coefficient (Fig. 5.21).

The same results are obtained by damping the edges of the plates comprising the vibrating system. If the edges of the system are firmly fixed (Fig. 5.22), the coefficient of absorption is small and the resonance peak is obtuse (curves 1, 3). Where the edges of the system are damped the frequency characteristic of absorption rises higher and the peak becomes sharper and moves towards the lower frequencies (curves 2, 4).

In the acoustic treatment of enclosures, apart from panels in the form of plane vibrating systems, usually on the lower parts of the walls, frequent use is also made of cylindrical panels in the form of half columns. Cylindrical plywood panels have a maximum of absorption in the low frequency region and an area determined by the length of the chord *d* and the height of the arc *h* of the panel (Fig. 5.23). The wide use of these panels is recommended; apart from absorption of sound energy they also help in scattering.

Bekesha screens are another form of rigid resonant system. These are constructions in the form of frames, on one side of which is stretched a wire grid, and on the other oil cloth. The gap between the surfaces is filled with a soft porous material. This construction gives a high effectiveness of absorption, allowing the attainment of an absorption coefficient of 0·8 in the range 100 to 400 Hz.

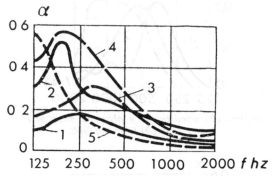

Fig. 5.22. Frequency curves of sound absorption of plywood with the edges rigidly fixed (1, 2) and fixed with damping material (3, 4, 5). Curves 1, 3, for plywood with a thickness of 4 mm, 2, 4, for plywood with a thickness of 3 mm.

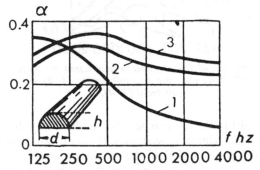

Fig. 5.23. Curves of sound absorption by cylindrical panels: (1) d = 113 cm, h = 40 cm; (2) d = 90 cm, h = 35 cm; (3) d = 70 cm, h = 32 cm.

Thus we come to the following conclusions:

1. The frequency characteristic of a rigid vibrating system is represented in the form of a resonance curve.

2. The absorbing capacity of these systems depends on their elasticity, on the specific weight, dimensions and the method of attachment of the panels, their position relative to the hard surface (the wall) and on the filling of the gap between the system and the wall.

166

3. An increase in m, d and l_0, a reduction in s, damping of the edges, and the filling of the gaps with soft materials results in an increase in the absorption coefficient and a shift to the left of the maximum of its frequency characteristic.

4. To obtain the most regular frequency characteristic of absorption possible from a rigid vibrating system, it should be made up of elements having differing characteristic frequencies as far as possible evenly distributed in the relevant frequency range. This can be achieved by varying the resilience and the specific weight of the panels or their distance from the wall.

5.8 Absorption of sound by porous-vibratory systems

A porous vibratory system consists of a rigid layer or layers, non permeable to air, standing off a little distance from a hard surface, and having a number of apertures covered by air-permeable cloth (Fig. 5.24). Such layers may be made of plywood, metallic or plastic sheets.

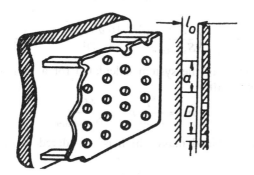

Fig. 5.24. A porous-vibratory system.

The design of these systems is directed towards the problem of increasing the absorption coefficient, which became particularly critical in connection with the planning of the enormous auditoria of the Palace of the Soviets in Moscow. The treatment was carried out by S. N. Rzhevkin, V. S. Nesterov, M. S. Antisferov, and G. D. Malyuzhents. Each cell of this system, consisting of an aperture and an air space behind it, despite the absence of any division between the cells, was regarded as a resonator. Such an approach to the solution of the problem is correct only under conditions of normal incidence of the sound wave.

In order to find the coefficient of absorption for the systems under consideration, it is essential, as follows from expression 5.14, to determine the resistance (R) and the reactance (X) components of specific acoustic impedance.

The resistive component of impedance, caused by friction, should be largest at those points where the highest velocity of air motion is set up,

167

i.e. near the apertures covered in air-permeable fabric (see Fig. 5.24). The reactive component of impedance may be considered as a loss of external sound pressure in overcoming the inertia of air in the apertures and in its compression in the cavity of every cell.

Let us define both elements of the specific acoustic impedance for a single layer system.

Each of the cells, which have a breadth and a height equal to a can be compared in action to the pre-horn chamber of a loudspeaker, and so the resistive component of impedance for a unit of space of such a system is represented as:

$$R = \frac{r_{fab}}{S_{cell}} \cdot n^2$$

where r_{fab} is the active impedance of the fabric, S_{cell} is the area of one cell the lengths of whose sides are equal to a, n is the acoustic transformer-ratio:

$$n = \frac{S_{cell}}{S_{aperture}} = \frac{a^2}{\frac{1}{4}\pi D^2}$$

Thus

$$R = \frac{r_{fab}}{S_{cell}} \cdot \left(\frac{S_{cell}}{S_{aperture}}\right)^2 = \frac{16}{\pi^2} \frac{a^2 r_{fab}}{D^4} = 1 \cdot 64 \frac{a^2}{D^4} r_{fab} \qquad 5.59$$

Similarly the mass of the system calculated for a unit of space can be expressed by the equation:

$$m = \frac{M}{S_{cell}} n^2 = \frac{MS_{cell}}{S^2_{aperture}}$$

where M is the combined mass which is considered to be approximately equal to:

$$M = \frac{S^2_{aperture}}{D} \rho_0$$

If we insert this value into the preceding formula, we obtain:

$$m = \frac{S^2_{ap}}{D} \rho_0 \frac{S_{cell}}{S^2_{ap}} = \frac{a^2}{D} \rho_0 \qquad 5.60$$

For a low frequency the resilience of the air space in the cell, per unit of area is:

$$s' = \frac{s}{S_{cell}} = \frac{K_0 S^2_{cell}}{V_{cell} S_{cell}} = \frac{K_0 a^2}{a^2 l_0} = \frac{K_0}{l_0} \qquad 5.61$$

where

$$K_0 = c_0^2 \rho_0$$

On the basis of equations 1.60 and 1.61 we get the full reactive impedance:

$$X = \omega m - \frac{s'}{\omega} = \frac{\omega a^2}{D} \rho_0 - \frac{K_0}{\omega l_0} \qquad 5.62$$

168

If we introduce the values of the active and reactive components of impedance into expression 5.14, we obtain after some calculations:

$$\alpha_\perp = \frac{6 \cdot 56 a^2 r_{\text{fab}} c_0 \rho_0}{D^4 \left(1 \cdot 64 \dfrac{a^2}{D^2} r_{\text{fab}} + c_0 \rho_0\right)^2 + \left(\dfrac{\omega a^2}{D} \rho_0 - \dfrac{K_0}{\omega l_0}\right)^2} \qquad 5.63$$

For high frequencies, when the distance between the vibratory system and the wall becomes large in relationship to the wavelength, the stiffness of the air space in the cell should be considered as distributed. The reaction of the air in the cell is the same as in the channels of a porous material with a rigid skeleton. As the cell ends in a hard wall, its stiffness corresponds to the impedance of porous material where $z_2 = \infty$. From equation 5.44 this impedance is:

$$z \coth bl = - jz \cot jbl = jc_0\rho_0 \cot \omega l_0 \sqrt{\frac{\rho_0}{K_0}}$$

Thus the coefficient of absorption for any frequencies can be expressed by equation 5.63 in which the stiffness $\dfrac{K_0}{\omega l_0}$ is replaced by the value $c_0 \rho_0 \cot \omega l_0 \sqrt{\dfrac{\rho_0}{K_0}}$.

From equation 5.63 we can easily determine the resonant frequency of the system. It is clear that $\alpha_\perp$ will be largest when the reactive component of impedance,

$$\frac{\omega a^2}{D} \rho_0 - \frac{K_0}{\omega l_0} = 0$$

whence

$$\omega p = \frac{1}{a} \sqrt{\frac{K_0 D}{l_0 \rho_0}} = \frac{c_0}{a} \sqrt{\frac{D}{l_0}} \qquad 5.64$$

In the present case, assuming that R is constant (see Fig. 5.5) it is easy to obtain a frequency characteristic for $\alpha_\perp$ by the use of the circles of equal absorption (Fig. 5.5). If we choose values for the active impedance of R_1', R_2', and R_3', etc. and draw through these points on the axis of the abscisses straight lines vertical to it, these straight lines intersect the circles at points with varying X'. As X is a function of frequency (see 5.62) then its value can be used to find the value of ω which corresponds to the given coefficient of absorption.

If we construct curves $\alpha = f(X')$ for different values of R' (Fig. 5.25), it will be seen that for low values of R', absorption at the resonant frequency ($X' = 0$) is high, and the frequency characteristic is too narrow. An increase in R' results in a reduction in the maximum of absorption and a levelling out of the frequency characteristic.

As formulae 5.63 and 5.64 show, a porous vibrating system can be calculated to ensure the required absorption in a given frequency range.

In this lies its great superiority over other absorbers. Another great advantage of this system is the constancy of the parameters of the system, its strength and its economy.

Porous vibratory systems have been put into practical use with the space between the wall and the perforated system filled with soft porous material. Usually the fillings are of glass, mineral or organic fibre, prepared in the form of mats. The use of a filling results in the stiffness of the air space in the cell, represented by equation 5.61 being increased by as many times as

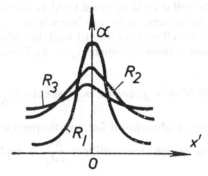

Fig. 5.25. Curves $\alpha = f\ (X')$ for varying values of the active component of acoustic impedance of the system.

the volume of air is decreased. The reduction in the volume of air in a perforated system depends on the porosity of the material, and so for a given case equation 5.61 should be re-written in the form:

$$s' = \frac{K_0}{Pl_0} = \frac{c_0{}^2 \rho_0}{Pl_0}$$

Unlike normal porous vibratory systems, the covering perforated layer is frequently made quite thick and heavy, for instance from plaster, brick, etc. As a result of this the mass per unit area of the covering layer m^2 is greater than the mass of a system with an air gap.

In the analogous circuit of the system these masses are placed in parallel. The total mass of the system with a filling is determined by the equation:

$$\frac{1}{m_\sigma} = \frac{1}{m} + \frac{1}{m_2}$$

If we consider that $m_2 \gg m$, the influence of the mass of the perforated layer may be ignored, and so for the reactance of a filled system we may write, similarly to equation 5.62:

$$X = \frac{\omega a^2}{D} \rho_0 - \frac{c_0{}^2 \rho_0}{\omega Pl_0} \qquad\qquad 5.65$$

This expression enables us to describe the role of the porous filling of the system.

170

In fact an increase in the resilient impedance of the system as a result of the porosity of the filler P results in a reduction of the reactive impedance X. As a result the coefficient of absorption α increases at low frequencies. As the frequency is increased the reactive impedance increases, resulting in the gradual reduction of α.

Fig. 5.26 shows characteristics of absorption for a number of porous vibratory systems with an air space, for which D and a are identical and respectively equal to 0·7 and 3 cm, and l_0 varies.

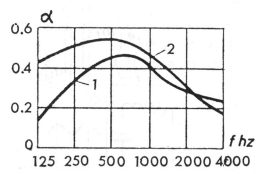

Fig. 5.26. The absorption of sound by porous vibratory systems with distances from the wall of 5 cm (1) and 10 cm (2).

As a result of the analysis of the work of porous vibratory systems we have the following conclusions:

1. The coefficient of absorption for these systems is defined by the inertia and resistance of the air in the passages of the porous material and by the stiffness of the air in the cells between the material and the wall.

2. The absorption characteristic of a single layer construction has a clearly defined maximum.

3. The frequency of maximum absorption increases with an increase in the diameters of the apertures D and a reduction in the distance between apertures a or between a perforated layer and the wall l_0.

4. Where K is increased the coefficient of absorption at the resonant frequency decreases, but the absorption of the system becomes less selective.

5. The system can be calculated for given parameters, allowing the necessary sound absorption to be achieved within the limits of the specified frequency bands.

6. The system is very efficient, economical, and stable in operation.

7. A serious shortcoming of a layer construction is that it has a comparatively narrow and peaky absorption characteristic.

5.9 The absorption of sound by a two-layer porous vibratory system

A more regular frequency characteristic of absorption can be obtained if a series or parallel combination of resonance constructions is used. The

171

plan of a two layer construction working in series, and its equivalent electrical circuit are shown in Fig. 5.27.

For a two layer construction, just as for a single layer, it is essential in the first place to determine the coefficient of absorption, expressing it in terms of the basic parameters of the system. This can easily be done if, as can be seen from formula 5.14, the values for this system of the resistive and reactive components of acoustic impedance are found.

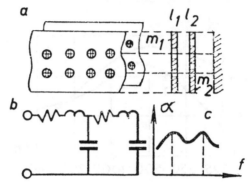

Fig. 5.27. A double porous vibratory system (a), its equivalent circuit (b) and its frequency characteristic (c).

By solving the equation of motion we can find the specific acoustic impedance of the whole system for normal incidence. This impedance is given by:

$$z = z_1 + \frac{z_2}{\sin^2 kl_1}$$

and the resistive and reactive components of it may be determined from equations:

$$R = R_1 + \frac{R_2}{\sin^2 kl_1(R_2{}^2 + X_2{}^2)} = R_1 + \sigma R_2 \qquad 5.66a$$

$$X = X_1 + \frac{X_2}{\sin^2 kl, (R_2{}^2 + X_2{}^2)} = X_1 - \sigma X_2 \qquad 5.66b$$

where $k = \dfrac{\omega}{c_0}$ the wave number, l_1 is the distance between the layers of the construction and

$$\sigma = \frac{1}{\sin^2 kl_1 (R_2{}^2 + X_2{}^2)}$$

For calculations in a two layer system, it is first regarded as a single-layer system with lumped constants, though this gives a true picture only for low frequencies. In this case, given that the necessary coefficient of absorption is α_1, and the extreme frequencies of the range of maximum

172

absorption are f_1 and f_2, the total thickness of the system is given by the formula

$$l = 2704 \frac{\alpha_1}{f_1\sqrt{1-\alpha_1}} \qquad 5.67$$

and also the distance from the hard wall to the first layer and between the first and second layers:

$$l_2 = l \cdot \frac{f_2{}^2 - f_1{}^2}{f_2{}^2 + f_1{}^2} \quad \text{and} \quad l_1 = l - l_2 \qquad 5.68$$

The impedance components of the two parts of the system can be found on the basis of the equations

$$m_1 = \frac{c_0{}^2}{\omega_0{}^2 l} \quad \text{and} \quad m_2 = m_1 \left(\frac{f_2{}^2 + f_1{}^2}{f_2{}^2 - f_1{}^2}\right)^2 \qquad 5.69$$

$$R_1 = \frac{2-\alpha_1}{\alpha_1} - R_2 \left(\frac{l_2}{l}\right)^2 \quad \text{and} \quad R_2 = \sqrt{\frac{m_2}{l_1}} \qquad 5.70$$

where

$$\omega_0 = 2\pi \sqrt{\tfrac{1}{2}(f_1{}^2 + f_2{}^2)}$$

As at high frequencies, where the dimensions of the system become commensurable with the wavelength, the system should be considered as one with distributed parameters, and corrections must be made in the calculations. Dimensions l_1 and l_2 can be found by the formulae

$$\cot \frac{\omega_0}{c_0} l_1 = \frac{\omega_0}{c_0} m_1 \quad \text{and} \quad \cot \frac{\omega_0}{c_0} l_2 = \frac{\omega_0}{c_0}(m_1 - m_2) \qquad 5.71$$

and the active components of impedance by formulae

$$R_2 = \frac{1}{\sin k_0 l_1} \sqrt{\frac{l_1 + l_2 + m_2 \sin^2 k_0 l_1}{l_1 + m_1 \sin^2 k_0 l_1}} \qquad 5.72$$

and

$$R_1 = \frac{2-\alpha_1}{\alpha_1} - \sigma R_2$$

After calculating R_1, R_2, m_1 and m_2 it is easy to find the constructional dimensions of a two layer system, which is done in the following order:

The coefficient of perforation for each layer is defined as

$$C_{1,2} = \frac{r_{\text{fab}}}{R_{1,2} c_0 \rho_0} \qquad 5.73$$

Then the diameter D of the apertures and the distance d between them are found by formulae

$$D_{1,2} = 1 \cdot 27\phi(\sqrt{c_{1,2}})(m_{1,2}c_{1,2} - d_{1,2}) \quad \text{and} \quad a_{1,2} = \frac{0 \cdot 89\, D_{1,2}}{\sqrt{c_{1,2}}} \qquad 5.74$$

173

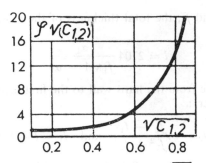

Fig. 5.28. Graph to determine the value of $\phi(\sqrt{C_{1,2}})$ in formula 5.74.

Fig. 5.29 shows calculated and measured frequency characteristics of sound absorption in a two layer system. The constructional parameters of the system are: breadth of the air spaces l_1 and l_2, 1·7 and 3 cm, thickness of the panels d, 0·1 and 0·2 cm, diameter of the apertures D, 1·2 and 0·35 cm, distance between the apertures a, 2 and 0·9 cm, the specific impedance of the material in the apertures r_{fab}, 3 and 3·6 mechanical ohms/cm².

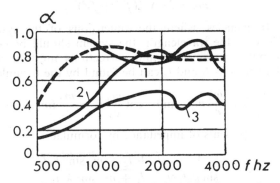

Fig. 5.29. Frequency characteristics of sound absorption for two layer systems: 1 for vertical incidence of sound, 2 and 3 for a diffuse field. The broken curve is one obtained by calculation.

Curve 1 is obtained for vertical incidence of sound, curves 2 and 3 for a diffuse field, and curve 3 is for a case where the fabric covering the apertures is stretched on a frame, while curve 2 is when this fabric is glued to each layer of the system.

A comparison of the curves demonstrates the importance of placing the fabric in close contact with the apertures in the panels. Moreover the correct choice of fabric and an accurate measurement of its acoustic impedance have an important effect on the value of the sound absorption coefficient.

The use of a two-layer construction results in an increase in the coefficient of absorption by 0·2 over a one layer system, while on going to a three layer system only an 0·05 increase in α results.

174

This analysis of a two layer porous vibratory system yields the following conclusions:

1. An increase in the number of layers in the system from one to two considerably broadens the limits of the frequencies for which the coefficient of absorption remains comparatively large.

2. The formulae for calculation allow a sufficiently accurate agreement identity between calculated and measured values of the coefficient of absorption.

3. To increase the breadth of the frequency characteristic of absorption, the distances between layers should not be increased as they come nearer the hard wall, i.e. they should tend towards $l_1 < l_2 < \ldots < l_n$. A similar trend is best with the resistive and inertial components of impedance.

4. To avoid gaps in the frequency characteristic, the air spaces should not only not be equal, but should not be multiples of one another.

5. The sound transparent material should be very carefully chosen and well glued to the panels, as this has a marked effect on the coefficient of absorption.

6. Resonant one-, two-, and three-layer systems have small dimensions and are robust, allowing their use for the deadening of ventilation trunking with strong air flows.

7. In making panels of metal or plastic, the perforations can easily be punched in the sheet material.

5.10 The use of sound absorbent materials and systems in practical conditions

The sound absorbing materials and the sound reflecting areas of the interior treatment of enclosures should be chosen not only to ensure the necessary acoustic conditions, but also to further the architectural and interior design.

Porous materials are used in the form of sound absorbing plasters with a granular or fibrous structure, of fabrics or mats made of organic mineral or artificial wool, or of acoustic tiles and blocks compressed from fibres with the addition of binders. The random-incidence coefficients of a number of such materials are given in Appendix 1.

Sound absorbing plasters are the strongest, and are easy to mount on suitably prepared surfaces. However, when they are made up on building sites they do not always possess uniform coefficients of absorption.

Where fabrics and mats of fibrous materials are used, a number of conditions have to be met. First the recommendations on the thickness of the layer and the distance between it and the wall have to be followed. In fitting drapes to doors, windows, etc., it should be remembered that the absorption characteristic varies with the distance between the fabric and the rigid surface. The absorption of energy is proportional to the square of the particle velocity, and therefore if the fabric is placed at the distance of a quarter wavelength from the wall ($l = \lambda_{1/4}$), where this velocity is a maximum, α reaches a peak at the corresponding frequency ($f_1 = c_0/\lambda_1$).

175

The thickness of mat should be chosen in accordance with the data in Appendix 1. Choosing too thin a mat results in a reduction of the absorption coefficient at low frequencies, while the use of too thick a material is uneconomic and wasteful.

Basing ourselves on theoretical data, the thickness is considered to be adequate if $l = 100/\sqrt{R}$, where R is the acoustic impedance of the material in grams/cm^3. sec.

Practically adequate thicknesses of some porous materials are given in Table 1.

TABLE 1

Materials	Thickness in cm
Cotton wool	40–60
Woollen felt	18
Compressed felt	12
Mineral wool	8–10
Cork tile	7–8
Dry plaster	3–4
'Wet' plaster	0·9–1
Cardboard	2
Insulite	0·8–1
Porous plaster	0·6

Fibrous materials are made up into mats, quilted with thin fabric. The mats are fastened to wooden supports fixed at a distance of 5–10 cm from a rigid wall and 1·5–2 metres from one another. On their outer surfaces they are protected by a metal grid. Sometimes fibrous absorbers are placed in the cells of a rigid frame assembled against the wall. The depth of the cells should be equal to the thickness of the layer, and the height, in order to prevent the material from sagging should be not more than 0·5–0·6 m.

In practice the use of fibrous absorbent materials is associated with various perforated coverings which have circular or slit-shaped or decorative apertures. These coverings meet the demands of contemporary architectural and interior decorating styles. They have the advantage that when a particular construction using a perforated covering has been chosen, the frequency characteristic of the absorption of the basic material can be corrected.

Coverings are usually made of plywood, fibreboard, metal or plasterboard. They can easily be painted and they can be given any decorative shape, which allows the design of acoustic constructions to be combined with architectural design of the bearing surfaces (Fig. 5.30).

Compressed porous tiles are the most convenient form of absorber from the point of view of the ease of fixing. The tiles can be given any shape and they are easily fastened to supporting battens.

176

Arborite, insulite and similar materials are made from paper waste, sawdust and wood pulp. They are quite strong and fire resistant. Vermiculite, made from heated mica, has one further quality—it is non-hydroscopic.

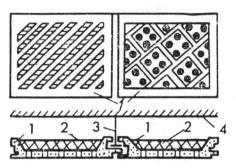

Fig. 5.30. Method of fastening acoustic tiles (1) with fibrous material (2) to the ceiling (4) by use of metal supports (3).

Like other porous materials, tiles suffer a change of absorption characteristic depending on the distance between them and the rigid wall. If they are fastened directly to the wall (curves 1, 2, 3, 4 in Fig. 5.31) their absorption characteristic at low frequencies is lower than when they are fastened at a distance of from 4–6 cm from the wall (curves 1′, 2′, 3′ in Fig. 5.31).

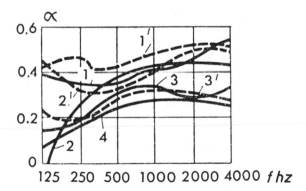

Fig. 5.31. Frequency characteristics of sound absorption for ROSNIIMS tile (1) arborite (2) insulite (3) and vermiculite (4). Curves 1′, 2′, 3′ are for the same tiles when they are fastened 5 cm away from the wall.

Acoustic tiles may be painted with paints which do not cover the pores (water solutions of dyes) or by paints which produce a thin resilient film which transmits the vibrations sufficiently well into the internal pores of the material.

177

Hard panel systems have a number of constructional and artistic advantages compared with porous materials, including damp resistance, durability, and the ability to accept any kind of surface treatment and to be redecorated. They can be painted, varnished or polished.

From the decorative point of view the only drawback of panel systems is that their sizes must be varied in order to broaden the frequency range over which the sound absorption is adequate. This variation results in the appearance of joints between individual panels and in projections in the surface at points where the distance between the wall and the system changes.

As the absorption of each element of the system is determined by the constructional data such as the material, the dimensions of the system, the

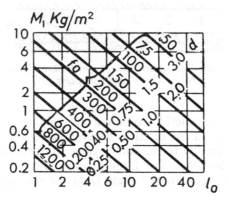

Fig. 5.32. Graph to determine the resonance frequency f_0 for panels of weight m, thickness d with an air space l_0.

way in which it is secured and the distance it is placed from the wall, great attention must be paid during assembly to the strict observance of all its calculated parameters. To check the accuracy of the solution of the problem, the graph in Fig. 5.32 may be used. This graph, for a given mass of panel *m* allows the choice to be made of dimensions for the panel *d* and for the distance from the wall l_0, to ensure the maximum absorption at the given frequency f_0.

For example, using the graph and knowing that *m* for six-ply and three-ply panels respectively is 5 and 2·2 kg/m², we find that if six-ply panels have dimensions 0·75, 1·0; 1·5 m and the same length, then where l_0 is respectively equal to 2, 4 and 8 cm, the systems will have maximum absorption at frequencies of 200, 150 and 100 Hz. The same resonance frequencies will be obtained for three ply panels if, having the same dimensions, they are placed respectively at 4, 7 and 18 cm from the wall.

Rigid panel systems, apart from their use in correcting the absorption at low frequencies also create in the enclosure a more diffuse sound field, as a plane wave reflected from a vibrating surface loses its directional properties. The sound scattering properties of rigid systems are enhanced if the distance *d* between neighbouring panels is greater than the width d_1 of each

178

system and the sum of $d + d_1$ is greater than the length of the sound wave (Fig. 5.33).

For the attachment of rigid panels, it is preferable to use damping washers, which will not lose their elastic properties with time. These washers should not be too much compressed during assemblies.

Perforated panel systems like solid panel systems, can easily be included in the general architectural design plan, as the dimensions of the systems and their outer decoration can be selected at will.

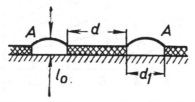

Fig. 5.33. Diagram of the correct disposition of rigid panel A.

The diameter of the apertures and their shape (Fig. 5.34) can also be varied to suit the designer. The systems are durable, and so their expense is economically justifiable. In treating an enclosure with perforated panel systems care should be taken to ensure that all the calculated elements of the construction are realized in practice, as their acoustic properties depend primarily on this.

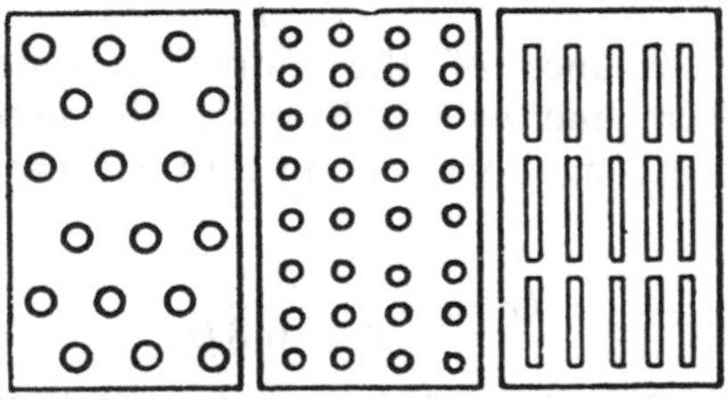

Fig. 5.34. Different designs of perforated panel systems.

For the calculation of perforated panel absorbers, formulae 5.63 and 5.64 may be used, and also the nomogram produced by Tsvikker and Kosten[90]. In the nomogram (Fig. 5.35) the required frequency ratio f_1/f_2, within which the coefficient of absorption does not vary by more than two to one, is plotted along the horizontal axis, and an the vertical axis are the maximum coefficient of absorption α (on the left) and the ratio $\dfrac{R}{nc_0\rho_0}$ (on the right).

The parameters of the curves are the values $n\dfrac{V}{\lambda}$ and $n\sigma\lambda$, where n is the number of apertures per m^2, V is the volume per unit area, and σ is the conductance of one aperture.

Let us consider the method of determining the parameters of a perforated panel system. First, the data of a system for which the maximum coefficient

179

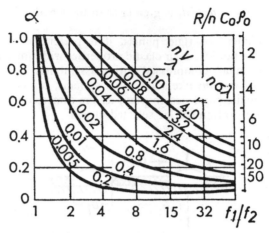

Fig. 5.35. Nomogram for the determination of the parameters of porous-resonant systems.

of sound absorption at a frequency of 300 Hz equals 0·6, and at frequencies of 100 and 900 Hz falls to 0·3. From Fig. 5.35 we find that for values of $\alpha = 0.6$ and $f_1/f_2 = \dfrac{900}{100} = 9$ the corresponding values are $n\dfrac{V}{\lambda} = 0.08$; $n\sigma\lambda = 3.2$ and $R/nc_0\rho_0 = 4.5$. As $\lambda = \dfrac{c_0}{f} = \dfrac{450}{300} = 1.5$, the volume per unit area, i.e. the distance of the vibratory system from the wall $l_0 = nV = 0.09 = 9$ cm and $n\sigma = 2.8$ cm^{-1}. The necessary value of $n\sigma$ can be found by various methods, taking into account that for a circular aperture

$$h\sigma = h\frac{S}{\delta + 0.8D},$$

where S is the area of section of the aperture, δ is the thickness of the system, and D is the diameter of the aperture.

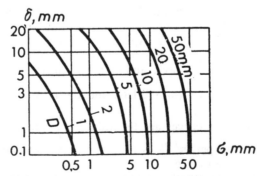

Fig. 5.36 Graph for the determination of the diameter D of the apertures and the thickness δ of a porous vibratory system.

180

Fig. 5.36 is drawn from this formula, showing the curves $\sigma = f(\delta)$ for various values of the diameter D of the apertures.

If we take the number of apertures per m^2 to be 250, then $\sigma = 11\cdot2$, and from the curves in Fig. 5.36 we find that D is 14 mm for $\delta = 1$ mm, or 18 mm for $\delta = 5$ mm.

Perforated panel systems consist of separate panels, thus interfering with the impression of continuity of surface in the decorative treatment of the walls or ceiling of the enclosure. To mask the joints between individual panels a new method of design for the inner surfaces of enclosures is used.

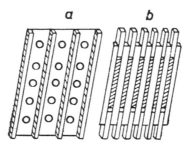

Fig. 5.37. Ways of fixing laths for the attachment of perforated panel (a) and porous systems (b).

It consists of fitting wooden or thin metal strips between rows of apertures (Fig. 5.37). Similar strips are often used in the treatment of surfaces covered by porous sound absorbent materials.

Fig. 5.38 shows ways of masking gaps, also by means of strips. Narrow gaps between the strips, particularly if they have a complex form, play the part of slit resonators, which give maximum absorption at a frequency

$$f_0 = \frac{c_0}{2\pi} \sqrt{\frac{b}{\left[1 + \frac{2b}{\pi}\left(1\cdot12 + \ln\frac{c_0}{\pi b f_0}\right)\right] S_{\text{trans}}}}$$

where l and b are the depth and width of the slit, and S_{trans} is the area of the transverse section of the slit gap.

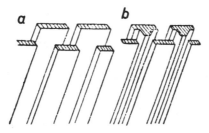

Fig. 5.38. Ways of masking gaps filled with sound absorbing materials.

6 Sound Insulation of Enclosures

6.1 Ways by which sound penetrates into the enclosure. Sound insulation as a natural property of the enclosure

By sound insulation is understood the protection of the enclosure against the penetration of interfering sounds or noises, the sources of which may be inside or outside the building.

To find ways of protecting the enclosure from noises, we must first establish the nature of these noises and the ways by which they enter the enclosure through its surfaces.

Noises can be divided into groups depending on the way in which they arise.

1. Continuous noise with a widely varied frequency spectrum: traffic noise, noise from neighbouring concert halls or music studios.

2. Continuous noise with an unchanging frequency spectrum, the source of which could be electric motors, machinery, falling water, etc.

3. Noise of short duration transmitted through the air, as, for instance, aeroplane noise, railway and ship siren, noise from a sawmill, etc.

4. Continuous and short noises emanating from mechanical vibrations of the building structure caused either by external excitation, as from a passing train, or by working machines or installations within the building.

There are several possible ways by which such noises can penetrate into the enclosure (Fig. 6.1).

Noises penetrate the enclosure by way of air transfer.

1. Through apertures and cracks in the walls or through ventilation trunking.

182

2. Through the pores in hard continuous walls.

3. By elastic vibrations of the wall separating the enclosure from the source of the interfering sound (bending vibrations).

4. By resilient longitudinal vibrations of non-adjacent walls. Vibrations propagated in the thickness of the walls are radiated into the enclosure by the side walls (flanking transmission).

Noises reach the enclosure, after generation in, and propagation through solid bodies.

5. As a result of vibrations of the material of the walls the material itself becoming a radiator of sound (impact sound transmission).

6. By vibrations caused by the moving traffic or by powerful machinery, transmitted through the ground, the foundation or other parts of the building structure.

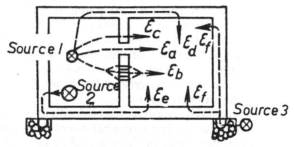

Fig. 6.1. Diagram of the penetration of noises into an enclosure. The indices of ε correspond to the items in the text .

To evaluate the insulation of each of the walls, the energy absorbed by them must be divided into two parts—the energy dispersed within the thickness of the wall and the energy which passes through it. Thus:

$$\varepsilon_{abs} = \varepsilon_{disp} + \varepsilon_{pass}$$

The energy which passes through is described by the value of the sound transmission coefficient, which is expressed by the ratio:

$$\alpha_{trans} = \frac{\varepsilon_{trans}}{\varepsilon_{inc}}$$

It is clear that the insulating capacity of the wall increases as this coefficient is reduced.

[*Editor's Note.*

The notation used by the author is different from that with which English-speaking readers will be familiar:

Transmission coefficient ratio of transmitted to incident energy is denoted by α_{trans}. This is more usually denoted by τ which he uses for the reciprocal.

This quantity τ he calls the coefficient of opacity and we have preserved this term which has no equivalent in English-speaking countries. Its logarithmic equivalent is the Sound reduction index.]

183

The amount of sound energy which is retained by the wall is described by the coefficient of opacity τ and is expressed as:

$$\tau = \frac{1}{\alpha_{\text{trans}}} = \frac{\varepsilon_{\text{inc}}}{\varepsilon_{\text{trans}}} = \frac{P_{a\ \text{inc}}}{P_{a\ \text{trans}}} \qquad 6.1$$

If this is expressed in decibels:

$$\tau_{\text{db}} = 10 \log \frac{1}{\alpha_{\text{trans}}} = 10 \log \frac{\varepsilon_{\text{inc}}}{\varepsilon_{\text{trans}}} \qquad 6.2$$

expresses the sound reduction index of the wall.

The total sound insulation of the enclosure is determined by the sound insulation of all the boundaries and depends both on the noise level behind these boundaries and on the level of noise permissible in the given enclosure. If, for example, the level of interfering sounds for each of the four walls of an enclosure comprises respectively 90 phon, (symphony orchestra) 80 phon (motor horn) 70 phon (noisy street) and 50 phon (quiet street), and the level of permitted noise in the enclosure (e.g. a cinema) is taken to be 40 phon, then the sound insulation of these walls should be no less than 50, 40, 30 and 10 dB respectively. The permitted noise level of an enclosure is the term used to describe the noise created by the apparatus and people normally working in the given enclosure.

In determining the practical questions connected with the sound insulation of an enclosure, it is necessary to know to what extent the sound insulation depends on the physical properties of the material of the wall and on the character of the interfering sound. It is especially important to know the frequency dependence of the sound insulation, and not only because the sound transmission of various materials is not the same at all frequencies, but also because aural perception also depends on frequency.

If we bear in mind that according to the equal loudness curves the sensitivity of the ear at low and high frequencies becomes progressively less as the level of sound pressure is reduced, then uniform reduction in this pressure will lead to a particularly noticeable reduction in the levels of loudness of low-frequency and high-frequency components of the noise. Consequently the greater the value of sound insulation, the more lacking in high and low frequencies the frequency spectrum of the penetrating sound appears to be.

Such a change in the subjective make-up of the spectrum is extremely desirable, as a significant suppression of low frequencies reduces the masking action of the noise, and suppression of high frequencies leads to an improvement when the interfering noise is speech, which loses its clarity with the loss of high frequency components.

6.2 Transmission of sound through apertures

If a plane sound wave impinges on a large rigid wall with an aperture, the wave which passes through the aperture is propagated onwards as if the aperture itself had become a source of sound. The form of the wave front is preserved or changes according to the dimensions of the aperture. If the diameter of the aperture is less than the wavelength of the sound falling on the wall, the sound which passes through the aperture is propagated in all directions and the wave front is approximately in the form of a hemisphere (phenomenon of diffraction).

If the diameter of the aperture is large compared with the wave length, the wave front retains its shape for a short distance from the aperture, and only at greater distance ($d > \lambda$) does it gradually become transformed into the spherical form.

Thus one and the same aperture may behave at low frequencies as a 'small' aperture, creating a wave of spherical form, while for high frequencies the aperture is 'large', and the wave passing through it retains its plane form.

To determine the sound insulation of a wall with an aperture, it is necessary to discover the value of sound energy falling on this aperture and passing through it.

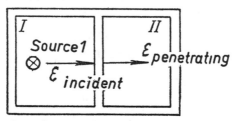

Fig. 6.2. Sketch for the derivation of formula for transmission of sound by a partition.

The sound energy falling on the wall during the time Δt is equal to the product of the total level of the stored energy in the enclosure where the sound source is situated (1 in Fig. 6.2) ε_1, and the probability of impact of the wave on this wall in that period of time:

$$\Delta \varepsilon_{inc} = \varepsilon_1 \omega_4$$

If we insert the value of ω_4 derived from expression 2.13 we find that:

$$\Delta \varepsilon_{inc} = \varepsilon_1 \frac{c_0 S}{4V} \Delta t$$

The sound energy falling on the same area S in a unit of time is:

$$\varepsilon_{inc} = \frac{\Delta \varepsilon_{inc}}{\Delta t} = \varepsilon_1 \frac{c_0 S}{4V} = \frac{c_0 S}{4} E_1$$

The energy density may be expressed in terms of sound pressure, as:

$$E_1 = \frac{p_1^2}{\rho_0 c_0^2}$$

so,

$$\varepsilon_{inc} = \frac{c_0 S}{4} \cdot \frac{p_1^2}{\rho_0 c_0^2} = \frac{p_1^2 S}{4 \rho_0 c_0} = \frac{\pi D^2 p_1^2}{16 \rho_0 c_0} \qquad 6.3$$

where S and D represent the area and the diameter of the aperture.

The sound energy passing through the aperture, is expressed by the equation:

$$\varepsilon_{trans} = \frac{p_1^2 \, R_{ap} S^2}{[2R_{ap} + j\omega(2M + M_{ap})^2]} \qquad 6.4$$

In this expression R_{ap} is the radiation impedance of the aperture, M_{ap} and M are the inertances of the air in the aperture and of the surrounding wall mass.

For small dimensions of the aperture compared with the wavelength, the radiation impedance of the aperture as given by the equation

$$R_{ap} = \frac{k^2 S^2}{2\pi} \rho_0 c_0 \qquad 6.5$$

is also small. If we insert the value of R_{ap} in the numerator and ignore this value where it occurs in the denominator, the expression for the energy passing through the aperture can be rewritten:

$$\varepsilon_{trans} = \frac{p_1^2 k^2 S^4 \rho_0 c_0}{2\pi |\omega(2M + M_{ap})|^2} \qquad 6.6$$

In this expression the inertance is

$$2M + M_{ap} = \frac{\rho_0 S^2}{G} \qquad 6.7$$

where G is the acoustic conductivity; if we consider that the length of the passage l is not very great, G can be written as an equation:

$$G = \frac{\pi D^2}{4l + \pi D} \qquad 6.8$$

For these values of inertance and conductivity, the equation 6.6 takes the form:

$$\varepsilon_{trans} = \frac{8\pi p_1^2 D^4}{(4l + \pi D)^2 \rho_0 c_0} \qquad 6.9$$

The sound reduction of a wall of thickness l, which has an aperture small compared to a wave length, is thus the ratio of the values defined by equations 6.3 and 6.9, and has the value:

$$\tau = \frac{\varepsilon_{inc}}{\varepsilon_{trans}} = \frac{\pi D^2 p_1^2}{16\rho_0 c_0} \cdot \frac{(4l + \pi D)^2 \rho_0 c_0}{8\pi p_1^2 D^4} = \frac{1}{8}\left(\frac{l}{D} + \frac{\pi}{4}\right)^2$$

or

$$\tau = 0{\cdot}12 \frac{l^2}{D^2} + 0{\cdot}19 \frac{l}{D} + 0{\cdot}08 \qquad 6.10$$

Thus the sound insulation of the wall depends both on the diameter of the aperture and on the length of the passage.

186

If the ratio $\dfrac{l}{D}$ is large, which may happen where the aperture is of small size of where the passage is long (where the wall is thick), then in accordance with expression 6.10 sound insulation becomes significant and increases almost in proportion to the square of the ratio l/D. In this case, for a fixed thickness of wall, sound insulation falls as the diameter of the aperture increases, and as this latter increases as the wavelength decreases, then sound insulation will also fall with a reduction in λ, or, which amounts to the same thing, with an increase in the frequency of the interfering sound.

These considerations give rise to the curves of sound insulation as a function of the diameter of a small aperture and of the frequency of the sound, shown in Fig. 6.3.

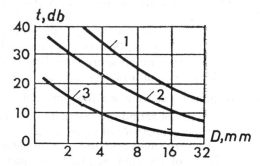

Fig. 6.3. Curves of the dependence of natural sound insulation of a wall on the diameter of an aperture for frequencies of 200 Hz (1), 1000 Hz (2) and 4000 Hz (3).

If the ratio l/D is small, which corresponds to a large aperture or to a thin wall, then the first two terms of the equation 6.10 can be disregarded, and the sound insulation will be a constant and comparatively small value.

Consequently apertures which are large compared to wave length in thin walls are comparatively good conductors of sounds. This situation is illustrated by curves in Fig. 6.3 which for high values of D fall towards the frequency axis, and come close together.

. For large apertures, such as door or window openings, or grids of ventilating systems, the sound energy which passes through is proportional to the area of the opening. Taking this into account and using the equal loudness curves, it is easy to establish that with a hundredfold decrease in the area of the aperture (e.g. from 1 m² to 1 dm² to 1 cm²), the loudness level at middle frequencies changes smoothly and comparatively slowly (at a frequency of 1000 Hz by 20 phon). At low and high frequencies the loudness level falls rapidly at first as the area of the aperture decreases, and although the fall slows down later, the absolute value of the loudness level is always less than at middle frequencies.

It follows that:

1. For apertures which are small in comparison to the wavelength the sound insulation of the wall decreases with increase in the diameter of the aperture.

2. The sound insulation of a wall with small apertures is greater at low frequencies and significantly decreases at high frequencies.

3. An increase in the thickness of a wall, which results in an increase in the length of the passage of the aperture, significantly increases its sound insulation.

4. If the wall contains apertures which are large in relation to the wavelength the sound insulation is relatively small and does not depend on the frequency of the interfering sound.

5. For very large apertures, a reduction in area results in a slow reduction of the loudness level at middle frequencies and a fairly rapid reduction at low and high frequencies.

6. A reduction in the area of very large apertures has the greatest effect on sounds of medium loudness. Protection of the enclosure from the penetration of strong sounds is much more complicated than protection from sounds with weak or medium loudness levels.

From these observations we can make a few practical conclusions:

1. From the point of view of sound insulation it is better to have several small apertures than one large aperture of equal area.

2. If the wall contains very small apertures, we should worry about the danger of the transmission of the high frequency components of noise; if the apertures are large then we should take steps relating to the middle frequencies.

3. To reduce the difficulties connected with protection against loud sounds, and to reduce the cost of sound insulation measures, a building should be designed to keep loud sound sources away from enclosures requiring protection.

4. If it is impossible to reduce the dimensions of apertures, e.g. of the grids of ventilating systems, it is essential to lengthen the trunking and to treat it with absorbent material, to increase the exposed area of the ducts by supplementary ribs or to introduce acoustic filters (see Chapter 6.10).

6.3 The transmission of sound through porous walls

Porous walls are those which have a large number of small apertures, the orientation and length of which differ one from another. The sound insulation of such walls is provided both by the reflecting of incident energy and by its internal attenuation resulting mainly from the viscosity of the air and friction at the walls of the pores.

It is clear that the energy reflected from the surface of the material and the energy which decays within its thickness depends on the porosity of the material. Moreover, the decay of energy caused by friction is connected with the length of the passages of the pores, i.e. with the thickness of the material.

188

To determine the sound insulation of a porous material we have to find an expression for the sound energy which passes through it. If we reckon that for a uniform distribution of pores the sound wave passing through has a plane form, we can write that

$$\varepsilon_{\text{trans}} = v^2 S \rho_0 c_0 = \frac{p_1{}^2}{|z|^2} S \rho_0 c_0 \qquad 6.11$$

Here S is the area of the whole wall.

If we limit ourselves to a consideration of comparatively low frequencies and remember that the mass of porous material per unit area is small, then in accordance with formula 5.47 the acoustic impedance of a material with rigid walled pores is:

$$z = \sqrt{\left(\frac{K}{\omega l P}\right)^2 + \left(\frac{lR}{3}\right)^2}$$

whence

$$\varepsilon_{\text{trans}} = \frac{p_1{}^2 S \rho_0 c_0}{\left(\dfrac{K}{\omega l P}\right)^2 + \left(\dfrac{lR}{3}\right)^2} = \frac{9 p_1{}^2 S \rho_0 c_0 \omega^2 l^2 P^2}{9 K^2 + l^4 R^2 \omega^2 P^2} \qquad 6.12$$

The sound opacity coefficient of the wall is determined as a ratio of expressions 6.3 and 6.12, i.e.

$$\tau = \frac{\varepsilon_{\text{inc}}}{\varepsilon_{\text{trans}}} = \frac{p_1{}^2 S}{4 \rho_0 c_0} \cdot \frac{9 K^2 + l^4 R^2 \omega^2 P^2}{9 p_1{}^2 S \rho_0 c_0 \omega^2 l^2 P^2} = \left(\frac{K}{2 \rho_0 c_0 \omega l P}\right)^2 + \left(\frac{lR}{6 \rho_0 c_0}\right)^2 \qquad 6.13$$

It follows from this last expression that the insulation of a porous wall increases with increase in flow resistance and with the thickness of the wall. The sound insulation also increases with a reduction in the porosity of the material, and this is confirmed by the experimental data shown in Fig. 6.4.

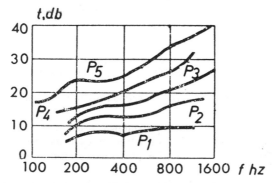

Fig. 6.4. The change of natural sound insulation of layers of felt having different degrees of porosity: $P_1 > P_2 > P_3 > P_4 > P_5$

With an increase in frequency the first part of equation 6.13, which results from the stiffness of the material, becomes less important and the sound insulation of the enclosure becomes largely determined by the second part of the formula 6.13 which expresses resistive losses. Following from the fact that losses as a result of friction of the air in the pores increase at

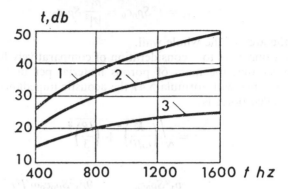

Fig. 6.5. The dependence of natural sound insulation of felt on frequency for different thicknesses of layer: (1) where l = 6 cm, (2) where l = 4·5 cm, and (3) where l = 3 cm.

middle and high frequencies, the relationship between sound insulation of porous material and frequency is shown in Fig. 6.5.

If we consider porous elastic materials, then on the basis of expression 5.53, supposing that $\rho_1 \rho_2$, $K_1 + K_2$ are constant values, we can conclude that z, and consequently τ increase with increase in frequency ω or with the thickness of the wall l.

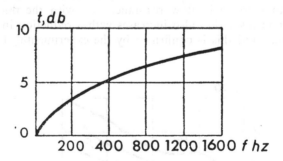

Fig. 6.6. The frequency characteristic of sound insulation for porous elastic material.

As coth $j\omega l$ asymptotically approaches unity for large values of the argument, the dependence of τ_dB on frequency is as shown by the curve of Fig. 6.6. The figure shows experimental measurements of attenuation in a layer of felt 1 cm thick.

It would seem that a reduction in sound insulation at low frequencies, characteristic for porous materials, might, as follows from expressions

190

5.45 or 5.53, be compensated for by an increase in the thickness of the material *l*; however, this requires such thick walls that their use is uneconomic.

The sound insulating properties of porous material which we have considered above apply only when the material is in isolation from other, non-porous materials, e.g. not fastened to a rigid surface.

Let us formulate some general conclusions about the sound insulating properties of walls made from porous materials.

1. Porous walls provide low sound insulation (10–20dB) which can be of practical significance only at high frequencies and where the material is very thick.

2. The sound insulation of these materials increases as their porosity decreases and the resistance increases.

3. These comments refer to the sound insulating properties of porous materials only when they are used as walls on their own.

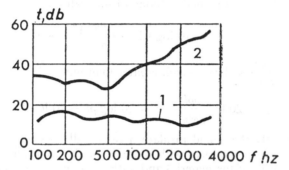

Fig. 6.7. The frequency characteristic of sound insulation for a porous wall without plaster (1) and with plaster (2).

4. Porous materials are not good enough insulators against airborne sounds, and as follows from Fig. 6.7, where curve 1 corresponds to an unplastered foam concrete wall 10 cm thick, they can hardly be used as sound insulating walls. Sandwiched between other materials they also give almost nothing. In this case, although sound insulation increases (see curve 2, obtained for the same wall after plastering), it does so only to such a degree that its total sound insulation still does not exceed that of a non-porous rigid wall of the same weight. However, porous materials can be usefully employed for complex and light weight sound insulating constructions if they used in conjunction with other materials but not in direct contact with them (see para. 6.5)

6.4 The transmission of sound via transverse vibrations of walls

Under the action of a sound wave, the wall dividing one enclosure from another may be brought into a state of vibration. Even very small

191

vibrations, with amplitude as little as 7.10^{-5} mm are transmitted into the adjoining enclosure and can be easily heard.

If we assume that the wall undergoes piston type vibrations, and is a system controlled by its mass, then where its movement is of small amplitude the energy passing through it can be expressed by the equation:

$$\varepsilon_{\text{trans}} = v^2 S \rho_0 c_0 = \frac{p_1{}^2}{\rho_0{}^2 c_0{}^2 + \omega^2 m^2} S \rho_0 c_0 \qquad 6.14$$

where m is the mass of the wall per unit area, equal to the product of the density of the material ρ and its thickness l.

If we divide expression 6.3 by 6.14, we can easily determine coefficient of opacity of a wall which is moving under the action of sound vibrations:

$$\tau = \frac{\varepsilon_{\text{inc}}}{\varepsilon_{\text{trans}}} = \frac{p_1{}^2 S}{4 \rho_0 c_0} \cdot \frac{\rho_0{}^2 c_0{}^2 + \omega^2 \rho^2 l^2}{p_1{}^2 S \rho_0 c_0} = \frac{1}{4} \left[1 + \left(\frac{\omega \rho l}{\rho_0 c_0} \right)^2 \right] \qquad 6.15$$

From this, the sound reduction index of the wall is:

$$\tau_{db} = 10 \log \left[1 + \left(\frac{\omega \rho l}{\rho_0 c_0} \right)^2 \right] - 6 \qquad 6.16$$

As the ratio $\dfrac{\omega \rho l}{\rho_0 c_0}$ is always much greater than unity, then

$$\tau_{db} = 20 \log \frac{\omega \rho l}{\rho_0 c_0} - 6 = 20 \log \rho l + 20 \log f - 43 \qquad 6.17$$

Consequently the sound insulation of a wall moving under vibrations as a single whole increases with increase of the frequency of the sound, with the density of the material and with the thickness of the wall itself.

If we bear in mind that an increase in sound insulation results in a disproportionately rapid reduction in the loudness level of low and high frequency components of the noise (see Chapter 6.1), it is sensible to calculate the mean value of sound insulation, taking a mean for the frequency range of from $f_1 = 100$ Hz to $f_2 = 3200$ Hz.

In this case formula 6.17 takes the form:

$$\tau_{db} = 20 \log \rho l + 20 \log \sqrt{f_1 f_2} - 43 = 20 \log \rho l + 12 \cdot 3 \qquad 6.18$$

This formula is correct only if the conditions set out above are met, that the frequency of the incident wave is high in comparison with the frequencies of the normal modes of vibration of the wall. These characteristic frequencies, which depend on the density of the material ρ and the thickness of the wall l, are spread over a wide range, lying, in practice, between 10–50 Hz for brick walls and 200–500 Hz for wooden walls (see shaded areas of curves in Fig. 6.8).

The presence of natural modes of vibration results in the fact that the theoretical curve 1, of the dependence of τ_{db} on the mass of the wall (Fig. 6.9) lies higher than the experimental curve 2. Curve 2 lies in the middle of the zone obtained by measuring the sound insulation of various walls and

192

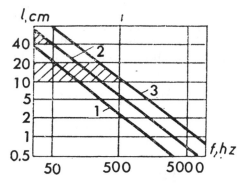

Fig. 6.8. Change of frequency of normal modes of vibration depending on the thickness of the wall for materials having different densities: 1 for glass, 2 for brick, 3 for wood.

results from the fact that sound insulation is a function not only of mass, but to a certain extent of the rigidity and the resistive impedance of the wall. Taking this into account, the following formulae are used for practical calculations:

$$\left.\begin{array}{llll} \tau_{db} = 13\cdot5 \log \rho l + 13 & \text{where} & \rho l \leqslant 200 \text{ kg/m}^2\\ \tau_{db} = 23 \quad \log \rho l - 9 & \text{where} & \rho l > 200 \text{ kg/m}^2 \end{array}\right\} \qquad 6.19$$

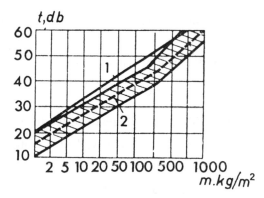

Fig. 6.9. Theoretical (1) and experimental (2) curves of the dependence of insulation on the mass of the wall.

These formulae enable practical problems of sound insulation to be solved sufficiently accurately, but, as they are empirical, they cannot provide an explanation of the mechanism of sound transmission through barriers.

In deriving formulae 6.17 and 6.18 it was supposed that the wall moved as a single whole. This may be true for a uniform action of the sound wave on all parts of the wall, i.e. where the wave is normally incident. But if the wave impinges at an oblique angle θ, then the forces acting on various

193

parts of the wall will have different phases, and, as experience shows, the wall executes bending vibrations (Fig. 6.10).

With this approach, the wall should be considered as a beam undergoing bending vibrations. The equation of motion of the beam is represented in the form:

$$\Delta p + j\frac{\varepsilon I}{\omega}\frac{\mathrm{d}^4 v}{\mathrm{d}x^4}\, j\omega m v \qquad\qquad 6.20$$

where Δp is the sound pressure acting on the wall, ε is the modulus of elasticity, and I is the moment of inertia.

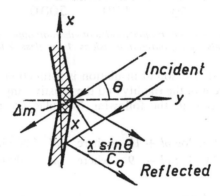

Fig. 6.10. Diagram of the action of a sound wave falling obliquely onto a wall.

If the sound acting on the wall has a sinusoidal form, we may write

$$v = e^{j\omega\left(t - \frac{x\sin\theta}{c_0}\right)}$$

where $\dfrac{x\sin\theta}{c_0}$ (see Fig. 6.10) is the difference in the time of arrival of two rays distance x apart impinging on the wall at an oblique angle θ of incidence, and

$$\frac{\mathrm{d}^4 v}{\mathrm{d}x^4} = j^4 \frac{\omega^4}{c_0^4}\sin^4\theta\cdot v$$

Equation 6.20 then takes the form:

$$\Delta p = j\omega\left(m - \frac{\varepsilon I}{c_0^4}\omega^3\sin^4\theta\right)v = j\omega m\left(1 - \frac{\varepsilon I}{m}\frac{\omega^3}{c_0^4}\sin^4\theta\right)v \qquad 6.21$$

For a uniform beam $I = \dfrac{l^3}{12}$ and $m = \rho l$, and so the acoustic impedance of the wall, on the basis of expression 6.21 is written as:

$$z = \frac{\Delta p}{v} = j\omega m\left(1 - \frac{\varepsilon l^2}{12\rho}\frac{\omega^3}{c_0^4}\sin^4\theta\right) \qquad\qquad 6.22$$

194

If we substitute this value of acoustic impedance into expression 6.11, we get:

$$\varepsilon_{\text{trans}} = \frac{p_1{}^2 S \rho_0 c_0}{\omega^2 \rho^2 l^2 \left(1 - \dfrac{\varepsilon l^2}{12\rho} \dfrac{\omega^3}{c_0{}^4} \sin^4 \theta \right)^2} \qquad 6.23$$

The ratio of the energies defined by equation 6.3 and 6.23 gives the expression for the coefficient of opacity:

$$\tau = \frac{\varepsilon_{\text{inc}}}{\varepsilon_{\text{trans}}} = \frac{p_1{}^2 S}{4\rho_0 c_0} \cdot \frac{\omega^2 \rho^2 l^2 \left(1 - \dfrac{\varepsilon l^2}{12\rho} \cdot \dfrac{\omega^3}{c_0{}^4} \sin^4 \theta \right)^2}{p_1{}^2 S \rho_0 c_0}$$

or

$$\tau = \frac{\omega^2 \rho^2 l^2}{4\rho_0 c_0} \left(1 - \frac{\varepsilon l^2 \omega^3 \sin^4 \theta}{12 \rho c_0{}^4 \rho_0} \right)^2 \qquad 6.24$$

From this equation it follows that the insulation of a wall undergoing bending vibration depends, not only on the frequency of the interfering sound, the thickness of the material and the elasticity of the wall, but also on the angle θ of incidence of the wave.

If we introduce the symbol

$$c_1{}^4 = \frac{\varepsilon l^2}{12\rho} \omega^2 \qquad 6.25$$

where c_1 is the speed of propagation of free waves along the wall. Then expression 6.22 for acoustic impedance takes the form:

$$z = j\omega\rho l \left[1 - \left(\frac{c_1 \sin \theta}{c_0}\right)^4 \omega \right]$$

and z becomes zero if

$$c_1 = \frac{c_0}{\sqrt[4]{\omega \sin \theta}} \qquad 6.26$$

i.e. if the speed of propagation of bending waves, c_1 coincides with the speed of forced vibrations, which depends both on the angle of incidence of the wave and on the frequency of the sound. In such a case τ has a minimum value.

It is clear that there are several such minima. Their frequencies can be determined from expression 6.25 in the form:

$$f_0 = \frac{c_1{}^2}{2\pi l} \sqrt{\frac{12\rho}{\varepsilon}}$$

or, taking equation 6.26 into account, in the form:

$$f_0 = \frac{c_0{}^2}{2\pi l \sin^2 \theta} \cdot \sqrt{\frac{12\rho}{\varepsilon \omega}} \qquad 6.27$$

Consequently for a wall able to sustain bending vibrations, the curve of sound insulation frequency as a function of frequency has a number of minima, i.e. is represented as shown in Fig. 6.11 by line *b*. Curve *a* represents the relationship for a wall executing piston type vibrations.

In real conditions, a significant part of a wall or floor behaves as a resilient wall, and has dips in the curve of sound insulation against frequency, reaching 10 or 15 dB. The frequency of the basic resonance is comparatively low (see Fig. 6.3). Then comes a group of particular resonances resulting from the coincidence effect, the last of which depends on the material of the wall and increases in frequency as the thickness of the wall diminishes.

As is seen in expression 6.27 the coincidence effect is observed for specific values of the angle of incidence θ of the wave. Thus, when the wave incidence

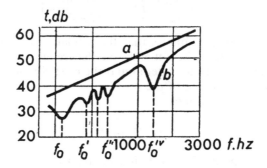

Fig. 6.11. The frequency dependence of sound insulation for walls which cannot sustain bending vibrations (a) and for those which do (b).

is random, i.e. when impact at any angle is equally probable, there is always the possibility of the appearance of a number of frequencies at which wave coincidence will take place.

In order to find the value of natural sound insulation under conditions of a diffuse field, the value τ_θ from expression 6.24 must first be substituted into the equation

$$\tau = \int_0^{2/\pi} \tau_\theta \sin 2\theta \, d\theta,$$

obtained by analogy with equation 6.10 and then the value of τ derived in decibels.

Such a solution of the problem, for the case $f > 2f_0$, enables us to obtain the following expression for sound insulation of a partition.

$$\tau_{db} = 20 \log \frac{\pi f_0 m}{\rho_0 c_0} + 30 \log \frac{f}{f_0} + 10 \log \eta - 3 \qquad 6.28$$

where η is a coefficient which takes into account the internal friction of the material of the barrier.

196

The product of the frequency of the basic resonance f_0 and the thickness of the wall l for each of the materials is almost constant; for wooden walls it amounts to 450–500, for concrete 600–650 and for brick 1000–1100. This is confirmed by Fig. 6.12, which shows the dependence of the value of

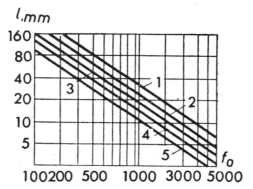

Fig. 6.12. The dependence of frequency f_0 on the thickness of the wall l for plaster of Paris, (1) for plexiglass (2) for concrete (3) cast iron (4) glass or deal (5).

the first characteristic frequency on the thickness of the wall for a number of materials.

It follows from Fig. 6.12 that the reduction in mean sound insulation of the wall in the neighbourhood of frequency f_0 can be avoided by moving this frequency beyond the limits of the standard range (100–3200 Hz).

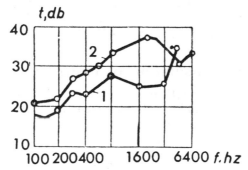

Fig. 6.13. Frequency characteristics of sound insulation for a solid panel (1) and a slotted panel (2).

With thick walls frequency f_0 may be reduced by *increasing* the rigidity of the walls, while with thin walls it is reduced by *reducing* the rigidity of the wall.

The influence of the change in rigidity is shown in Fig. 6.13; curve 1 represents the frequency characteristic of sound insulation for a solid wooden panel, curve 2 the characteristic for the same panel, whose rigidity has been reduced by narrow slots cut into it in two directions. As can be

197

seen from a comparison of curves 1 and 2, the reduction in the rigidity of the wall results in a shift of frequency f_0 outside the limits of the standard range of frequencies.

We can vary the rigidity of walls by adding to their surfaces two rows of crossing ribs or grooves.

Thus:

1. Sound penetrates into an enclosure as a result of the vibrations it excites in the walls.

2. We can consider that brick walls 40–50 cm thick, the normal modes of which occur at low frequencies, carry out piston-type vibrations and that their sound insulation is proportional to the mass of the wall and of the frequency of the vibrations. An increase in sound insulation of such walls is possible within very narrow limits by increasing their thickness or the density of the material.

3. For walls which sustain bending vibrations (if the characteristic frequency of the vibrations lies within the limits of the standard frequencies) sound insulation depends not only on the thickness of the wall and the density of the material but also on the elastic properties of the material. Owing to the coincidence effect such walls have a number of frequencies at which their natural sound insulation noticeably decreases (by 10–15 dB).

4. We can avoid the undesirable effects of coincidence if we increase the rigidity of thick walls and reduce it for thin.

5. Bearing in mind the disproportionately rapid fall in the sensitivity of the ear at low and high frequencies, if we have a uniform reduction in the sound level as a result of sound insulation, we need only consider the values at middle frequencies of sound insulation, defining the mean value by:

$$f_{\text{mean}} = \sqrt{f_1 f_2} = \sqrt{100.3200} \simeq 550 \text{ Hz}$$

6.5 The sound insulation of multi-layer walls

It follows from formulae 6.19 and from Fig. 6.9 that to increase the sound insulation of a single wall by 10 dB we have to increase its weight ten times, which is extremely uneconomic. So the possibility of increasing the effectiveness of the wall, by dividing it into a number of thinner leaves so that the sound is reduced in stages is of great interest. In this case, grossly oversimplifying, we can say that each leaf works independently, and if one leaf of mass m is split into two with masses m_1 and m_2, then the total sound insulation of this double leaf is shown in the form:

$$\tau_{db2} = 20(\log m_1 + \log m_2) = 20 \log m_1 m_2 \qquad 6.29$$

From this the gain in sound insulation is expressed by the equation:

$$\Delta\tau_{db} = \tau_{db2} - \tau_{db} = 20 \log m_1 m_2 - 20 \log (m_1 + m_2) = 20 \log \frac{m_1 m_2}{m_1 + m_2}$$

$$6.30$$

198

In fact sound insulation is a result not only of the inertial impedances of the two screens, but also of the elasticity of the air in the space between them (Fig. 6.14). If each screen is supposed to represent only inertial impedance, the sound pressures acting on the first and second leaves can be respectively represented by the equations:

$$p_1 = j\omega m_1 v_1 \quad \text{and} \quad p_2 = j\omega m_2 v_2 \qquad 6.31$$

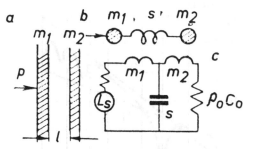

Fig. 6.14. A section of a double-leaf wall (a) and its equivalent mechanical (b) and electrical (c) circuits.

The vibration velocities of the leaves, taking into consideration the presence of the air space between them, are expressed as:

$$v_1 = j\frac{\omega l}{\rho_0 c_0^2} \quad \text{and} \quad v_2 = \frac{1}{\rho_0 c_0} p_{\text{trans}} \qquad 6.32$$

If we solve these two pairs of equations 6.31 and 6.32 together, we find that:

$$\tau = \frac{p_1}{p_{\text{trans}}} = -j\frac{\omega^3 m_1 m_2 l}{\rho_0^2 c_0^3}$$

The sound insulation of a double leaf wall is represented in the form:

$$\tau_{db} = 20\log\frac{\Gamma_1}{P_{\text{trans}}} = 60\log\omega + 20\log m_1 m_2 + 20\log l - 20\log \rho_0^2 c_0^3$$
$$6.33$$

The presence of an elastic coupling between the leaves results in the curve of sound insulation of the system against frequency having a number of dips due to resonances, and the frequency of the basic resonance has the value:

$$f_0 = \frac{c_0}{\pi}\sqrt{\frac{\rho_0}{2l m_1 m_2}} \qquad 6.34$$

With the aim of removing this undesirable drop in the frequency characteristic, it will be seen from equation 6.34 that we must increase the mass of each of the leaves and the width of the air space between them. In this case the resonance frequency can be reduced in such a way that it passes outside the limits of the sound range.

It follows from Fig. 6.15 that below the frequency of the basic resonance, when the system is controlled by mass, this curve coincides with the curve for a single layer wall. At higher frequencies, the elastic properties of the air space begin to have an effect, and the curve turns steeply upwards.

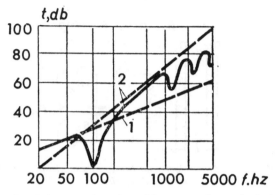

Fig. 6.15. Sound insulation as a function of frequency for a single layer (1) and a double layer (2) wall.

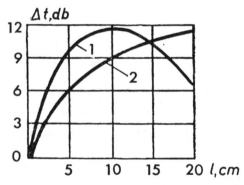

Fig. 6.16. An increase in sound insulation resulting from a change in the breadth of the air space in double walls, (1) of wood and (2) of glass.

Thus the improvement of sound insulation $\Delta \tau_{db}$ shown by the difference of the ordinates of the curves for double- and single-layer walls, is caused by the action of the air space. This improvement (Fig. 6.16) increases noticeably with increase of the width of the air space to between 8 and 12 cm. A further increase in the breadth of the space has little effect on the sound insulation of the complex wall.

The drop in sound insulation at resonance frequencies can be partly avoided if measures are taken to reduce the coupling between the leaves. This can be achieved by widening the air space l or by placing porous

200

material within the space in such a way that it is not in direct contact with the leaves.

These measures, and also constructing the wall in various materials and at various thicknesses, results in the variation of phases between the vibrations of the two parts of the wall, and reduces the possibility of the appearance of resonance dips. Apart from improving the frequency characteristic of sound insulation, these measures result in a general increase of the insulation of the double wall.

Thus:

1. Walls made up of several layers and having no rigid connection, behave as complex vibratory systems which reach a significantly greater value of sound insulation than can be obtained from a single wall of the same mass. By simple theory one would expect a doubling of the mass of a single wall to result in a maximum increase in sound insulation of 6 dB, while two leaves of the same mass would give double the value of sound insulation.

2. Air spaces between the leaves increase the sound insulation of the construction, as they play the part of dampers, the action of which becomes more effective as the width of the gap is increased and it is filled with absorbent material which does not touch the basic wall.

3. The dips in the frequency characteristic which arise as a result of resonances can be partly removed by increasing the mass of the leaves or the gap between them, or by altering the thickness, density and construction of each leaf.

6.6 The transmission of sound by way of longitudinal vibrations in walls

Noises may penetrate from a number of neighbouring enclosures into an enclosure which is to be isolated, and not only through a common wall. All the other walls of the enclosure in which the sound source is located, and which are connected by flanking walls with the room to be insulated, transfer sounds as a result of longitudinal vibrations of their walls (see Fig. 6.1).

Such an indirect transfer (see Fig. 6.10) arises from the fact that if a vibrating element of a wall Δm is connected with other elements in direction x, these neighbouring elements are brought into a state of vibration and the vibrations are propagated along the wall. Longitudinal vibrations along the side walls may reach the isolated room.

The speed of propagation of longitudinal compression waves in the wall is defined by the equation:

$$c_x = \sqrt{\frac{\varepsilon}{\rho}}$$

and depends on the density and elastic properties of the material of the wall. The parameters of the wall and its construction should be such as to avoid the coincidence of the wave velocity of obliquely incident sound in the air with the velocity of propagation of bending waves in the wall. In this case the flanking transmission of noise can be considerably reduced.

An increase in the sound insulation from flanking transmission can be achieved by an increase in the attenuation of sound in the material, which can be obtained by constructing the transverse walls of materials with different densities and elasticities. The same effect can be obtained if gaps are made in the wall constructions and filled with materials of a rigidity different from that of the basic material. Finally, complete insulation from the flanking transmission of sound waves can be accomplished by the maximum possible constructional separation of the elements making up the enclosure.

Frequently, where there is a high noise level in an enclosure, no heed is paid to the flanking transmission of sounds, and attempts are made only to increase the sound insulation of common walls. This does not always result in success, as in removing one of the causes of sound penetration, the other is allowed to remain.

6.7 The propagation of vibration and impact noise

Mechanical shocks caused directly in the structure of a building by such sources as rotating machinery or moving traffic are propagated through the walls in the form of compression, bending, and even torsional waves. The compressional waves are propagated through the earth, the foundation and the structural elements of the building in all directions, gradually becoming attenuated with distance from the source of the shock or vibration. The rate of attenuation and the distance to which percussive waves are propagated depend to a significant degree on the properties of the material through which they travel.

In some materials, e.g. in earth, the vibrations decay rapidly and their energy decreases in inverse proportion to the square of the distance; in others, e.g. in metal constructional frameworks and pipes, attenuation proceeds very slowly and the energy is distributed to great distances, reaching rooms remote from the source of the vibrations.

It is clear that there may be several ways of protecting an enclosure from vibration. In the first place we should try to reduce the amplitude of the vibrations at the point where they arise. Enclosures of which the noise level should be low should be placed as far as possible away from the source of the noise. Walls lying along the line connecting the source of the vibrations and the room to be protected should be constructed of materials having good attenuation. And finally, in structures made of materials which have good sound conducting properties, paddings of materials which do not conduct these vibrations should be provided.

Let the mass of the machine M_1 together with the mass M_2 and the elasticity s_2 of the mounting be a resonant system, on which is acting a driving force caused by the movements of the machine. The mechanical and equivalent electrical diagrams are shown in Fig. 6.17.

For such a system, in accordance with expression 6.1, we can write that:

$$\tau' = \frac{\varepsilon_{\text{inc}}}{\varepsilon_{\text{trans}}} = \left|\frac{z_{\text{trans}}}{z_{\text{inc}}}\right|^2 = \frac{\rho_0^2 c_0^2 + \left[j\omega(M_1 + M_2) - \dfrac{S_2}{j\omega}\right]^2}{\rho_0^2 c_0^2}$$

If we divide the numerator by the denominator, and then take $\left(\dfrac{S_2}{\omega}\right)^2$ outside the square brackets, we get:

$$\tau' = 1 - \left\{\frac{\dfrac{S_2}{\omega}\left[\left(\dfrac{\omega}{\omega_0}\right)^2 - 1\right]}{\rho_0 c_0}\right\}^2 \qquad 6.35$$

where $\omega_0 = \sqrt{\dfrac{S_2}{M_1 + M_2}}$, the natural frequency of the system.

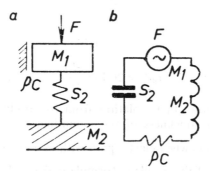

Fig. 6.17. Mechanical (a) and electrical (b) plans of a vibratory system representing a machine placed at the foundations.

As follows from this equation, isolation will be a minimum where $\omega = \omega_0$ i.e. at resonance. In order to ensure sufficient isolation, the natural frequency of the system should be as low as possible. It is desirable that the ratio $\dfrac{\omega}{\omega_0}$ should be greater than five.

As the compliance of the antivibration mounting or pad, s_2, is equal to:

$$\frac{\varepsilon S_2}{l}$$

then

$$\omega_0^2 = \frac{\varepsilon S_2}{(M_1 + M_2)l} = \frac{\varepsilon}{\rho ml} \qquad 6.36$$

Consequently, the characteristic frequency of the system can be lowered by increasing the mass of the baseplate under the machine, the thickness of the elastic layer, and by choosing an elastic material which has the necessary elastic properties and can retain them long enough. Such properties can be found in rubber and cork, which, for example, where $l = 2\cdot5$ cm and $m = 10$ kg/cm^2, has $f_0 = 16$ Hz.

If pads are used which have a thickness l_1 and a modulus of elasticity ε_1, then for longitudinal oscillations distributed in the material of the walls, the waves do not penetrate from the air into the wall, as was considered earlier, but from the wall into the pad. Consequently the role of the air medium in this case is played by the wall, and that of the wall by the pad.

Bearing that in mind, with some simplicity we can suppose that the longitudinal vibration is represented by formula 6.17, we can replace the density of the air and the material of the wall (ρ_0 and ρ) in accordance with the considerations above by values inverse to the moduli of elasticity of the wall and the pad (ε and ε_1). The formula of isolation against longitudinal waves is then:

$$\tau_{db}' = 20 \log \frac{\omega l_1 \rho_1}{2\rho c} = 20 \log \frac{\omega l_1 \varepsilon}{2c\varepsilon_1}$$

or, as

$$c = \frac{\omega\lambda}{2\pi} :$$

$$\tau_{db}' = 20 \log \frac{\pi l_1 \varepsilon}{\lambda \varepsilon_1} \qquad\qquad 6.37$$

Formula 6.37 shows that isolation from longitudinal waves increases with increase in the thickness of the mounting, of frequency, or in the ratio $\dfrac{\varepsilon}{\varepsilon_1}$, which occurs when elastic pads with $\varepsilon_1 < \varepsilon$ are used. Despite the simplified solution of the problem, formula 6.37 can be used for practical calculations. Isolation from transverse vibrations is significantly lower for the same materials than it is for longitudinal vibrations, particularly at low frequencies. This can be seen from Fig. 6.18 where curve 1 shows the insulation from longitudinal and curve 2 from transverse waves, for a concrete wall 10 cm thick with a cork layer 3 cm thick.

The insulation of an enclosure from vibrations caused by blows on a boundary is of great importance. This is particularly so in the case of floors in houses, which are subject to impacts due to footsteps, moving furniture, etc.

The isolation of an enclosure from impact sound propagated in the material in both longitudinal and transverse directions is determined basically as for airborne sound by the mass of the floor and by its elasticity. In this case, to increase the insulation we should try to increase both the mass and the flexibility of the floor. Where the floor is of large mass the amplitude of bending vibrations is negligible, while if the floor is flexible enough impacts are smoothed out by the elastic deformation of the floor surface.

For impacts on the boundary surface of an enclosure, separation of the boundary into two (preferably unequal) parts, with air or a flexible lining, between them produces in general the same effect as does a multi-layer wall on the transmission of airborne sound.

Consequently to reduce the transmission of vibratory or percussive sound, it is necessary to:

1. Take measures to eliminate the possibility of such sounds arising in the building, and for this reason to place sound conductive building elements (e.g. metal frameworks, pipes) far away from the sources of the noise.

2. Reduce vibrations at the source where they arise; this can be accomplished by putting machines on special foundations of sufficiently large mass, separated from building by elastic pads of the necessary thickness and flexibility. The best quality pads are of cork or rubber.

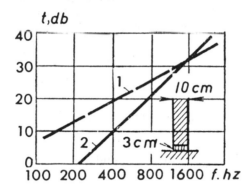

Fig. 6.18. Frequency curves of vibration isolation for longitudinal (1) and transverse (2) vibrations of a 10 cm thick concrete wall mounted on a 3 cm cork.

3. Weaken the transmission of longitudinal vibrations propagated through the construction of the building, by insertion of discontinuities (for example at corners) and by filling the discontinuities with inserts of materials which have lower moduli of elasticity (lead, rubber, cork).

4. Make it difficult for impacts to penetrate, by ensuring the requisite thickness of the floors and by making them of a multi-layer construction, alternating rigid and flexible materials (glass wool, mineral wool, etc.). Footstep impacts should be damped at source by covering floors with resilient materials (carpet, felt backed linoleum, etc.).

5. Isolate the room to be protected from the load bearing structures in the building by suspending the room on sprung supports. This applies in particular to inter-storey floors. An enclosure within another enclosure connected with it by a resilient support forms a 'floating' system, which allows for a high degree of insulation from vibration and impacts.

6.8 The total level of noise in an enclosure

The establishment of the necessary degree of sound insulation does not mean the total prevention of the breakthrough into the enclosure. A

205

demand for total sound insulation, as can be seen from a consideration of the mechanism by which various sounds penetrate into the enclosure, involves the use of a great quantity of expensive materials and is not necessary. It is important to aim at sound insulation of such an order that the total level of penetrating noise in the enclosure is permissible for the given enclosure.

The permissible noise level in an enclosure depends on the purpose to which the enclosure is put. For enclosures used for sound recording or broadcasting (sound studios, radio studios) the permissible noise level is set at 25–30 dB. This noise is caused by the equipment and the people working on sound recording or on a radio broadcast. The permissible noise level in a film stage or in a television studio is somewhat higher (30–35 dB) and is caused mainly by the operation of lighting and film or television cameras. The permissible noise level in cinemas and theatres, caused by the large number of people assembled in them, is 40 dB, in living quarters 35 dB, and in offices 40–45 dB.

Consequently in order to decide what is an acceptable degree of sound insulation, we must define the total level of noise penetrating into the enclosure, and compare this with the permissible noise level for the enclosure.

The total noise level in the enclosure should be determined from the noise level beyond its boundaries and the value of the sound insulation of these boundaries.

Let there be behind one of the boundaries of the enclosure a source whose noise level is:

$$N_n = 10 \log \frac{E_n}{E_{\text{thresh}}}$$

where E_{thresh} is the density of energy at the threshold of hearing.

The sound energy density created by the source is defined thus:

$$E_n = E_{\text{thresh}} 10^{0.1 N_n} \qquad 6.38$$

This value can be found from expression 2.21 if we bear in mind that we need not calculate the mean absorption coefficient α for a wave impinging on a wall. And consequently:

$$E_n = \frac{4 P_{a_n}}{c_0 S_n} \qquad 6.39$$

from whence

$$P_{a_n} = \frac{c_0 S_n}{4} E_n$$

If we substitute for E_n in this expression its value 6.38, this last equation can be rewritten as:

$$P_{a_n} = \frac{c_0 S_n}{4} E_{\text{thresh}} 10^{0.1 N_n} \qquad 6.40$$

According to equation 6.1 the power of the noise transmitted through the wall is:

$$P'_{a_n} = P_{a_n} \cdot \alpha_{\text{trans}} = \frac{c_0 E_{\text{thresh}}}{4} S_n 10^{0 \cdot 1 N_n} \cdot \alpha_{\text{trans}}$$

On the basis of expression 6.2 the coefficient of sound conductivity α_{trans} can be replaced by the following value:

$$\alpha_{\text{trans}} = 10^{-0 \cdot 1 \tau_{db}}$$

and the power of the noise transmitted through the wall is expressed by the equation:

$$P'_{a_n} = \frac{c_0 E_{\text{thresh}}}{4} S_n 10^{0 \cdot 1 (N_n - \tau_{db})} \qquad 6.41$$

The power of the noises penetrating into the enclosure through all the partitions is the sum of the separate powers penetrating through each of them:

$$P_{a\Sigma} = \frac{c_0 E_{\text{thresh}}}{4} \sum_{n=1}^{n} S_n 10^{0 \cdot 1 (N_n - \tau_{db})} \qquad 6.42$$

The energy density caused in the enclosure by the penetrating sound is, from expression 2.16 and 6.42:

$$E_0 = \frac{4 P_{a\Sigma}}{c_0 \alpha S} = \frac{E_{\text{thresh}}}{\alpha S} \sum_{n=1}^{n} S_n 10^{0 \cdot 1 (N_n - \tau_{db})} \qquad 6.43$$

and the total noise level penetrating through all the partitions, and corresponding to this energy density, is:

$$N = 10 \log \frac{E_0}{E_{\text{thresh}}} = 10 \log \sum_n S_n 10^{0 \cdot 1 (N_n - \tau_{db})} - 10 \log \alpha S \qquad 6.44$$

Expression 6.44 shows that the level of noise coming into the enclosure depends on the sound insulation of the boundaries, their area, and the level of noise outside the boundaries.

Moreover, the level of noise in an enclosure decreases as the absorption coefficient α of the inner surfaces of the enclosure is increased, as indicated by the presence of the second term in expression 6.44. If we construct curves for the relationship between noise level $\nabla N = 10 \log \alpha S$ and the absorption coefficient (Fig. 6.19) we can see that for any dimensions of the enclosure, satisfactory results may be obtained by treating the enclosure with materials having $\alpha \geqslant 0\cdot 4$.

It should be noted that N_n and τ_{db} are included in expression 6.44 as power indices, and that a small change in these quantities can have a significant influence on the noise level. This is why, if the sound insulation is low for even one small area of wall, or, alternatively, if the external noise level is high, then the term representing that area may become large corresponding in many cases to a noticeable increase in noise level in the enclosure.

A large reduction in insulation of this kind can happen if a hole for a door or a window is cut in a wall, the insulation capacity τ_{db_2} of the hole being significantly lower than the insulation capacity τ_{db_1} of the wall. In this case it is useful to make a preparatory determination of the value $\nabla\tau_{db}$, the reduction in overall sound insulation of the enclosure due to cutting the space. The total energy impinging on the wall under these conditions is equal to the sum of the energy falling on the part of the wall without the space, and on the space:

$$\varepsilon_{inc} = \varepsilon_{inc_1} + \varepsilon_{inc_2} \tag{6.45}$$

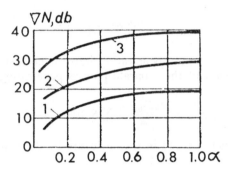

Fig. 6.19. Effect of sound absorption in decreasing the noise level in an enclosure for $S = 100$ m² (1) 1000 m² (2) 10,000 m² (3).

Formula 6.1 enables us to represent the parts of expression 6.45 in the form:

$$\varepsilon_{inc_1} = 10^{0 \cdot 1 \tau_{db_1}} \cdot \frac{S_1}{S} \cdot \varepsilon_{trans} \tag{6.46}$$

and

$$\varepsilon_{inc_2} = 10^{0 \cdot 1 \tau_{db_2}} \frac{S_2}{S} \varepsilon_{trans}$$

where S_1 and S_2 are respectively the areas of the surrounding wall and of the space.

Substitution of values from equations 6.46 for the energies falling on each part of the wall gives the following expression:

$$\varepsilon_{inc} = \left(\frac{S_2}{S} 10^{0 \cdot 1 \tau_{db_2}} + \frac{S_1}{S} 10^{0 \cdot 1 \tau_{db_1}} \right) \varepsilon_{trans}$$

The total sound insulation of a wall with a space is:

$$\tau_{db} = 10 \log \frac{\varepsilon_{inc}}{\varepsilon_{trans}} = 10 \log \left(\frac{S_2}{S} 10^{0 \cdot 1 \tau_{db_2}} + \frac{S_1}{S} 10^{0 \cdot 1 \tau_{db_1}} \right) \tag{6.47}$$

If we symbolize the difference in insulation of the wall and the space by

$$\tau' = \tau_{db_1} - \tau_{db_2}$$

208

the reduction in sound insulation resulting from the cutting out of the space can be expressed, on the basis of expression 6.47 by the equation:

$$\nabla\tau_{db} = \tau_{db_1} - \tau_{db} = \tau_{db_1} - 10 \log \left(\frac{S_2}{S} 10^{-0.1(\tau' - \tau_{db_1})} + \frac{S_1}{S} 10^{0.1\tau_{db_1}} \right)$$

or, as

$$\frac{S_1}{S} = 1 - \frac{S_2}{S}$$

$$- \nabla\tau_{db} = \tau_{db_1} - 10 \log 10^{0.1\tau_{db_1}} + 10 \log \left[\frac{S_2}{S} (10^{-0.1\tau'} - 1) + 1 \right]$$

and finally

$$\nabla\tau_{db} = 10 \log \left[\frac{S_2}{S} (10^{0.1\tau'} - 1) + 1 \right] \tag{6.48}$$

Fig. 6.20 shows the results of calculations by means of this formula. It represents the loss of sound insulation $\nabla\tau_{db}$ as a function of the relative area occupied by the space $\frac{S_2}{S}$ for a series of values of the differences in the insulation capacity of the uniform wall and that of the window or door filling the cut out space τ'. It shows that where the relative area $\frac{S_2}{S}$ is equal to 0·05–0·5 which includes most practical appreciations, the reduction in sound insulation of the wall is comparatively large and reaches 50–85 per cent of the difference of the transmissions losses of the surrounding wall and the space τ'. A small reduction in the sound insulation can be achieved only by reducing the area of the cut out space, or if this is impossible, by increasing the transmission loss of the door or window opening.

For differences in the insulation of the wall and the filling $\tau' > 15$ dB the loss of total isolation $\nabla\tau_{db}$ increases very quickly if the relative area of the cut out $\frac{S_2}{S}$ is increased. This is particularly noticeable where $\frac{S_2}{S} < 0·2$. Thus, for example, where $\frac{S_2}{S} = 0·1$ and the insulation capacity of the material filling the door space is $\tau_{db_2} = 20$, an increase in the sound insulation of the wall τ_{db}, from 30 to 60 dB, which is the equivalent of an increase in τ' from 10 to 40 dB (see Fig. 6.20), results in an increase of $\nabla\tau$ from 3 to 30 dB, i.e. to a reduction in total sound insulation of only 3 dB. This shows that it is irrational to increase the insulation capacity of the wall by more than 10–20 dB compared with the insulation of the windows or doors, as this produces almost no increase in the total sound insulation.

If there are cracks in the door or window fittings, the sound reduction index of the construction diminishes still further to a value

$$\nabla\tau_{db} = 10 \log \left(1 - n \frac{S_0}{S} 10^{0.1\tau_{db}} \right)$$

209

14

where S_0 is the area of the cracks, S and τ_{db} are the area and sound reduction of the whole construction, and n is the coefficient of effective increase of the area of the crack. According to the experimental data of the Institute of Building Physics, this coefficient has the following values: 3, 7·7, 9·8, 6·1 and 4·2 respectively for a breadth of the crack of 1 mm, 4 mm, 5 mm, 10 mm and 20 mm for length 1 m.

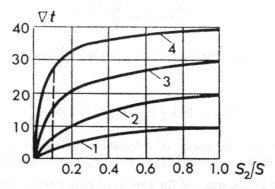

Fig. 6.20. Reduction in sound insulation Δ_τ as a function of the relative area of the cut out space with differences in the insulation of the wall and the construction filling the cut out space of $\tau' = 10$ dB (1) $\tau' = 20$ dB (2) $\tau' = 30$ dB (3) $\tau' = 40$ dB (4).

An analysis of the results we have obtained allows us to form the following conclusions:

1. The total level of noise should not exceed the level of noise permitted in an enclosure of the given type (see Appendix 4).

2. The total level of noise penetrating into an enclosure can be reduced by reducing the level of external noise and by increasing the sound insulation of the boundaries of the enclosure. Enclosures with low permissible noise levels should be placed remote from the sources of noise (noisy streets, factories), and steps should be taken wherever possible (separate foundations, baffles, etc.) to reduce the noise of the sources themselves.

We should not allow differences in the amounts of sound penetrating into the enclosure through different boundaries. For this purpose, during the period the enclosure is being built, walls beyond which there is a high level of noise should be made from materials having a high sound insulation capacity, while for existing enclosures one should take steps to ensure that beyond walls which have low sound insulation capacity should be enclosures with a low permissible noise level.

3. Treatment of the inner surfaces of the enclosure with material with good sound absorbing properties helps to reduce the level of noise.

4. The construction of windows or doors in the walls has a considerable effect on the level of the noise which penetrates into the enclosure. With the aim of reducing this, the transmission loss of doors and windows should be increased by strengthening their construction, to minimize the

210

difference between the insulation of the main wall and the door or window filling the cut-out space. Unnecessary increases in the area of such insertions is not recommended as this greatly increases the level of the transmitted noise.

5. It is essential to prevent the appearance of cracks in the fittings of doors and windows as this increases the level of noise in the enclosure.

6.9 Standardization of the value of transmission loss and the method of calculation

The sound transmission loss of a partition between two enclosures is normally described by a mean value determined over a standardized frequency range 100–3200 Hz. To find this value, empirical formulae (6.19) or curve 2 of Fig. 6.9 are used.

However, data recently obtained in this way have come to be regarded as merely indicative. The present norms[77] (*Building Norms and Rules*, vol. 2, Moscow, 1960) demand that the value of transmission loss should be determined by measurement or calculation for a number of frequencies from 100 to 3200 Hz.

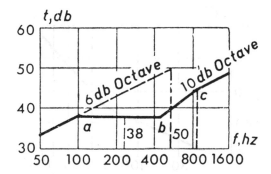

Fig. 6.21. Construction of the frequency characteristic transmission loss for a concrete wall where l = 12 cm.

Let us illustrate the simplest method of obtaining the frequency characteristic of sound insulation which can be recommended for walls having a comparatively low (and below 50 Hz) first resonance frequency.

A logarithmic scale of frequencies is laid out along the horizontal axis on squared paper and along the ordinate axis is marked the value of sound insulation. A transmission loss characteristic can then be constructed in four sections (Fig. 6.21). If, for example, we are asked to construct the characteristic for a reinforced concrete panel 12 cm thick, the weight of 1 m² of the panel is worked out from the table reproduced on page 212.

$$m = 2300 \cdot \frac{12}{100} = 276 \text{ kg/m}^3$$

TABLE 2

Material	Density kg/m³	τ_{db} on the horizontal section	f_b/f_a
1	2	3	4
Aluminium	2700	29	11
Ferroconcrete	2300	38	4·5
Brickwork	2100	37	4·5
Steel	7850	40	11
Glass	2600	27	10
Plywood	600	19	6·5

From the curve in Fig. 6.22 the mean value of sound insulation ($\tau_{db} = $ 50dB) corresponding to the value of m is next found, and remembering that the curve is constructed for a mean frequency of $f_0 = 550$ Hz an ordinate 50 dB high is drawn on the grid in Fig. 6.21 at that value of frequency.

Through the end of the ordinate a straight line is drawn to the left at an angle of 6 dB/octave. Then, according to the data in column 3 of Table 2,

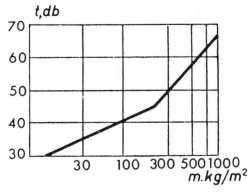

Fig. 6.22. The dependence of the sound insulation of a wall on its mass.

a straight line is drawn on the grid in Fig. 6.21 parallel to the horizontal axis and at 38 dB. This line, and the left hand part of the inclined line are two parts of the characteristic.

The point of their intersection a (Fig. 6.21) corresponds to a frequency of 100 Hz. Hence, as from Table 2, $f_b/f_a = 4·5$, the value of the frequency is determined for the end of the parallel section of the characteristic $f_b = 100.4·5 = 450$ Hz.

The characteristic should run from point b for the next octave (to $f_c = 900$ Hz) at an upward angle of 10 dB/octave, and from point c onwards with an inclination of 6 dB/octave.

A similar method can be used to construct a transmission loss/frequency characteristic for walls of any material and any thickness.

212

To judge whether the sound insulation provided by a wall is adequate, comparison is made of the measured or calculated frequency characteristic of sound insulation with the standard recommended characteristic (Fig. 6.23). The performance of a partition against airborne and impact noise respectively is characterized by indices C_A and C_p. This index is the whole number of

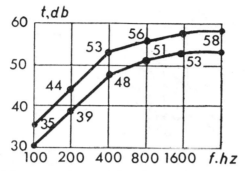

Fig. 6.23. Standard frequency characteristics of sound insulation: the upper with and the lower without departure paths for noise.

decibels by which it is permissible to reduce the mean value of sound insulation of an existing partition in order that its frequency characteristic should differ by no more than 2 dB from the standard characteristic.

The index of sound insulation from airborne noise for the walls and floors of living quarters is established at 1 dB; for impact noise penetrating

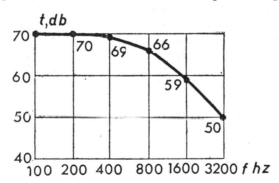

Fig. 6.24. Standard frequency characteristic for sound insulation from percussive noise.

through the floors and ceilings it is 0 dB. The standard curve for impact noise is shown in Fig. 6.24.

The procedure for comparison of the frequency characteristic of sound isolation from airborne noise for a real panel with a standardized characteristic can be illustrated on Fig. 6.25 where curve 1 is the characteristic calculated for a ferroconcrete panel and curve 2 is the standardized

213

characteristic. Values of sound insulation are indicated at various points on both characteristics.

First, the mean variation of the measured characteristic from the norm is measured at those frequencies where the former lies below the latter. These unfavourable divergences (see Fig. 6.25) occur at a number of

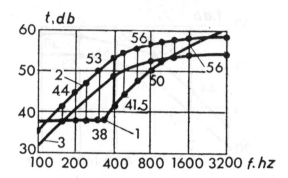

Fig. 6.25. To find the index of sound insulation for airborne noise.

frequencies from 125–1600 Hz, and have the following values: 3, 6, 9, 12, 11·5, 9, 7, 6, 5, 4 and 2 dB.

The mean of these values is determined by dividing the sum of the unfavourable divergences by the number of all the frequencies (15) and in the case quoted has a value of approximately 5 dB.

In this case, as in a number of others, when the mean value of the unfavourable divergences is greater than 2 dB, the standardized curve is moved vertically downwards by a whole number of decibels arranged so that if the calculation of the mean value of unfavourable divergences is repeated it will be no greater than 2 dB. If this is achieved, then the value by which the standardized curve has been moved is the index of sound insulation.

In the case under consideration the standardized curve was moved by 5 dB, (curve 3) and the unfavourable divergences (beginning from a frequency of 200 Hz) are 1, 4, 7, 6, 4, 2, and 1 dB. The mean unfavourable divergence will be $25 \div 15 = 1\cdot7$, i.e. just less than 2 dB. Consequently the index of sound insulation for the given ferroconcrete partition is equal to 5 dB, which is significantly less than the standard value $C_A = -1$. Thus this panel does not meet the demands of sound insulation from airborne noise.

The question of the adequacy of the sound insulation from impact noise transmitted through the floors is answered in the same way by comparing the measured and standard frequency characteristics (see Fig. 6.26).

Unlike the determination of the index of sound insulation from airborne noise, divergences in impact noise insulation lying above the standard

214

curve are considered unfavourable (Section *ab* in Fig. 6.26). For this section they comprise 1, 3, 4, 4, 5, 4, 4 and 2 dB. The mean unfavourable divergence is $27 \div 15 = 1 \cdot 8$ dB. As the mean divergence obtained is less than 2 dB, the index of sound insulation C_p is 0 dB, and consequently the floor meets the demands in regard to sound insulation from impact noise.

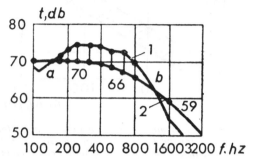

Fig. 6.26. To determine the index of sound insulation from impact noise.

6.10 Sound insulating materials and constructions

An important factor in the choice of material for main walls with regard to their sound insulation is the mass of the wall per square metre of surface. If walls are made of uniform material, then the mass per square metre for the required insulation can be determined by formula 6.19 or by the graph corresponding to this formula in Fig. 6.22.

It follows from the graph that to increase the sound insulation, walls have to be built thick enough, choosing the densest materials (stone, brick) and bonding materials. As there is a logarithmic relationship between the sound transmission loss and the mass of the wall, a small increase in the former necessarily entails a significant increase in the latter. Moreover, only for walls having a weight greater than 200 kg per square metre does a doubling of the weight increase sound insulation by 6 dB. As can be seen from Fig. 6.22, doubling the mass of walls weighing less than 200 kg per square metre results in an increase in sound insulation of only 4 dB.

We cannot omit to mention the other properties of the materials of which walls are made on which sound insulation capacity also depends. This is particularly important for light walls (Fig. 6.27) where sound insulation can be doubled according to whether the upper or lower limit of the zone is used to determine this value. The upper limit corresponds to materials with low sound conductivity, and the lower to materials with high sound conductivity. Practice shows that, for equal mass, wooden walls are less sound conductive than brick, which in turn gives better results than uniform concrete. From the point of view of air borne noise, the main walls and the carcass of the building which are made heavy and thick for reasons of strength, can ensure sufficient sound insulation, (50–55 dB). However, these very walls most often turn out to transmit structure-borne vibration.

215

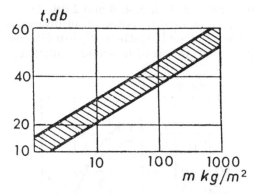

Fig. 6.27. The dependence of mean sound insulation on mass for walls of various materials. The upper limit is for the less sound conductive materials (wood) and the lower limit is for materials which are better sound conductors (concrete).

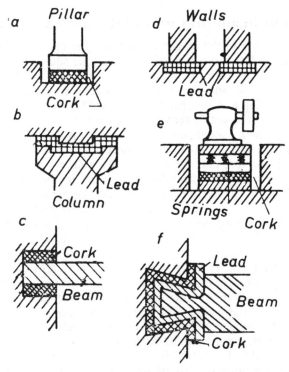

Fig. 6 28. Various examples of the use of sound insulating packings.

The simplest methods of defence are the building of the structure using dissimilar materials and the use of washers or paddings of heavy or plastic materials (lead or cork). The first method reduces the possibility of indirect transmission of noises via longitudinal vibrations, the second hinders the distribution both of longitudinal vibrations and bending vibrations. The thickness of the lining or pad can be chosen to be between 5 to 20 mm.

Fig. 6.28 shows variants of the use of sound insulating packings. Sometimes special metal springs are used as resilient packings.

Breaks in constructional elements should be complete, and should not have even small bridge pieces of rigid materials.

A more effective method of combating noises penetrating through the constructional elements of a building is to have total break between the

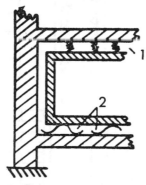

Fig. 6.29. Diagram of the suspension of a 'floating' system. (1) coil springs, (2) springs.

structure of the building and the wall of the enclosure to be protected from noise. This method is used where very stringent demands are made on the sound insulation of the enclosure, and in particular in the case of radio studios, music studios, etc.

Such a system of protection, known as 'floating', has the enclosure totally isolated from the bearing elements of the construction and resting on resilient devices (metal springs, resilient packings) or suspended from them (Fig. 6.29). The attenuation introduced by the sprung connections results in a great reduction of airborne, percussive or vibratory noise.

If it is necessary to isolate the enclosure from the noise of machinery, it is recommended that apart from mounting the machine on a separate foundation (see Fig. 6.28e) it should be placed in a separate enclosure or in a box, the walls of which are made of a sound opaque material (brick) with an inner treatment of sound absorbent materials.

Interior walls cannot be massive and thick because of high cost and the impossibility of placing heavy loads on the load-bearing parts of the structure. Hence arises the need to increase the sound insulation of walls with comparatively small mass. This is accomplished by the use of multi-layer constructions and the use of porous materials.

It is desirable that two wall elements making up a complex construction should be if possible different from each other in mass, thickness and even material. Such measures reduce the possibility of the appearance of resonances causing dips in the frequency characteristic of sound insulation. If all bridges across the air space are resilient, the sound insulation is increased, and an increase in the gap to 10–12 cm is technically justified.

The sound insulation of a two layer lightweight wall can be increased to a level similar to that of large solid walls, by placing in the cavity a layer of soft porous material which is not allowed to touch the sides. Apart from

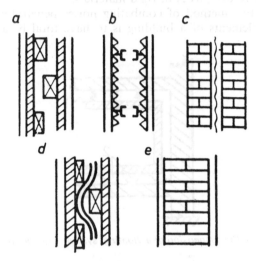

Fig. 6.30. Various sound insulating constructions: (a) a double wall of fibreboard with plaster on both sides (m = 64 kg/m²), (b) plaster on a metal lath, (m = 88 kg/m²), (c) two walls of hollow brick with a filling of porous material (m = 170 kg/m²), (d) a double wooden wall with plaster, with porous material in the gap, (m = 66 kg/m²), (e) a brick wall 200 mm thick with plaster (m = 340 kg/m²).

additional sound insulation this introduces a disorder into the system which reduces resonances, thereby greatly improving the operation of the system.

For the isolation of an enclosure from flanking transfer of sound energy, the internal walls which abut on to it should have breaks with packings of resilient materials. Fig. 6.30 shows the construction of double walls differing both in complexity and in the degree of sound insulation they offer. The transmission loss given by these walls is illustrated by the curves in Fig. 6.31. If we consider these two figures, we can see that the double wooden wall with plaster surfaces and with a layer of porous material in the cavity (see Fig. 6.30d) is five times lighter than the brick wall (Fig. 6.30e) and provides almost the same degree of sound insulation (50–60 dB).

Floors, which are just as much boundaries as are walls, are distinguished from them by the fact that they are far more exposed to impacts (footsteps, moving of furniture, etc.).

218

As regards airborne transfer of sound, floors are no different from walls, and their insulation increases with increase of mass. Normal floors with a weight of 200–300 kg/m² have transmission loss figures of from 45–50 dB. However, this insulation is insufficient to protect against impact sounds and further steps have to be taken.

A 'floating' ceiling, suspended on springs, can increase sound insulation if its mass is not very small, if the layer of air between the ceiling and the floor has a depth of 10 cm and if the edge of the floating ceiling has breaks filled with packings where it abuts on to the walls. In this case a normal two-layer partition is created, whose sound insulation can be increased still further by placing a layer of porous material in the air space (glass wool or felt).

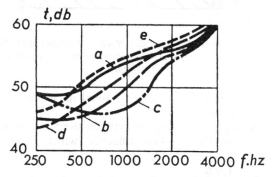

Fig. 6.31. Frequency dependence of the sound transmission loss of sound insulating constructions shown in Figure 6.30.

A particularly effective protection from impact noise is offered by a 'floating' floor, which must be fully isolated from the walls and the true, load-bearing floor. For this the false floor is placed on soft resilient packings (glass wool, cork or rubber) and is separated from the walls by similar packings. Thus with a floating floor and a floating ceiling below, the floor becomes a three-layer construction, the total sound insulation of which is much higher than that of any ordinary floor.

Finally the impact sound transmission of floor can be still further reduced if the surface of the floor is covered in soft materials. In particular[37] where the floor is covered in a rubber sponge carpet the additional insulation amounts to 14 dB, and with a pile carpet to 20 dB.

Fig. 6.32 is a diagram of a complex floor taking into account all ways of increasing its sound insulation.

Doors and windows are often the weakest elements of walls in respect to sound insulation. This is because they have comparatively low weight and do not always close tightly.

The basic factor influencing the sound insulation of windows is the thickness of the glass. Calculations and experiments have shown that the sound insulation of ordinary double-glazed windows made of glass of 2·5–3·0 mm is 25–27 dB. With glass 6 mm thick this increases to 32 dB.

Double windows have a higher sound insulation. Experience confirms the theoretical prediction that good results should be obtained by using glass of different thicknesses in double windows. Where the inner and outer frames are glazed with glass of differing thicknesses (3 and 6 mm) the

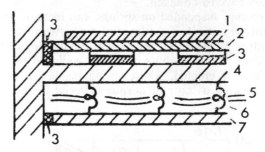

Fig. 6.32. Schematic section of a complex floor. (1) carpet, (2) 'floating' floor, (3) resilient packings, (4) concrete floor, (5) porous material, (6) springs, (7) suspended ceiling.

same sound insulation (33 dB) was obtained as when both frames were glazed with 6 mm glass.

As with double walls, the air space between the panes of windows has an influence of sound insulation. This influence, which is determined by the

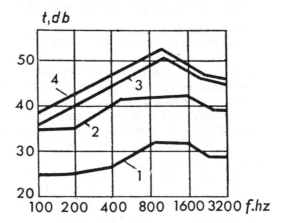

Fig. 6.33. Frequency characteristics of sound insulation for windows with glass 3 mm thick. (1) for a single pane, (2), (3), (4), for double windows with gaps of respectively 10, 20 and 30 cm.

depth of the air layer, is illustrated in Fig. 6.33 where curves 1, 2 and 3 relate to the use of double windows for a thickness of air space respectively 10, 20 and 30 mm. The small increase of additional insulation obtained by increasing the depth beyond 20 mm shows that this is the maximum depth which need be allowed for.

The use of packings in the window frames increases the sound insulation

220

by 6–7 dB, and the material of which the packings are made (felt, porous rubber, brass ribbon) has almost no influence on this value.

Experiments carried out by the Institute of Building Physics have shown that the method used to fix the panes in the frames and the deadening of the window frames between the inner and outer window have no influence on the sound insulation of the window.

These observations about the sound insulating properties of windows relate to a large degree also to doors. The material of which they are made,

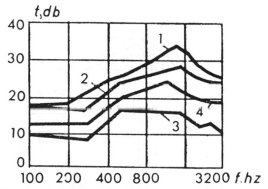

Fig. 6.34. Frequency characteristics for sound insulation of layered doors.

the method of construction, the presence of cracks round the door in the frame all have an effect on the value of sound insulation achieved. This can be seen in Fig. 6.34 which gives curves for sound-insulating doors. Curves 2 and 3 are for doors filled with glass wool and mineral felt without the use of door seals, and 1 and 4 are for the same with seals. It follows from Fig. 6.34 that the sealing of the cracks increases sound insulation by 5–7 dB.

The influence of the construction and the height of the door is clearly shown in Fig. 6.35. The normal panelled door gives sound insulation of

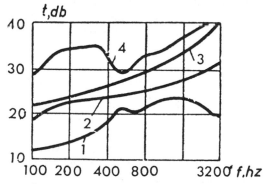

Fig. 6.35. Frequency characteristics of sound insulation for doors of different types. (1) panelled, (2) single slab, (3) two single slabs with a filling of mineral wool, (4) two single slabs in a common frame.

221

about 20 dB (curve 1), while a door made of whole blocks (curve 2) and one of blocks with a layer of mineral wool between them (curve 3) give respectively 23 and 28 dB. The sound insulation of two slab doors in one frame (curve 4) on average comprises 34 dB.

In Fig. 6.36, *a* and *b* show plans of the constructional method of door and window seals.

When we are faced with problems of sound insulation, we must not forget the network of pipes which act as a distributor of sound, particularly

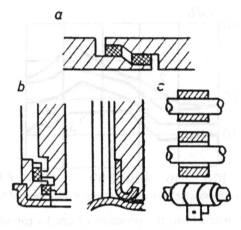

Fig. 6.36. Diagrams of the method of construction of door and window seals ((a) and (b)) and pipe packings (c). Dense shading shows resilient packings.

of impact noises. In installing pipes, we should try to see that as few as possible go into the enclosure we are dealing with. Pipes, radiators and taps must be firmly fastened to the walls with the help of resilient packings of felt, asbestos or rubber, and should be fed through the walls in such a way that there is a resilient sleeve between the wall and the pipe (Fig. 6.36c). If the pipes pass through double walls, any rigid contact between them and the elements of the wall must be avoided. For this purpose we may use either resilient sleeves at each of the walls or insert a flexible pipe section in the gap between the walls.

6.11 The reduction of noise from ventilating systems

Large enclosures, which by the nature of their work must house powerful lights (television and film studios) or where large numbers of people will assemble (theatres, cinemas), are usually equipped with ventilation or air conditioning systems. These systems are often sources of noise, which originates in the fans and is distributed throughout the building through the ventilation trunkings.

The noise from a fan or from the electric motor which drives it can be transmitted either as vibration or as airborne noise. Steps taken against the vibration from the ventilator are just the same as those used in any other case of this kind of sound (separate foundation or flexible mountings). The worst aspect of the airborne noise is the fact that the air passages of the ventilation system are excellent channels for the transmission of noise.

In trying to reduce ventilation noise, we should first turn our attention to the source of this noise. The level of noise created by the ventilation fans themselves depends on the diameter of the rotor D, and very much on the speed of rotation. At a speed of 10 to 20 m/sec the noise is proportional to the square or even to the fourth degree of the speed.

The noise level of the fan can be found if its diameter D, the number of revolutions per min n, and some constructional data are known.

To find this level we should use the formula:

$$N = N_0 + 80 \log D + 60 \log n - 83 \qquad 6.49$$

where the value N_0, known as the abstract noise level, is basically defined by the type of ventilation fan and the conditions under which it is working.

Fig. 6.37, can be used to find the value of N_0. This figure shows the dependence of N_0 on $\bar{Q}$, which is connected with the ventilator fan, its

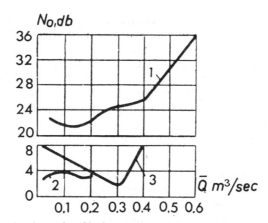

Fig. 6.37. Curves to determine N_0 from the diameter of the rotor and rev/min. for ventilation fans of various types.

number of rev/min and the flow Q in cubic metres per second, and can be found from the following formula:

$$\bar{Q} = 12 \cdot 4 \, \frac{Q}{D^3 n}$$

As different ventilators have different outputs, Fig. 6.37 shows several curves for $N_0 = f(\bar{Q})$. Graph 1 is for a centrifugal ventilator type with blades bent forward, 2 is for a centrifugal ventilator with blades bent backwards, 3 is for a single stage axial ventilator with the blades set at 45°.

223

The necessary air exchange can be achieved by using a large diameter fan at low rev/min. In this case the noise created is of low frequency and of comparatively low level. The same air exchange can be obtained by increasing the speed of the fan and reducing the diameter. The high frequency noise emitted as a result of this condition can be attenuated by the use of sound absorbent materials.

The transmission of air through the ventilation trunking, whether the system is an impeller or an extractor system, can be mathematically described by an equation similar to equation 4.9.

In actual fact a ventilation passage can be regarded as a rectangular 'enclosure' of extreme length and comparatively small section. The solution of this equation is similar to the solution of 4.9a, shown in equation 6.50, bearing in mind that the 'enclosure' to which a ventilation passage is being compared has infinite length ($l_z = \infty$) and the component of a wave reflected in this direction equals 0 $\left(\cos \dfrac{n_z \pi z}{l_z} = 1 \right)$ under these conditions:

$$p = \sum_{n=1}^{\infty} A_n \cos \frac{n_x \pi x}{l_x} \cdot \cos \frac{n_y \pi y}{l_y} \cdot e^{j(\omega t - \delta_r)} \qquad 6.50$$

As the section of the passage is comparatively small, waves falling on its walls undergo multiple reflections, as a result of which, if the walls have a high absorption coefficient the waves decay rapidly. The attenuation of waves during transmission along the axis of the duct is caused only by losses in air, and is consequently small. The attentuation of these waves can be increased by submitting them to reflections from the walls of the duct, which can be done in practice by making repeated changes in the direction of the duct.

Thus from formula 6.50 it follows that reduction in ventilation noise can be obtained by:

1.   An increase in the length of the ducts.
2.   The treatment of the inner surfaces of the air ducts with a good sound absorbent material.
3.   Increasing the number of bends in the duct.

The attenuation which can be achieved by these factors can be calculated from the formula:

$$\Delta \tau_{db} = 1 \cdot 1 \frac{Pl}{q} \phi(\alpha) \qquad 6.51$$

where P and q are the perimeter and cross-section of the duct in metre units, l is the length of the duct, and $\phi(\alpha)$ is a value connected with α by the relationship shown in the graph in Fig. 6.38.

The perimeter of a cross-section of the same area increases as the section of the duct becomes flatter, so the use of circular or square section trunking is to be avoided. The perimeter of the cross section can be still further increased or, which is the same thing, the decay of sound can be still further increased, by fitting the ducts with splitters (Fig. 6.39).

224

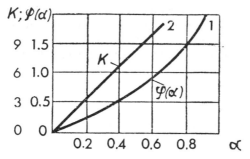

Fig. 6.38. Curves of K and $\phi(\alpha)$ as functions of the coefficient of sound absorption.

It is wise to treat the walls of ventilation trunkings with sound absorbent materials with a very high coefficient of absorption, because any increase in this coefficient is accompanied by a disproportionate increase in the

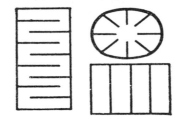

Fig. 6.39. Sections of ventilation trunkings with different forms of splitters.

attenuation. The material should be very firmly fastened in position, e.g. by a metal mesh or by perforated sheets, in order that it should not be torn away or otherwise disturbed by the strong airflow.

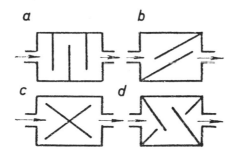

Fig. 6.40. Diagrams of various acoustic filters.

If there is a need to increase the attenuation of sound in the system still further, acoustic filters may be used. These are rectangular chambers with a greater cross section than the trunking, and with additional baffle walls treated with sound absorbent materials (Fig. 6.40).

225

As the sound wave moves from the trunking into the chamber and as it leaves the chamber, and also as it moves along the trunking, a loss of energy occurs between the additional baffle walls. This loss can be calculated. Thus the attenuation in each link in the filter (see Fig. 6.40a) can be calculated from:

$$\tau_{db} = \frac{\alpha S}{q} \qquad 6.52$$

For the filter indicated in Fig. 6.40b the sound decay can be calculated from the formula

$$\tau_{db} = 10 \log \frac{a}{b} \frac{\alpha}{0 \cdot 055 n (1 - \alpha)} + K \qquad 6.53$$

In these formulae: α is the coefficient of sound absorption, S is the area of the sound absorbent material, q is the cross section of the air passage, n is the mean number of sound reflections within the chamber, a/b is the ratio of the width of the passage and the width of the chamber, and K is a value depending on α and shown by curve 2 in Fig. 6.38.

7 The Acoustic Characteristics of Film, Radio, and Television Studios

7.1 Acoustic conditions in speech studios

Speech studios, or as they are called in radio and television, talks studios, are primarily used for the recording or broadcasting of an announcer's speech. In the film industry these studios are widely used for speech dubbing of the original soundtrack or for adding a sound track in a new language. The recording of speech is carried out while silent film is being shown on the screen. This silent film may have been shot earlier, or may be film in which the previous sound track was in another language. Such studios are used in television and radio for the transmission of talks, lectures, and other spoken word programmes, where it is not intended to give the listener the impression that the performers are in large enclosures.

The basic requirement of speech studios is that speech in them should be extremely clear and distinct and that the peculiarities of timbre of the voice of the performer should be preserved.

As can be seen from expression 2.77, good clarity of speech in sound transmission can be achieved on condition that each of the coefficients which express the dependence of articulation on the level of strength of the sound of speech k_L and on the reverberation time of the enclosure k_T should be as close as possible to unity. The first of these coefficients approaches unity for a sound level of from 50–80 dB (see Fig. 3.16) which in practice be achieved by bringing the performer close enough to the microphone. The second coefficient is close to unity for values of the reverberation time of the studio of less than one second (see Fig. 2.20).

227

If we bear in mind that in the types of transmission for which a talks studio is used the number of performers is not usually more than ten, the volume of these enclosures can be small, which makes it fairly easy to achieve a reverberation time, at middle frequencies, of 0·4–0·8 sec. A more accurate value for the optimum reverberation time for a talks studio, on the basis of its volume, can be found from the curve in Fig. 7.1.

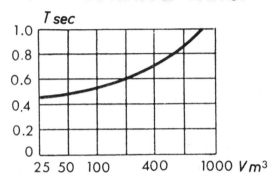

Fig. 7.1. The dependence of reverberation time on the volume of a dubbing studio.

At the present time no final answer has been arrived at to the question of the dependence of the optimum reverberation time of a talks studio on frequency. Some of the recommendations based on the propositions of Lifshits[47] and Knudsen[36] include (see Chapter 2.10) a lift in the frequency characteristic at low frequencies, and in this case the frequency characteristic of reverberation time for speech studios is represented by the graph in Fig. 7.2.

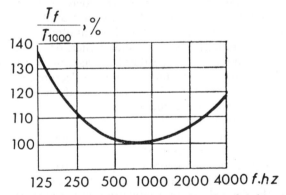

Fig. 7.2. The frequency dependence of optimum reverberation time for a speech studio.

Other recommendations[37] are based on the fact that in small-volume enclosures there is a danger of speech masking by low frequencies, and of extreme emphasis of low frequencies. To avoid the effect of masking and to achieve correct transmission of voice timbre it is found necessary to reduce the lift in the frequency characteristic (Fig. 7.3).

There is reason to think that neither of these recommendations is correct as they are both based on a fundamental proposition which is untenable. They originate from the idea that it is necessary to have a rise in the frequency characteristic at low frequencies, and this is based in turn on the proposition that the levels of all the frequency components of the signal should fade simultaneously to the threshold of hearing. Such a proposition would be valid for isolated signals following each other at long intervals, comparable with the reverberation time. But, in view of the comparatively high rate at which speech syllables or notes of music follow one another, the simultaneous decay of their various components is of no importance, since the end of the reverberation is masked by the basic signal. Because of this,

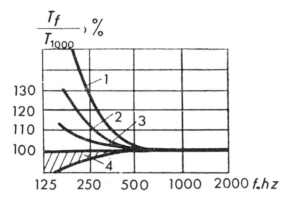

Fig. 7.3. The frequency characteristics of optimum reverberation time for speech studios of various volumes: (1) where volume V=30,000m³, (2) V=3000m³, (3) V=300m³, (4) zone of permissible changes in the characteristic,

it would be more correct to recommend for speech studios a frequency characteristic of reverberation in the form of a line parallel to the frequency axis.

The shaded area 4 in Fig. 7.3 shows the limits within which reverberation time lies for a large number of actual speech studios having good acoustics. Thus experimental data obtained in recent years confirm that intelligibility and a distortion-free reproduction of the timbre of the voice are possible only with a level frequency characteristic, or even by one falling slightly (by 10–15 per cent) at the lowest frequencies.

It is also important for speech studios to make sure that the frequency characteristic is level at high frequencies as well because, in these comparatively small studios, the absorption of sound in air effectively loses any significance. This is why in speech studios every effort should be made to ensure a horizontal frequency characteristic to the highest frequencies.

Bearing in mind that speech recorded for a film is often accompanied by music or various sound effects, it is important to establish what effect this accompaniment will have on the clarity of the speech on reproduction. If we return to formula 2.77, we can see that the influence of accompanying sounds is expressed in that formula by the coefficient k_N, which (Fig. 7.4) quickly falls with reduction in the difference ΔN in levels of the useful and interfering sounds, i.e. with a relative increase in the level of the latter.

As can be seen from curve 2, this is particularly noticeable when the listeners cannot see the performers, e.g. in radio broadcasting. If we bear in mind that in films and in radio and television productions accompanying sounds are often absolutely essential, we can say, on the basis of Fig. 7.4,

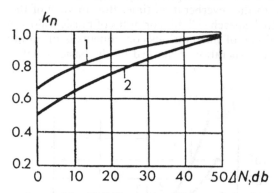

Fig. 7.4. The dependence of intelligibility on the difference in levels of speech and the accompanying sound; (1) where the sound source is visible to the listener and (2) where it cannot be seen by the listener.

that the difference in the levels of speech and accompanying sounds, particularly if these sounds are prolonged, should be not less than 20 or 30 dB. Thus to create optimum acoustic conditions for speech transmissions, the studio for these transmissions should have:

1.   Short reverberation time (0·4–0·8 sec).

2.   A frequency characteristic of reverberation time parallel to the frequency axis right along its length right through to the highest possible frequencies.

3.   A reverberation time at frequencies higher than 4000 Hz which is achievable by absorption at the boundaries of the enclosure and which differ as little as possible from reverberation time at low and middle frequencies ensured by absorption at the boundaries of the enclosure.

Apart from this it is essential that the level of the speech signal should be at least 50 dB, and that the level of accompanying sounds should be 20 or 30 dB below the speech level.

7.2 Acoustic conditions in a film dubbing studio

As in film studios the method of dubbing is mainly used for the recording of music, the stage intended for this purpose should have acoustic conditions in which music will sound well. It is clear that the same conditions should apply in music studios used for the transmission of all kinds of concert programmes by radio or television.

230

Because the character of musical works recorded or broadcast, and the size and nature of the orchestras which perform them, are very varied, to ensure optimum acoustic conditions for recording or transmission of music several specialized kinds of studio are used, or a single studio may be used in which the absorption can be varied.

As was pointed out in Chapter 2.10, the acoustic requirements of studios intended for music transmissions through an electro-acoustic system are not always the same. However, on the basis of the extensive and trustworthy experiments of Kuhl[45] we can say that the optimum conditions in large studios used for the recording or transmission of a symphony orchestra should be decided on the basis that the optimum reverberation time does not depend on volume if this is greater than 2000 m³.

Moreover in determining the optimum reverberation time appropriate for various kinds of music, we must start from a compromise solution, the reverberation time which a majority of experts agree gives reverberation too great for modern symphonic music and too small for romantic symphonic music.

If we allow that the number of expert evaluations is

$$N_B = N_M = 70$$

then, as can be seen from Fig. 2.26, for a studio used for the recording or transmission of symphony orchestras the optimum reverberation time should be 1·7 sec. Such a time can be recommended as the optimum if the studio is larger than 2000–3000 m³.

For music studios of smaller dimensions, the recommendations should be different, and for these we can probably, as before, regard the optimum reverberation time as dependent on volume. The character of this dependence at the present time is not yet agreed by different authorities (see Fig. 2.22). In this figure curve 1 relates to the recommendations of Lifshits, expressed by formula 2.78, curve 2 to the recommendations of R. Thiele, represented by the empirical formula:

$$T_{\text{opt}} = 0·09 V^{\frac{1}{3}} \qquad\qquad 7.1$$

and curve 3 represents the dependence in which we are interested, and which is represented by another empirical formula:

$$\log T_{\text{opt}} = -0·375 + \tfrac{1}{6} \log V \qquad\qquad 7.2$$

If formula 4.16 is presented in the form

$$T = 0·0018(nV)^{\frac{1}{3}} \qquad\qquad 7.3$$

it becomes similar to the two previous formulae.

From formula 7.3, optimum reverberation time can be interpreted as the time during which a strictly determined number n of reflected waves arise in the enclosure, a number which cannot reduce the quality of the basic signal by the addition of seriously delayed high order reflections.

In order to keep the number of reflected waves constant, as follows from equation 7.3, reverberation time in a music enclosure should be directly proportional to $V^{\frac{1}{3}}$.

If we use this dependence, expressed for example by equation 7.1, and the formula for reverberation time expression 2.41, we can find the optimum value of the mean coefficient of absorption for music studios.

Equating expressions 7.1 and 2.41, we can see that:

$$0.09 V^{\frac{1}{3}} = \frac{0.164V}{-S \ln (1 - \alpha)}$$

whence

$$- \ln (1 - \alpha) = 1.83 \frac{V^{\frac{1}{3}}}{S} \qquad\qquad 7.4$$

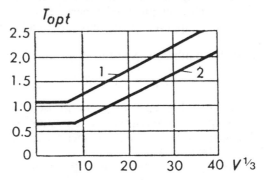

Fig. 7.5. The dependence of optimum reverberation time on the value of $V^{\frac{1}{3}}$; (1) for concert halls, (2) for cinemas.

The value $V^{\frac{1}{3}}/S$ for rectangular enclosures changes comparatively little. If the enclosure has a cubic shape, then

$$\frac{V^{\frac{1}{3}}}{S} = 0.17$$

For an enclosure in the form of an elongated rectangular parallele piped, with sides in a ratio of $2.5 : 1.5 : 1$.

$$\frac{V^{\frac{1}{3}}}{S} = 0.24$$

for enclosures of these forms we have respectively:

$$- \ln (1 - \alpha) = 0.312 \quad \text{and} \quad \alpha = 0.52$$
$$- \ln (1 - \alpha) = 0.44 \quad \text{and} \quad \alpha = 0.67$$

Consequently, optimum reverberation time in enclosures for the broadcasting or recording of music is achieved if the mean coefficient of absorption is 0·5–0·65. This last statement means that for a studio of a specific

232

shape the increase in optimum reverberation time with increase in its volume is in effect an attempt to keep the mean coefficient of absorption constant. A change in this coefficient within certain limits is connected only with a change in the shape of the enclosure.

It should be noted that the view that optimum conditions may be achieved simply from a fixed mean coefficient of absorption, can be used only for approximate checks.

The nature of the variation of optimum reverberation with frequency, as described above (Chapter 2.10), and based on theoretical propositions, is generally confirmed by experience. However, experimental findings relating to a music studio with good acoustic qualities have certain divergences. The results of studies of music broadcasting studios in a number of countries

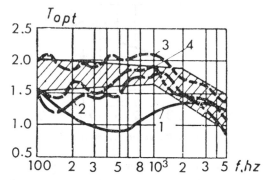

Fig. 7 6. The range of optimum reverberation time for large music studios, and curves for studios with (1) V = 1500 m³, (2) V = 2900 m³, (3) V = 6000 m³.

show that the frequency characteristic of optimum reverberation time can be expressed by the whole area shown as a shaded area in Fig. 7.6. This same figure shows curves for some actual studios. As can be judged from the diagram, the general tendency is that the best music studios have a small rise in the frequency characteristic at low frequencies, and a noticeable drop at high frequencies. This drop is explained by the high absorption of sound in air, which increases as the sound frequency and the size of the studio are increased.

Thus, a small rise in the frequency characteristic of optimum reverberation time at low frequencies (up to 20–30 per cent at a frequency of 125 Hz) can be considered desirable. This rise should be put down to the aesthetic tastes and traditions of listeners who prefer a certain emphasis of low frequencies in music transmissions.

An important factor determining quality in a music studio is a high degree of diffuseness of the sound field, measured by one of the methods indicated in Chap. 10.3. This state of diffusion can be achieved by an even distribution of sound absorbing materials within the studio, or by the use of good scattering surfaces which are of convex form or composed of elements set at an angle to one another. Scattering and absorbing surfaces should have dimensions comparable with the length of the sound wave, and should alternate along the walls of the enclosure.

It should be borne in mind that resonant sound absorbing systems, during vibration, deform the front of the sound waves reflected from them and thus contribute to an increase in the diffuseness of the field in the enclosure.

Summarizing the results of these views, the acoustic demands for a sound stage or for radio or television music studios can be summed up as follows:

1. The optimum reverberation time for studios of small and medium volume (up to 3000 m³) changes within comparatively small limits and may be selected, according to volume, from the graph in Fig. 7.7.

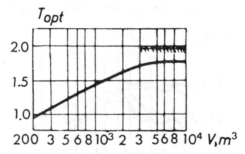

Fig. 7.7. The recommended curve $T_{opt} = f(V)$ for music studios.

2. Optimum reverberation time for large studios used for the transmission of symphonic music, depends little on the volume of the enclosure. On the basis of experimental data, an optimum reverberation time of 1·7 to 1·8 sec can be recommended. This can be increased, for example by the use of variable absorption in the enclosure, up to 2 sec, particularly for the performance of 'Romantic period' music (the right hand part of the curve in Fig. 7.2).

3. The frequency characteristic of optimum reverberation time may have a lift at low frequencies not exceeding 20–50 per cent above the reverberation time at middle frequencies.

4. The sound field in the studio should have a high degree of diffusion, and to achieve this the acoustic treatment of the studio should involve the most even possible distribution of sound absorbent materials on the walls of the enclosure alternating them with scattering surfaces. As resonant systems have the capacity both to absorb and diffract sound energy, they should be recommended for the treatment of the interior surfaces of music studios.

7.3 Acoustic conditions in synchronous shooting stages

On synchronous shooting stages sound is recorded simultaneously with the shooting of film. Filming conditions demand the presence of large sets, lights, cameras, and sound recording apparatus in the studio. Usually speech is recorded synchronously with the shooting of the film, but musical background and effects may be added afterwards.

In synchronous shooting the sound heard by the listeners when they watch the film must correspond acoustically with the scenic setting and the scale of the scene shown in the shot. As synchronous sound recording may take place in sets showing small or large enclosures with varying degrees of absorption, it would seem that the stage should be able to allow for reverberation times varying from a fraction of a second to several seconds. This is even more so because, as the scale of the shot changes and objects in shot change their dimensions, the characteristics of the sound should also alter to correspond to the difference in loudness when a sound source rapidly approaches the listener.

Television and radio studios intended for the transmission of plays and various types of television programmes are no different from film studios in the conditions of their use and in the demands placed on them in respect of their acoustic properties. Consequently the question of acoustic conditions in synchronous shooting stages and in studios for large television programmes can be considered in the same way.

As the recording and transmission of sound takes place, not in the stages and studios themselves, but in the large sets erected in these enclosures, the need arises to consider the possibility of establishing the necessary acoustic conditions by use of the sets themselves.

A set erected in a studio and occupying part of it is essentially an enclosure linked acoustically with the remaining part of the studio. If we suppose that this acoustic coupling exists only by way of gaps in the set, then on the basis of formula 2.52 we can say that the only circumstances in which acoustic conditions within the set are not influenced by the acoustic data of the studio are when the coupling coefficients Q_1 and Q_2 are close to zero. Keeping in mind the impossibility of any significant reduction in the size of the coupling 'aperture' s, such conditions can only be created, as can be seen from equation 2.67, by very heavy deadening of the studio resulting in

$$T_{\text{stage}} < T_{\text{set}}$$

Measurements made in various sets have shown that their reverberation times do not exceed 0·5–0·7 sec. Consequently to make these sets acoustically independent, the reverberation time of the studio must be less than 0·5 sec. However, acoustic independence of the sets under these conditions yields nothing, as in these conditions it is impossible to obtain a sufficiently long reverberation time within them. Thus, acoustic conditions in the sets depend mainly on the acoustic properties of the studio, and it is not possible to control reverberation time within wide limits by varying these properties. Sets change reverberation time only by introducing additional absorption.

As measurements made in the Moscow Television Centre have shown, this absorption A_{set} is proportional, for practical purposes, to the floor area S_{set} occupied by the set. Dependence of the mean coefficient of absorption $\alpha_{\text{set}} = A_{\text{set}}/S_{\text{set}}$ on frequency (Fig. 7.8) shows that the influence of large sets on the total absorption of studios may be significant.

The circumstances noted above result in the need to create in the enclosure under consideration conditions close to the optimum for the accompaniment of speech, using a group of microphones variously arranged relative to the sound source to control the acoustic characteristics within small limits, and special equipment, described in Chapter 7.6 below, to vary the reverberation more widely, up to several seconds.

To obtain optimum acoustic conditions in speech recording studios, it is necessary to have a reverberation time of 0·5–0·8 sec, which can be achieved in such large enclosures as film studios only by the use of highly efficient absorbing materials which allow the attainment of a mean coefficient of

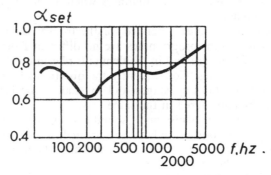

Fig. 7.8. The frequency characteristic of the mean coefficient of sound absorption of a set.

absorption of 0·8–0·7. If we consider that a large part, (up to 30–40 per cent) of the interior surface of a film studio is usually taken up by electrical, lighting and ventilation equipment, and surfaces such as the floor, rostra, etc. which cannot be treated with efficient materials, it becomes clear that it is extremely difficult to obtain the reverberation time indicated above in large studios. Consequently, for synchronous shooting stages and television studios, it is necessary to limit ourselves to the practical minimum of reverberation time obtainable, which, by covering all available areas and efficient absorbent materials, will be approximately 0·8–1·0 sec. This time is somewhat reduced by the absorption contributed by the set.

A further reason for the desirability of a high degree of deadening in film studios and television drama studios is that in order to avoid the possibility of microphones appearing in shot, they are placed a fairly long distance (1·5 to 4 m) from the source of sound, as a result of which they become relatively more sensitive to incidental noise than they are at close range. Heavy deadening enclosures results, as follows from formula 6.44, in a noticeable reduction in the level of noise received by the microphone. This same reason also makes it necessary to insist on the highest possible standards of insulation of film and television studios from outside noise.

The frequency characteristic of reverberation for these enclosures should be appropriate to speech transmission, i.e. should be represented by a line parallel throughout its whole length to the frequency axis.

236

Thus to establish good acoustic conditions in synchronous shooting stages and television studios intended for drama productions, the following demands must be met:

1. The reverberation time should be short, and no longer than 0·8–1·0 sec, irrespective of the volume of the enclosure.
2. Deadening of the enclosure should be carried out by the use of efficient sound absorbent materials with a mean coefficient of absorption of 0·7–0·8, using as much as possible of the walls and ceiling for these materials.
3. The frequency characteristic of reverberation time should be close to a straight line, and should not rise either in the low or high frequency regions.
4. The sound insulation of these enclosures should be high enough to ensure that noise which penetrates into them remains below the permissible level.
5. The aural and visual impression of the enclosure represented by the set shown in each shot should be made to coincide, and the sound perspective indicated by the visual image on the screen achieved by the use of supplementary microphones placed remote from the sound source and by the use of special acoustic and electro-acoustic devices providing artificial reverberation.

7.4 Acoustic conditions in a studio for stereophonic transmission

Acoustic conditions in studios for stereophonic recording or transmission, apart from satisfying a number of general conditions, should contribute towards the creation of natural sound and to the correct location of the apparent sound source in the reproduction.

These factors are of equal importance in sound filming. For moving shots, usually connected with speech accompaniment, the second is of more importance, while for static shots, which are usually involved with the recording of music, it is more important to ensure that the former requirement is met.

The naturalness of sound and the creation of a spatial 'sound picture' are of primary importance in stereophonic broadcasting.

These two factors are mutually contradictory, as naturalness and fullness of sound are caused by a certain reverberation time in the enclosure, and good accuracy of localization of the source is better achieved with reverberation time reduced to zero.

Probably the best way of overcoming this contradiction is to reduce the reverberation time as far as possible in studios for speech transmissions, where localization is of major importance, and to preserve or perhaps even increase the reverberation time for stereophonic transmission of music, in the attempt to give the music more naturalness and fullness.

The use of stereophonic transmissions of several microphones leads to a change in the ratio of reflected and direct sound, i.e. to a change in

237

equivalent reverberation time. In fact, the reflected energy received by the microphones of a stereophonic system is proportional to the number of microphones. Direct energy also increases, but more slowly, as a result of the fact that the microphones of the system are placed at different distances to and at different angles from the source of sound.

As it is usual to evaluate acoustic conditions in a studio on the basis of its use for single channel sound recording, it is necessary to determine the way in which the acoustic conditions in an enclosure should be changed when a multi-channel stereo system is in operation, in order to achieve the reverberation effect similar to that in single channel operation.[56]

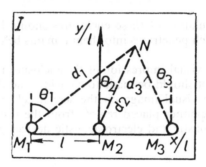

Fig. 7.9. The position of the performer N relative to microphones M_1, M_2, and M_3.

If in a studio (Fig. 7.9) m microphones, each connected to the corresponding channel of a stereophonic system, are arranged in front of a sound source N, the sound pressures acting on each of the microphones is expressed by the following equations:

$$p_1 = p_0\Phi(\theta_1) \cdot y/d_1 = p_0\Phi(\theta_1) \cos \theta_1;$$
$$p_2 = p_0\Phi(\theta_2) \cdot y/d_2 = p_0\Phi(\theta_2) \cos \theta_2; \qquad 7.5$$
$$. \; . \; . \; . \; . \; . \; . \; . \; . \; . \; . \; . \; . \; . \; . \; . \; . \; . \; .$$
$$p_m = p_0\Phi(\theta_m) \cdot y/d_m = p_0\Phi(\theta_m) \cos \theta_m$$

where p_0 is the sound pressure created by the source if it is placed on the axis of the microphone; $\Phi(\theta_1)$, $\Phi(\theta_2)$, . . ., $\Phi(\theta_m)$ are functions expressing the directional qualities of the microphones.

By calculating equations 7.5, the power developed by the microphone of a single channel as a result of direct sound, can be expressed:

$$P'_{\text{dir}} = kE_{\text{dir}} = \frac{p_1{}^2}{\rho_0 c_0{}^2} = \frac{p_0{}^2}{\rho_0 c_0{}^2} \Phi^2(\theta) \cos^2 \theta_1, \qquad 7.6$$

and the electrical power developed by the microphones of a multi-channel system:

$$P'_{\text{dir}_\Sigma} = kE_{\text{dir}_\Sigma} = \frac{\sum\limits_{m=1}^{m} p_m{}^2}{\rho_0 c_0{}^2} = \frac{p_0{}^2}{\rho_0 c_0{}^2} \sum_m \Phi^2(\theta_m) \cos^2 \theta_m \qquad 7.7$$

238

The power developed by a directional microphone and caused by reverberant energy alone, with the microphone working in a single channel system is:

$$P'_{\text{ref}} = kE_{\text{ref}} = \frac{16\pi r^2 p_0{}^2}{\rho_0 c_0{}^2 S\Omega} \cdot \frac{1-\alpha}{\alpha} \qquad 7.8$$

where Ω is the coefficient of directionality of the microphone, expressed by the equation:

$$\Omega = \frac{2}{S\Phi^2(\theta)^2 \sin\theta \, d\theta}$$

The power developed by the microphones of an m-channel stereophonic system when reflections are acting on them is m times greater than for a single channel system, and is represented as:

$$P'_{\text{ref}\Sigma} = kE_{\text{ref}\Sigma} = m \frac{16\pi r^2 p_0{}^2}{\rho_0 c_0{}^2 S\Omega} \cdot \frac{1-\alpha}{\alpha} \qquad 7.9$$

On the basis of expressions 7.6 and 7.8 we can determine that the acoustic ratio for a single channel system is:

$$R = \frac{P'_{\text{ref}}}{P'_{\text{dir}}} = \frac{16\pi r^2}{S\Omega\Phi^2(\theta_1)\cos^2\theta_1} \cdot \frac{1-\alpha}{\alpha} \qquad 7.10$$

while for a multi-channel system, from expressions 7.7 and 7.9, this ratio is expressed as:

$$R_m = \frac{P_{\text{ref}\Sigma}}{P_{\text{dir}\Sigma}} = \frac{16\pi r^2 m}{S\Omega \sum\limits_m \Phi^2(\theta_m)\cos^2\theta_m} \cdot \frac{1-\alpha}{\alpha} \qquad 7.11$$

If the last two expressions are equated, we can find the mean coefficient of absorption α' necessary to ensure that the acoustic conditions in an enclosure for a multi-channel system are the same as the conditions for single-channel operation. After some transformations, this coefficient is shown by the equation:

$$\alpha' = \frac{m\alpha\Phi^2(\theta_1)\cos^2\theta_1}{(1-\alpha)\sum\limits_m \Phi^2(\theta_m)\cos^2\theta_m + m\alpha\Phi^2(\theta_1)\cos^2\theta_1} \qquad 7.12$$

As in a single channel sound transmission the microphone is usually placed in front of a stationary source, or moves with the source, then in the last expression we can consider $\cos\theta_1 = 1$ and $\Phi(\theta_1) = 1$ then:

$$\alpha' = \frac{m\alpha}{(1-\alpha)\sum\limits_m \Phi^2(\theta_m)\cos^2\theta_m + m\alpha} \qquad 7.13$$

Therefore, taking formula 2.42 also into consideration, the reverberation time for multi-channel transmission T' can be related to reverberation time for single-channel transmission T in the same studio in the following way

$$T' = kT$$

where

$$k = \frac{\log(1 - \alpha)}{\log(1 - \alpha')} \qquad 7.14$$

is a coefficient characterizing the change in reverberation time when changing from single-channel to multi-channel transmission.

By using expressions 7.13 and 7.14 we can construct curves for $k = f(\alpha)$ for various types of microphones in systems with various numbers of channels.

Fig. 7.10 shows curves $k = f(\alpha)$ for the use of non-directional microphones; curves 1, 2 and 3 were obtained respectively for 2-, 3- and 5-channel systems with the relative distance y/l from the line of the microphones to the line along which the performer moves (see Fig. 7.9) equal to 0·5; curves 1′, 2′ and 3′ and 1″, 2″ and 3″ correspond to the same number of channels but are drawn for values of y/l of 1 and 2 respectively.

Fig. 7.11 shows the function $k = f(\alpha)$ for use in a 5-channel stereophonic system of microphones of various kinds. The symbols shown by each curve show the shape of the directional characteristic of the microphones; solid lines are curves corresponding to $y/l = 0·5$ and broken lines $y/l = 1$.

Considering Figs. 7.10 and 7.11 we can form some general conclusions. If we are trying to ensure that the reverberation effect for multi-channel stereophonic recording or transmission is similar to that for a single channel system, these conclusions can be summed up as follows:

1. The coefficient k always remains less than unity when the coefficient of absorption α, the number of channels m and the distances of the performer from the line of the microphones y/l change. Thus T' is always less than T.

2. This coefficient decreases with decrease of α. Thus the difference between T' and T is greater for lightly damped enclosures.

3. As the number of channels is increased, the coefficient k at first grows rapidly, and then where $m \geqslant 5$ changes more slowly. Thus, in an enclosure equipped for the operation of five or more channels, acoustic conditions may not change.

4. The range of values of k dependent as it is on α and m, grows narrower as the relative distance between the line of the microphones and the performer y/l increases. Where $y/l = 2$, the coefficient k becomes almost constant.

5. The greatest change in reverberation time in a studio for stereophonic operation is required for figure-of-eight microphones, and the least change by omni-directional microphones.

6. Bearing in mind that in practice stereophonic recording never uses fewer than three channels, and that the distance y/l is always between 0·5 and 1, coefficient k may be given some value between 0·6–0·85. This means that the average coefficient of absorption for speech studios should be increased for stereophonic recording to between 0·8 and 0·85.

7. As the use in stereophonic transmission of a large group of microphones results in an increase in equivalent reverberation, which gives greater

240

naturalness to music, acoustic conditions in music studios should not be changed while stereophonic transmission is taking place. This recommendation should be checked in practice.

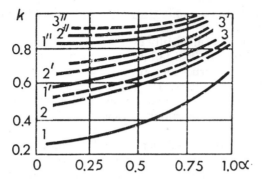

Fig. 7.10. Curves $k = f(\alpha)$ for a stereophonic system using non-directional microphones. 1, 2, 3, for $m = 2, 3, 5$ and $y/l = 0{\cdot}5$; 1', 2', 3' and 1'', 2'', 3'' for the same values of m, but with $y/l = 1$ and $y/l = 2$.

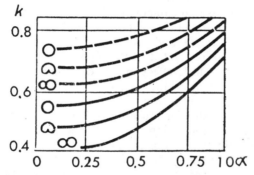

Fig. 7.11. Curves $k = f(\alpha)$ for a five channel system with various microphones: continuous lines for $y/l = 0{\cdot}5$ broken lines for $y/l = 1$.

7.5 The dimensions, shape and acoustic treatment of film and radio studios

The dimensions and shape of an enclosure have a distinct effect on its acoustic properties. Incorrect choice of dimensions for an enclosure may result not only in an inefficient use of its volume and inconvenience in use, but also in an impairment of the uniformity of the sound field. Thus, judging from formula 6.18, the number of characteristic frequencies per unit bandwidth is significantly less in small enclosures than in large ones, and this has an effect on the uniformity of the sound field.

No less important is the question of the ratio of the dimensions of an enclosure, because if they are equal to, or even multiples of one another (as

241

can be seen from formula 4.10) the number of modal frequencies of the enclosure is reduced as a result of the excitation of a number of coincident frequencies. A poor choice of the geometrical shape of the enclosure often leads to distinct acoustic deficiencies. In enclosures of good shape, optimum reverberation time can be significantly increased without noticeable damage to clarity of sound.

In order to achieve a sufficiently even sound field in a small talks studio, and also to avoid the effect of fatigue on the announcer, which is a noticeable effect of a short reverberation time, these studios should have a volume not less than 20 cubic metres. The shape usually chosen is a rectangular parallelepiped, which is convenient in use and which does not introduce any noticeable acoustic distortions in view of the heavy deadening of these enclosures.

As the volumes of these studios are small and an attempt must be made to create a uniform sound field, absorbent materials meeting the required reverberation times in efficiency and dimensions may be distributed evenly on all the bounding surfaces of the enclosure.

The dimensions of speech dubbing studios should be chosen with regard to the convenience of working of the group of performers, taking into account the need to have a screen in front of the performers in dubbing, and free space for lights and cameras in the case of a television broadcast. For these reasons, the dimensions of speech dubbing studios are selected to give volumes between 300 and 800 m^3.

As a speech studio is treated with highly efficient sound absorbent materials, its shape has little effect on the distribution of the sound field, and consequently it may be arbitrarily chosen. However, rectangular speech studios are more convenient in use.

The volume for a studio for music recording or transmission is chosen primarily by considerations connected with the need to create optimum acoustic conditions in the studio. It is important from this point of view that sufficient loudness should be maintained in the studio, demanding a specific relationship between the power of the source and the volume of the studio.

A level of loudness in a studio high enough to cause overloading is extremely undesirable, because it disorientates the performers and the conductor and upsets the balance between the sounds of various instruments of the ensemble. As the loudness level of musical ensembles depends on the number of players, the volume of an enclosure for music recording or transmission is related to their number.

There is another reason why the volume of a music studio is determined according to the number of performers. To remove any emphasis of individual instruments or of groups of instruments against the background of the sound produced by the ensemble as a whole, it is necessary to move the microphones away from the performers to a distance commensurate with the mean free path of a sound wave and consequently the area and volume of the enclosure have to be increased if it is necessary to accommodate more performers.

242

If we remember that, to maintain a constant level of loudness in the enclosure, the power of the source must be increased in direct proportion to the volume of the enclosure to the power of 2/3, we may introduce a coefficient equal to the ratio of the acoustic power of a symphony orchestra to the number of performers, we can obtain the empirical formula:

$$N = 0{\cdot}125V^{\frac{2}{3}} \log V \qquad\qquad 7.15$$

Another empirical formula establishes the following connection between the number of performers and the volume of the studio:

$$V = 21N + 55 \qquad\qquad 7.16$$

Both formulae, illustrated respectively by curves 1 and 2 in Fig. 7.12 for an ensemble of the same number of performers, need studios of different volumes.

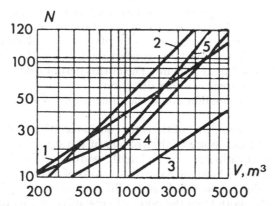

Fig. 7.12. The dependence of the number of performers on the volume of the studio: (1) and (2) according to formulae 7.15 and 7.16: (3) for concert halls; (4) for film studios; (5) for radio studios.

Formulae 7.15 and 7.16, which are frequently found in the literature of acoustics, are, as experience shows, out of date, and do not meet modern needs. Calculation according to these formulae leads to volumes for music studios which are clearly too small; this happens, apparently, because of the fact that when these formulae were composed no allowance was made for the additional floor area necessary solely to enable the microphones to be hung a certain minimum distance from the performers.

Another extreme view of the dependence of the volume of studios on the number of performers is expressed by curve 3 in Fig. 7.12. This curve is constructed from the mean data for a number of good concert halls, and is recommended[34] for radio studios, with a warning that the number of performers in them should not be even twice as many as the number of performers in the corresponding halls. This formula requires excessively high values for the volume of music studios.

Practice[44] shows that every modern film studio must have two or three music recording studios. One should be used for the recording of chamber ensembles of 25 or 30 performers, a second for orchestras of 75 or 100 players, the third for recording large orchestras with a chorus.

If we base ourselves merely on the norm of two square metres per performer, the first studio must have an area of 50 m². However, bearing in mind that microphone placing requires an area twice as large as that for the performers, the total area of the studio will be 150 m². In this case the height of the studio will be 6·5 to 7 m and the volume 1000 m³ (see Appendix 5).

Similar calculations enable us to demonstrate that the floor areas of studios for 75, 100 and 120 performers, should be respectively 450, 600 and 720 m², and the volume 4000, 5700 and 7200 m³ (curve 4 on Fig. 7.12).

The standards for radio studio centres[80] recommend that such centres should have four music studios, the volumes of which (curve 5 in Fig. 7.12) are close to the volumes recommended for film music studios. In accordance with these standards, the volume of chamber broadcasting studios for 10 or 15 performers should be between 1000 and 1400 m³; concert broadcasting studios intended for orchestras of 55–75 players should be between 2000 and 2500 m³ in volume, and large concert studios for 115–140 performers can have a volume of up to 5000 m³. The norms permit a departure of 10 per cent from the given areas in either direction.

More detailed information about the dimensions of studios are given in Appendix 5.

The creation of suitable conditions for music recordings or transmissions does not end with the choice of the volume of the studio, but includes also finding the correct ratio between its length, width and height.

It is not desirable to have the basic dimensions of a studio of equal length, or to have one dimension significantly different from the other two. This should be avoided both from the point of view of convenience in arranging the performers, and also from that of acoustic conditions. Where the dimensions are equal, it is difficult to place the performers because of insufficient floor area, and acoustic conditions are worsened because the ratio of the area of the boundary surfaces S and the volume V is at its lowest.

A reduction of the ratio S/V, as can be seen from formula 2.15, results in a reduction of the mean number of reflections per second, and makes the sound field less uniform. A great difference between the basic dimensions of a studio, on the other hand, is undesirable because it is difficult to fit in the performers and the microphones, and the danger also arises that an uneven field will be created because of the great differences in the time of arrival at the microphone of reflections of the same order.

Experience has shown that the wisest course is to select the dimensions of music studios according to the ratios in Table 3.

More precise ratios of dimensions can be found using Fig. 7.13.

If the shape of speech studios is not a matter of great importance because of the weak influence which reflections have on the sound field within them, this cannot be said of music studios. Their considerably greater reverberation time, and their greater dimensions can result in reflections between

224

parallel walls which set up a slowly-decaying system and lead to major unevenness in the field.

If we set the walls a little out of parallel, a large number of reflections is formed by more than two walls. In other words, putting walls out of parallel results in a reduction in the number of axial waves, which have the most orderly nature and which contribute greatly to the unevenness of the field.

TABLE 3

Studio volume m^3	Length l	Breadth b	Height h
Up to 250	1·6	1·3	1
From 650–1250	2·5	1·5	1
From 2000–4000	3	2	1
From 4000 upwards	3·3	2·2	1

These considerations cause us to resort to trapezoidal and other non-rectangular shapes for music studios. The angle of skew of the walls of such enclosures is usually up to about 10°. For the same reason, the walls, and sometimes the ceiling, of music studios are inclined at an angle of about 5°. In view of the difficulty of planning large studios with inclined and skewed walls, these are often made ribbed, or a non-rectangular shape is given only to that part of the enclosure where the performers are placed.

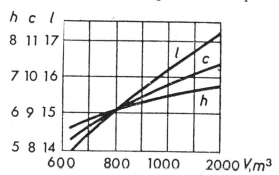

Fig. 7.13. Graph for the determination of the relative length, breadth and height of a studio of medium volume.

Greater uniformity of the sound field may be obtained by the even distribution of sound absorbent materials on the surfaces of the studio, and also by placing convex sound reflectors along the walls and in the ceiling. Such constructions, made of plywood, work as rigid resonant systems, providing additional absorption at low frequencies.

Sometimes, in the attempt to create good conditions for the performers, efforts are made to redistribute sound dispersing and absorbing surfaces, concentrating the former in that part of the studio where the orchestra is placed.

One single studio, or indeed several studios, cannot meet the demands of

245

broadcasting or recording in view of the fact that optimum acoustic conditions do not depend only on the number of performers, but also on the character of the programmes performed. This consideration leads us to the thought of the control of the acoustic properties of a studio by means of setting up some type of variable absorption.

For variable absorption in studios, absorbent materials with very different coefficients of absorption are mounted on movable screens. The screens can be moved along the walls parallel to each other, or, if mounted on hinges, can be opened and closed. In some studios one comes across remote control of acoustic characteristics, achieved by turning revolving columns which project only half way into the studio. As each half of the outer surface of the columns is treated in absorbing material of different coefficient, then, by revolving these columns, one can increase or decrease the areas occupied by the more effective absorbing materials and thus change the reverberation time.

The volumes of sound stages or television drama studios are not arrived at from acoustic considerations, but on the basis of the use to which they are to be put. The need to house large sets, and to find room for lighting, filming or television transmission apparatus makes it necessary to build stages of several thousand, or indeed, tens of thousands of cubic metres.

The ratios of the dimensions and the shape of these enclosures are also chosen to allow for the correct and convenient use of the technology of making pictures, either on film or in the television camera. However the dimensions ratios chosen purely for the above reasons, vary between 2·5 : 1·5 : 1 and 4 : 2 : 1, and are thus close to those which one might have selected in the hope of creating good acoustic conditions.

With the convenience of the film or television camera in mind, these enclosures are rectangular in shape; this cannot have any real effect on the quality of sound transmission as the studios are heavily damped. Moreover, the sets and the technical equipment help in the establishment of a diffuse field, which compensates to a certain extent for the worsening of acoustic conditions which results from the rectangular shape of the studio.

The walls and ceilings of these studios are evenly covered in efficient sound absorbing materials, and only for the lower part of the walls to a height of 3 m is the use recommended of materials or constructions with a rigid surface which cannot be damaged by the movement of scenery or equipment. The most frequently used absorbing materials for the upper part of the studio are mineral wool or asbestos wool 10 cm thick, held against the wall by a thin fabric cover. For the treatment of the lower part of the studio, porous resonant absorbers or resonant panels are recommended with the spaces between the system and the wall filled with mineral wool.

7.6 Artificial means of regulating acoustic conditions in enclosures

We have considered above the mechanical method of regulating acoustic conditions by means of variable absorption, sometimes used in music studios, where the amount of adjustment provided is comparatively small. It is clear that in film stages or television drama studios such a method cannot be used, because the necessary amount of adjustment in these cases is very large. This is why, in order to create a complete correspondence between the visual and aural perception of the film or television viewer, electro-acoustic methods are used during the process of recording or transmission to create and adjust the acoustic characteristics of the sound.

The idea of using artificial reverberation, enunciated in 1929 by V. S. Kazanski, involves the addition, during the process of recording or transmission, of a reverberant signal, produced by artificial means, to the direct

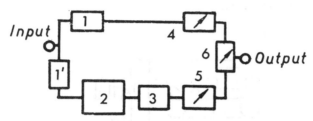

Fig. 7.14. Scheme of a system for the control of artificial reverberation. 1, 1' and 3 amplifiers; 4, 5, and 6 attenuators; 2, the reverberation producing equipment.

signal. The total effect in this case depends on the ratio of the levels of these signals and on the rate of decay of the reverberant signal. From this it is clear that it may be possible to control the reverberation by a relative change in the levels of the direct and the reverberant signals or by altering the steepness of the decay curve of the reverberant signal.

The block diagram of equipment for controlling artificial reverberation by the former method is shown in Fig. 7.14. The original signal, when it reaches the input of the device, is divided into two channels, in one of which is included the equipment, 2, which creates the reverberant signal. At the output of the device, both signals, amplified by amplifiers 1 and 3, are varied in level by faders 4 and 5, so that the resulting signal has the desired degree of reverberation.

The second method of regulation depends on the method of creating reverberant signals. Existing systems can be divided into three groups according to the method of operation. The first group includes systems which create a reverberant signal by electro-acoustic means, the second group consists of systems in which these signals are obtained purely acoustically, and finally, the third group includes signals obtained by a mixture of both methods.

In the first group we should include so called magnetic reverberators.

System with a magnetic reverberator

This system (Fig. 7.15a) is a tape recorder with tape loop 1 moved by spools 2. Apart from the usual record and erase heads 3 and 4, the reverberator has a series of reproducing heads 5, in the chain of each of which is included a volume control 6. These controls are adjusted so that the signal level produced by each successive head in the series diminishes according to a linear law (Fig. 7.15b). The last head in the reproducing series is switched to the input of the recording amplifier 8 and provides, by feedback, a continuous series of recordings and reproductions of a signal at successively reduced level.

The signal level in the chain of head 7 is adjusted with the aid of fader 9 so that the successive repetitions caused by the feedback decrease with the same exponential law as do the levels of the signals first reproduced by the heads of the reproduction block, and, as it were, arrange themselves in the series shown in Fig. 7.15b.

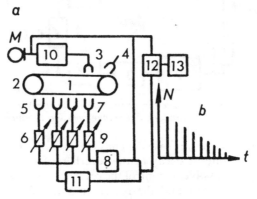

Fig. 7.15. Block diagram of a magnetic reverberator. 1, magnetic tape; 2, spools; 3, 4, records and rase heads; 5, 7, reproducing heads; 6, 9, levels adjusters; 8, 10, 11, 12 amplifiers; 13, sound recording equipment.

A simultaneous change in the position of attenuators 6 and 9 results in a relative growth or fall in the levels of the neighbouring signals, which changes the steepness of the decay curve. Thus the control of reverberation can be accomplished by the use of the attenuators in the chains of the reproducing heads.

Systems which allow a reverberant signal to be obtained acoustically include echo chambers and echo plates.

Echo chambers

In this case the system providing the reverberant signal is an enclosure with non-parallel surfaces and of volume not less than 100 m³, equipped with one or several loudspeakers and microphones. Part of the original signal from the studio is fed to the loudspeakers. As the interior surfaces of the echo-chamber are treated with good reflecting materials (e.g. concrete), the omnidirectional microphone placed fairly remote from the loudspeakers will receive a very reverberant signal, which is mixed with the original signal at the output of the system.

248

The character of decay of the reverberant signal in the echo chamber depends basically on the acoustic data of the chamber, and the influence of the studio which is acoustically coupled to it is extremely small. This can easily be realized if it is remembered that the combined reverberation time in enclosures coupled by an electro-acoustic channel is determined by the reverberation time of the more reverberant enclosure—in this case the echo chamber. This last statement is also derived from formula 2.74, if we consider that reverberation time in echo chambers reaches 5–6 sec for a reverberation time close to 1 sec for studios or shooting stages.

In view of what has been said above, we should try to see that an adequately uniform sound field is created in the echo chamber. This is why it is given a non-rectangular shape. In order to avoid these enclosures being frequency selective, a high density of modal frequencies should be aimed at, both in the high frequency and especially in the low frequency regions of the frequency range. This can be achieved with a comparatively large echo chamber of volume 100–200 m^3.

Adjustment of reverberation time by changing the average steepness of the decay curve is possible in this case within small limits by varying the distance between the microphone and the loudspeaker, by the use of a microphone with variable directional properties, and finally by the use of echo chambers of differing volumes.

Echo plates

The basis of this comparatively new method is a two-dimensional variant of the method of creating a reverberation process. The reverberant element is not a volume of air, but a flat steel sheet 2 m^2 in area and about 0·5 mm

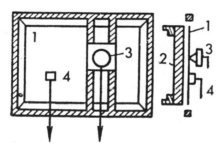

Fig. 7.16. Diagram of an echo plate.

thick. The sheet is spring mounted at its corners to a metal framework which is fixed vertically (see Fig. 7.16). On one side of the sheet 1 is mounted, also vertically, a second frame bearing a microporous material 2 which moves freely relatively to the first. On the other side there is an electro-dynamic loudspeaker 3 with a metal cone welded to the sheet in place of its usual cone. On the same side of the sheet some distance from the loudspeaker is fastened a piezo-electric sound pickup.

The reverberator is connected to a side-chain in parallel with the main circuit, by which a proportion of the signal is fed to the loudspeaker and thus brings the sheet into vibration. The transverse waves which are produced travel to the edges of the sheet, are reflected from them, and at each reflection lose part of their energy. The sound pickup receives these progressively decaying signals, whose delay, due to the low velocity of distribution of transverse waves, is close to the delay times characteristic of large enclosures. The spectrum of modal frequencies of the sheet, as in closed enclosures, grows more dense towards the high frequency end, and is accompanied by increasing density of reflected signals in time. In addition,

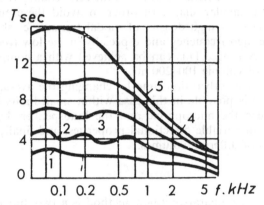

Fig. 7.17. The frequency characteristics of reverberation of an echo plate at distances between the sheet and the sound absorbent material of 0·5 cm (1), 1 cm (2), 2 cm (3), 3 cm (4) and 10 cm (5).

the law of decay of the reverberant signal at the output of the pickup is very little different from an exponential curve. All this helps to ensure the similarity of perception of an artificially obtained and a natural reverberation effect.

If the frame carrying the absorbing material is brought nearer to the sheet this speeds the decay of the transverse vibrations of the sheet. Thus by altering the distance between them (usually between 20 and 0·5 cm), reverberation time can be controlled between fairly wide limits (from between 6 and 10 sec to 0·6 sec).

As can be seen from Fig. 7.17 an increase in the distance between the vibrating system and the absorbing material, while resulting in a significant increase in reverberation time, also produces a noticeable worsening of the frequency characteristic of the reverberation process. Whereas for small distances the damping of the system results mainly from the sound absorbing material, for great distances it is mainly due to internal friction in the sheet, which causes losses, increasing rapidly as the frequency is increased. This defect in the system can be removed by the use of additional damping which introduces decay inversely proportionate to frequency.

250

Spring reverberator

As with plate echo, the basis of this system also is a device which imitates vibratory processes similar to those which take place in the volume of air within an enclosure. The imitation is performed with the aid of steel springs which undergo torsional vibrations under the action of mechanical vibrations produced by the sound.

By choice of the parameters of the springs the time of transmission from the transmitter attached to one end of the spring to the receiver fastened to the other end can be strictly controlled. Waves which are propagated along the spring are reflected and return to the receiver with a fixed delay time and

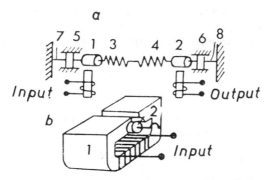

Fig. 7.18. Schematic diagram of a reverberation spring (a) and of one of the electromagnetic transducers (b).

with decreasing amplitude. This process is repeated many times until the reflected vibrations die away.

To increase the density of reflected signals, two or three springs with different parameters are used in parallel.

A diagram of a spring reverberator is shown in Fig. 7.18a, where 1 and 2 are an electromagnetic transmitter and receiver, the construction of which is shown at Fig. 7.18b; 3 and 4 are the transmitting springs, which have opposite directions of winding; 5 and 6 are dampers; 7 and 8 are supporting wires. The springs are composed of two elements with opposite windings in order to diminish mechanical interference and to avoid unwinding resulting from fatigue.

The audio frequency signal taken from the sound recording amplifier, is fed to the exciter coil 1 of an electromagnet (Fig. 7.18b), whose field acts on the rotor 2. The rotor begins to rotate, and the angle of its rotation is proportional to the exciting current. Torsional vibrations of the rotor are transmitted along the spring and after a certain delay bring the rotor of the receiver into motion. The delayed signals which are induced in the coil of the receiver are mixed with the direct signal in the mixing desk.

The delay time of the reflected signals depends on the velocity with which the wave is distributed along the spring. This velocity can be calculated in a similar manner to that for the velocity of propagation of electromagnetic

251

waves in a lossless transmission line. In both cases the velocity is expressed by the equation:

$$v = \frac{1}{\sqrt{Ic}}$$

7.17

where I is the moment of inertia of a turn of the spring, and c is the flexibility of the coil.

The moment of inertia I may be found if the coil is regarded as a toroid whose volume is V. In this case

$$I = V . R^2\rho = 2\pi R . \pi r^2 . R^2\rho = 2\pi^2 R^3 r^2 \rho$$

7.18

where r is the radius of a cross section of the wire, R is the radius of the coil, and ρ is the specific gravity of the material.

The toroidal displacement Φ of a spring of n turns under the action of a moment M is expressed by equation:

$$\Phi = \frac{8MRn}{\varepsilon r^4}$$

7.19

where ε is Young's modulus of elasticity. In this case the flexibility of one turn ($n = 1$) can be represented as:

$$c = \frac{\Phi}{M} = \frac{8R}{\varepsilon r^4}$$

7.20

If we substitute the values of I and c from equations 7.18 and 7.20 into equation 7.17 we obtain:

$$v = \sqrt{\frac{\varepsilon r^4}{2\pi^2 R^3 r^2 \rho . 8R}} = \frac{r}{4\pi R^2} \sqrt{\frac{\varepsilon}{\rho}}$$

7.21

This formula shows that the velocity of propagation of a wave for springs of similar material and with similar r and R, is constant and does not depend on the frequency of the signal. This formula enables us to determine the number of turns of a spring reverberator for a given delay time τ. In fact the number of turns is

$$n = v . \tau = \frac{r\tau}{4\pi R^2} \sqrt{\frac{\varepsilon}{\rho}}$$

7.22

A calculation from formula 7.22 leads to the conclusion that if the spring is made of steel wire for which $r = 0.6$ mm and it is wound with radius $R = 30$ mm, then for $\tau = 40$ msec the spring must have roughly 550–600 turns.

Industrially produced spring reverberators have two springs with different numbers of turns, giving delay times of 29 and 38 msec.

The speed of decay of a signal undergoing multiple reflections from the ends of the spring depends on the frequency of the signal. As a result of this the frequency characteristic of reverberation time drops at low and high frequencies. To improve the frequency characteristic, use is sometimes

252

made of the mechanical resonance of the support of the transmitter and receiver coils. The best results are obtained when the resonant frequency of the transmitter is in the low frequency region (100–200 Hz) and that of the receiver in the high frequency region (4000–5000 Hz). Under these conditions the reverberation time of the reverberator will vary from 2 sec at a frequency of 200 Hz to 1 sec at high frequencies.

A spring reverberator, like other similar devices, may be used in stereophonic systems. In this case the reverberant signal is fed to each channel of the system.

Ambiophonic system

This system, unlike those we have described previously, uses both acoustic and electro-acoustic means to create artificial reverberation. Its second distinguishing feature is that the mixing of the artificially induced reverberant

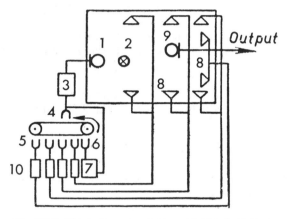

Fig. 7.19 Block diagram of an ambiophonic system.

signal with the original signal takes place directly in the studio or other primary enclosure, and this makes it possible to create the necessary acoustic optimum not only in the reproduced sound, but also for the performers who are in the studio.

The diagram of an ambiophonic system is shown in Fig. 7.19. A highly directional microphone 1 receives from the primary sound source 2 a direct signal, which having been amplified by amplifier 3 is fed to the recording head of a special reverberator. Each of the sound-reproducing heads 5 reproduces the signal with a fixed delay. A special head 6 feeds a feedback chain which includes an additional amplifier 7 to create a dense series of delayed signals. These signals are fed back into the studio by specially situated groups of loudspeakers thus creating additional reverberation which is received by the programme microphone 9 connected to the input of the sound recording or transmitting equipment. In order that the delayed signal levels may decay according to a linear law, additional amplifiers 10 are included in the chain of the heads 5, allowing the necessary control over these levels.

Thus the reverberation time created by an ambiophonic system can be controlled either by altering the levels of amplification in the reproducing head circuits, or by altering the amplitude of the feedback by means of amplifier 7. Simultaneously with the control of reverberation time a change is effected in the degree of diffusion of the sound field as a result of the additional series of delayed signals introduced by the artificial reverberation into the enclosure, which have various levels and which are distributed in various directions.

The process of the decay of sound energy in the enclosure with the ambiophonic system is similar to the process of decay in coupled enclosures. From this similarity, using the analogy with equation 2.56, the equation for a studio with ambiophonic equipment may be written as:

$$V \frac{dE}{dt} - P_a - P_{am} + P_{absorp} = 0 \qquad 7.23$$

When the primary sound source is switched off, i.e. when $P_a = 0$, the power of the loudspeakers declines exponentially and is represented as:

$$P_{am} = E_o . \delta . e^{-\delta_n t} \qquad 7.24$$

where δ_n is the index of decay of the ambiophonic system.

On the basis of this equation and formula 2.57, equation 7.24 can be presented as:

$$\frac{dE}{dt} - E_0 \delta e^{-\delta_n t} + E_0 \delta = 0 \qquad 7.25$$

As this expression is absolutely identical with expression 2.58, its solution is identical with solution 2.67:

$$E = \frac{E_0}{\delta - \delta_n} (\delta e^{-\delta_n t} - \delta_n e^{-\delta t}) \qquad 7.26$$

In order to find the value of the energy density E_0, in the studio in the steady state, for substitution into expression 7.26, we must solve equation 7.23, reckoning that $dE/dt = 0$:

$$P_a + P_{am} = P_{absorp} \qquad 7.27$$

In such a state the acoustic power of an ambiophonic system P_{am} is proportional to the power of the source P_a and to a constant K, which is connected with the index of decay of the system δ_n in the following way:

$$K = \frac{1}{1 - e^{-\delta_n t}}$$

Consequently:

$$P_{am} = kKP_a \qquad 7.28$$

where k is the coefficient of proportionality determining the amplification in the channels of the system.

254

If now, using expressions 7.28 and 2.18, P_{am} and P_{absorp} are replaced in equation 7.27 by their values given above, the equation takes the form:

$$P_a(1 + kK) = E_o V \delta$$

whence:

$$E_o = \frac{P_a}{\delta V}(1 + kK) \qquad\qquad 7.29$$

As follows from equation 2.70, $\delta = 13\cdot8/T$ and $\delta_n = 13\cdot8/T_n$. If we insert these values and E_o from expression 7.2 into the formula for the density of energy in the decay process (7.26), it can be presented in the form:

$$E = \frac{13\cdot8 P_a T}{V(T_n - T)}(1 + kK)\left(T e^{-\frac{13.8}{T_n}} - T_n e^{-\frac{13.8}{T}}\right) \qquad 7.30$$

This expression shows that decay of sound in an enclosure with an ambiophonic system does not follow an exponential curve. Moreover, the total reverberation time T_Σ depends largely on the ratio of T_n and T, and also on

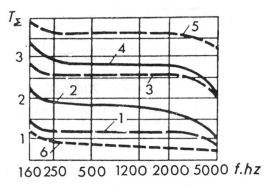

Fig. 7.20. *Frequency curves of total reverberation in a studio for different values of artificial reverberation time T_n with acoustic feedback (continuous lines) and without it (broken lines); 1 and 2 for $T_n = 1$ sec; 3 and 4 for $T_n = 2$ sec; 5 $T_n = 3$ sec and 6 $T_n = 0$ secs.*

kK. If $0\cdot5$ is selected as the value of kK, then total reverberation time T_Σ will be little different from artificial reverberation time T_n. In this case the ambiophonic system has a real influence on the acoustic conditions in the enclosure, and, which is very important, cannot be detected by the ear.

The resultant reverberation time can also be increased by increasing K for a constant value of T_n. This can be done by increasing the speed of the magnetic tape in the reverberator, as a result of which τ is diminished.

By way of illustration of these considerations, Fig. 7.20 and Fig. 7.21 show the reverberation frequency characteristics for a studio equipped with an ambiophonic system. The first of these diagrams shows how the total reverberation time T_Σ changes as a result of changes in artificial reverberation time T_n with acoustic feedback (continuous lines) and without it

255

(broken lines). Curves 1 and 2 were obtained for $T_n = 1$ sec, curves 3 and 4 for $T_n = 2$ sec, and curve 5 for $T_n = 3$ sec. The lower dotted curve 6 shows the reverberation characteristic of the studio when the ambiophonic system is not in operation.

The curves in Fig. 7.20, apart from giving quantitative information on the connection between total and artificial reverberation, show that reductions in the latter may have a distinct effect on acoustic conditions in the enclosure, increasing reverberation time by 3 to 3·5 times. Fig. 7.20 also shows that for small values of T_n, acoustic feedback can be used, but within reasonable limits which ensure the stability and inaudibility of the system.

Fig. 7.21 shows curves of total reverberation time against frequency obtained for a constant value of T_n and for two different values of τ caused by

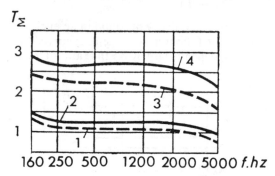

Fig. 7.21. Frequency curves of total reverberation time for various values of T_n and for tape speeds of 76 cm/sec (solid lines) and 38 cm/sec (broken lines); 1 and 2 for $T_n = 1$ sec; 3 and 4 for $T_n = 2$ secs.

a change in the speed of movement of the magnetic tape. It follows from this figure that a change in the speed of the tape has a greater effect for high values of artificial reverberation time than for low.

Judging by experimental data, the ambiophonic system also produces an increase in the sound level of 4–6 dB and increases the diffusion of the sound field in an enclosure.

The use of experimental systems in the studios of the Moscow Broadcasting and Sound Recording Centre, in the concert hall of the Gnesini Institute and of a special system in the hall of the Kremlin Palace of Congresses have shown convincingly that these systems give a great richness and fullness to music and have a good effect on the quality of sound, without impairing the ability to localize the source.

7.7 The methods of acoustic planning of film and radio studios

The aim of the acoustic planning of a studio is to specify the basic data which, considering the intended use and selected technical equipment, will provide the optimum acoustic conditions for the recording of sound, or of radio or television programmes.

The basis of planning is the technical data, by which is usually meant the purpose of the studio, the number of performers, and the use of the enclosures adjoining the studio. In planning a number of studios, the site of the studio block is indicated and the characteristics of sound sources near the site are indicated.

The acoustic plan includes the planning of studios or a studio complex with the subsequent specification of the shape and geometric dimensions of each enclosure, calculation of the necessary sound absorption, specification of optimum reverberation time and the drawing up of a sketch plan of the distribution of sound absorbing materials. Moreover, the plan includes a

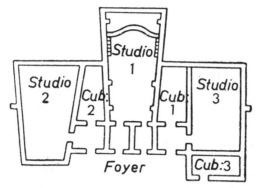

Fig. 7.22. Plan of the positioning of studios in the Moscow Broadcasting and Sound Recording Centre.

calculation of sound insulation from noises penetrating through the boundaries or by the ventilating system, and recommendations for the use of special sound insulating constructions and about additional measures to be taken to deaden water pipes and the whole ventilating system.

In planning separate studios or a studio complex, it is well to aim at the maximum direct access between the enclosures being planned and the enclosures involved in the technical processes of recording or broadcasting. And it should not be forgotten that this problem must be solved without any adverse effect on sound insulation.

Convenience of use of studios might suggest that they should be made adjacent to one another, but in view of the fact that this would make the problem of mutual sound insulation much more complicated, a plan is preferable in which the studios are separated by comparatively quiet areas (see Fig. 7.22). It will be seen from this diagram that the studios are separated from one another and from the noisy foyer by lobbies and control cubicles.

In planning the areas it is necessary to keep in mind that alongside music stages and broadcasting studios there must be rooms where the transmission is controlled by means of programme meters and loudspeakers,

257

and where the levels of the various signals, the dynamic range and the frequency characteristic of the recording or transmission are monitored.

In order to ensure a correct impression of the sound quality for aural control, the dimensions of these rooms, and particularly their reverberation time, should be close to mean values for rooms in which sound reproduction will take place (cinemas in the case of film sound recording, and domestic living rooms in the case of radio studios). The control rooms are visually linked with the studio by means of a window, which consists of three layers of plate glass with felt and rubber packings.

In large studios there should be two or three exits through lobbies with an area of three or four square metres each. Large television studios and sound stages should have carefully fitted scenery-doors through which sets can be carried. Near large studios and stages it is useful to have a foyer where performers can relax.

If the acoustic plan is part of an existing building adapted for broadcasting and sound recording, it is only necessary to calculate additional absorption to create optimum acoustic conditions in the studio, and to check the adequacy of the sound insulation.

For studios and sound stages which are being newly built, the following order of operations and acoustic planning may be suggested:

1. The volume of the enclosure is determined. For music studios and stages this can be decided from the number of performers, by using the data in Chapter 7.6 or in Appendix 5. For speech stages, studios and large television studios, the enclosure volume is selected and then on the basis of this decision the requirements for the enclosure, both from the point of view of acoustics and from that of the usage are considered, this term covering the possibility of the erection of essential sets, the housing of equipment, etc.

2. The shape of the enclosure is next decided. The rectangular shape, although convenient to use, is acceptable only if the optimum reverberation time of the enclosure is small and if the shape of the room cannot influence the uniformity of distribution of sound energy within it. Thus, a rectangular shape is usually chosen for speech studios, synchronous shooting stages, drama studios and small music studios. Music studios of large volume are usually designed in a trapezoidal shape with slightly non-parallel walls and ceiling.

3. The main dimensions of the studio are selected. For speech studios and sync. stages, the ratio of length to breadth to height may be chosen from $2 \cdot 5 : 1 \cdot 5 : 1$ to $4 : 2 : 1$. The choice of dimensions for music stages and studios must be decided with greater care. It should be done with the help of Table 3 or from the graphs in Fig. 7.12.

4. The area of all the surfaces bounding the enclosure are calculated. This area, for a rectangular enclosure is found from the equation:

$$S = 2(lb + lh + bh)$$

5. The necessary reverberation time for mean frequencies appropriate to the volume of the enclosure is found. For speech or music stages it is

found from Fig. 7.1 and 7.7, while for sync. shooting stages or large tele-vision studios on the basis of the recommendations set out in Chapter 7.3.

6. Using the graphs in Figs. 7.3 or 7.6, the reverberation time is worked out for the series of frequencies: 125, 250, 500, 1000, 2000, 4000 Hz, taking into consideration the purpose of the enclosure and the acoustic treatment.

7. Using the calculation, formula 2.42, the total absorption necessary to create the optimum reverberation time is worked out. For this purpose, we first find the value of $-\log_{10}(1 - \alpha)$ according to the formula:

$$-\log_{10}(1 - \alpha) = \frac{0 \cdot 071 V}{ST}$$

and then the necessary mean coefficient of absorption α is worked out, and finally the total absorption is found as a product of αS. Calculations are carried out for all the frequencies indicated under point 6, and the results are arranged in a table.

8. A calculation is made of the total absorption in the enclosure when its surfaces are covered with ordinary building materials. To do this, first the names of the materials to be used, their areas, and their coefficients of absorption for all the frequencies indicated above are set out in a table. These last data are obtained from Table 1. Then the total absorption of each material and of all materials for all frequencies are calculated and the results compared with the results of the calculations under point 7. This comparison will reveal at different frequencies a larger or smaller difference between the necessary total absorption and the total absorption which exists.

9. A selection is made of such highly efficient sound absorbent materials, (two or three) and in such quantity as to reduce to a minimum the differences under point 8. The results of the selection and the calculations of additional absorption are entered in a table similar to that made up under point 8.

10. Adding up the basic and the additional absorption, a mean coeffi-cient of absorption is found with which the acoustic conditions in the enclo-sure will be close to the optimum. Then the values of $-\log(1 - \alpha)$ at all frequencies are found, and the reverberation time in the enclosure, when it has been treated with additional absorbent materials, is calculated. The results of the calculations are entered in a table which also includes the re-quired reverberation time. By comparing the values of the calculated and the required reverberation time at all frequencies, the correctness of the choice of additional sound absorbing material is checked. If these values differ from one another by no more than ± 10 per cent, the calculation is taken as correct.

11. A decision is made about the positioning of the special absorbing materials or constructions, in accordance with the principles set out in Chapter 7.5.

12. The noise level penetrating into the enclosure is calculated, for which purpose, in accordance with formula 6.44 a table is constructed having columns: name of the boundary, their area S, the noise level beyond

boundary N, the sound insulation of the boundary τ_{db}, $N - \tau_{db}$, $10^{0.1(N-\tau_{db})}$ and $S \cdot 10^{0.1(N-\tau_{db})}$.

13. The indicated columns of the table are completed. The first and second are filled in from the drawings of the enclosure, the third and fourth from knowledge of the spectrum of the external noise and of the construction of the boundary itself. The numerical values of these last two items are found from the tables (see Appendices 2 and 3).

14. After calculations according to the last columns, the data in the last column are totalled, the logarithm to base 10 of this total is found and from this value is calculated the value $10 \log \alpha S$ defined from the data in point 10.

15. By comparing the results obtained with the data in Appendix 4, it can be seen to what extent the sound insulation of the enclosure meets the requirement. The noise level penetrating from outside should be equal to or less than the permissible noise level for the given enclosure.

7.8 An example of the acoustic planning of a music studio

For practice in mastering the method of acoustic planning of primary enclosures let us look at a concrete example of the planning of a large music studio.

The task is to carry out the acoustic planning of a music studio intended for recording a choir and orchestra with up to 120 performers. The outer walls of the studio overlook a quiet courtyard with a noise level of 75 dB. On the inner side the studio adjoins ancillary enclosures with a noise level of 50 dB and control rooms with a noise level of 85 dB.

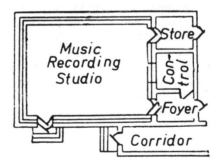

Fig. 7.23. An example of the planning of a studio and subsidiary enclosures.

1. *Planning and geometrical calculation of the studio.* A rectangular shape is chosen for the enclosure; let us site it in that part of the building which overlooks the quiet courtyard. The subsidiary enclosures which adjoin the studio are situated as shown in Fig. 7.23. On the first floor there is a quiet corridor, a control room and a musical instrument store, and on the second a speech studio and an announcer's room.

260

According to the recommendations in Appendix 5, a sound studio for the given number of performers should have

$$V = 7200 \text{ m}^3, \qquad S_{\text{floor}} = 720 \text{ m}^2, \qquad \text{and} \qquad h = 10 \text{ m}$$

The length and the breadth of the studio are determined in accordance with Table 3 by the equations:

$$l = 10 \cdot 3{\cdot}3 = 33 \text{ m} \qquad \text{and} \qquad b = 10 \cdot 2{\cdot}2 = 22 \text{ m}$$

The total area of the bounding surfaces of the studio will be:

$$S = 2(lb + lh + bh) = 2(726 + 330 + 220) = 2552 \text{ m}^2$$

and the exact volume:

$$V = 33 \cdot 22 \cdot 10 = 7260 \text{ m}^3$$

2. *Calculation of sound absorption.* On the basis of the recommendations in 7.3, the optimum reverberation time for studios at 500 Hz is 1·7 sec. If we consider that the lift in the frequency characteristic at 125 Hz should be of the order of 20–25 per cent, the frequency characteristic of the studio under planning will be expressed by Table 4.

TABLE 4

Frequency f Hz	125	250	500	1000	2000	4000
Time T sec	2·0	1·8	1·7	1·7	1·7	1·7

To determine the total absorption which would enable the provision of this reverberation time, on the basis of formula 2.41 we first find the values of $-\ln(1 - \alpha)$ and α.

For frequency $f = 125$ Hz these values are found to be:

$$-\ln(1 - \alpha) = \frac{0{\cdot}164V}{ST} = \frac{0{\cdot}164 \cdot 7260}{2552 \cdot 2} = 0{\cdot}232 \quad \text{and} \quad \alpha = 0{\cdot}21$$

Therefore the total absorption at this frequency should be:

$$A = \alpha S = 0{\cdot}21 \cdot 2552 = 536$$

If we carry out similar calculations for the other values of frequency and bear in mind that at frequencies from 2000 to 4000 Hz we must take into account the absorption of sound in air, using formula 2.40 for the calculations, we can find the total absorption for all frequencies. The data of such calculations are set out in Table 5.

TABLE 5

Frequency Hz	125	250	500	1000	2000	4000
T sec	2·0	1·8	1·7	1·7	1·7	1·7
$-\ln(1 - \alpha)$	0·232	0·258	0·273	0·273	0·248	0·230
$1 - \alpha$	0·79	0·77	0·76	0·76	0·78	0·80
α	0·21	0·23	0·24	0·24	0·22	0·20
A	536	585	612	612	560	510

TABLE 6

Description	Area or number	Sound absorption at a frequency of 125		250		500		1000		2000		4000 Hz	
		α	αS	α	αS	α	αS	α	αS	α	αS	α	αS
Walls—plaster on brick	1100 m^3	0·012	13·2	0·013	14·3	0·017	18·7	0·02	22·0	0·023	25·3	0·025	27·5
Ceiling—concrete slabs	726 m^3	0·01	7·3	0·012	8·7	0·015	10·9	0·019	13·8	0·023	16·7	0·035	25·4
Floor—carpeted	150 m^3	0·11	16·5	0·13	19·5	0·28	42·0	0·45	67·5	0·29	43·5	0·29	43·5
Floor—free. PVA	576 m^3	0·02	11·5	0·025	14·4	0·03	17·3	0·035	20·2	0·04	23·1	0·04	23·1
Doors—draped	20 m^3	0·08	1·6	0·29	5·8	0·44	8·8	0·50	10.0	0·40	8·0	0·35	7·0
Windows to control room	3 m^3	0·035	0·1	0·025	0·1	0·019	0·1	0·012	0·1	0·07	0·2	0·04	0·1
Ventilation grids	10 m^3	0·30	3·0	0·42	4·2	0·50	5·0	0·50	5·0	0·50	5·0	0·51	5·1
Performers	120	0·36	43·2	0·43	51·6	0·44	52·8	0·47	56·4	0·49	58·8	0·89	58·8
Instruments	120	0·23	27·6	0·26	31·2	0·26	31·2	0·29	34·8	0·32	38·4	0·36	43·2
Basic store of absorption A_0			124·0		149·8		186·8		229·8		218·0		233·7

Let us determine the reverberation time for the case when the inner surfaces of the enclosure have not been given any treatment with special sound absorbent materials. In calculating this sound absorption, which is often called the 'basic store' of sound absorption, we take into account not merely the absorption of the materials but also that of the people filling the enclosure. The coefficients of absorption of the materials are given in Appendix 1. The results of the calculation are given in Table 6.

Noting down from Table 5 the necessary total sound absorption and from Table 6 the basic store of absorption, we find that the difference between these two values is the additional absorption which is needed to ensure the attainment of optimum reverberation time in the studio.

TABLE 7

f Hz	125	250	500	1000	2000	4000
A	536	585	612	612	560	510
A_0	124	149·8	186·8	229·8	218·0	233·7
$A_{supp} = A - A_0$	412	435·2	425·2	382·2	342	276·3

If 60 per cent of the area of the walls and ceiling is treated with the additional sound absorbent materials, the total area will be:

$$S_{supp} = (S - S_{floor})\, 0{\cdot}6 = 0{\cdot}6(2552 - 726) \cong 1100 \ m^2$$

In this case materials will be needed which have a mean coefficient of sound absorption close to those shown in Table 8.

TABLE 8

f Hz	125	250	500	1000	2000	4000
α	0·37	0·4	0·38	0·35	0·31	0·25

As follows from the table, the additional material should have an almost straight line frequency characteristic falling off at frequencies of 2000 and 4000 Hz. As none of the existing absorbing materials offer such a characteristic, we shall have to select a number of materials and constructions with such a ratio of their areas that the mean coefficient of sound absorption, or total absorption will be close to those values of α and A_{supp}, which are shown in Tables 7 and 8.

By using Appendix 1, we select for the treatment of the inner surfaces of the enclosure rigid and porous vibratory systems which have high absorption at low frequencies and acoustic tiles which have superior absorption at high frequencies.

The results of the calculation of the additional sound absorption offered by the selected materials is shown in Table 9. The value of each of these materials in the creation of this additional absorption is shown in Fig. 7.24.

If we take into account that the additional sound absorbent materials should be placed on the walls and ceiling of the studio, their absorption

263

TABLE 9

Description	Area m²	125		250		Sound absorption at a frequency of 500		1000		2000		4000 Hz	
		α	αS	α	αS	α	αS	α	αS	α	αS	α	αS
Plywood panel height 1·5 m	160	0·32	51·2	0·35	56·0	0·19	32·0	0·13	20·8	0·11	17·6	0·10	16·0
Semi-cylindrical constructions	100	0·35	35·0	0·29	29·0	0·26	26·0	0·11	11·0	0·08	8·0	0·07	7·0
Semi-cylindrical constructions	300	0·30	90·0	0·34	102·0	0·35	105·0	0·32	96·0	0·28	84·0	0·26	78·0
Bekesy screens	200	0·8	160·0	0·81	162·0	0·73	146·0	0·58	116·0	0·46	92·0	0·45	90·0
Porous-vib systs. (4 mm thick $d=5$ mm, $D=100$ mm with filling PP-80)	100	0·77	77·0	0·64	64·0	0·30	30·0	0·15	15·0	0·15	15·0	0·10	10·0
Porous-vib slot form	70	0·10	7·0	0·19	13·3	0·28	19·6	0·26	18·2	0·24	16·8	0·23	16·1
Acoustic tile 20 mm	130	0·05	6·5	0·21	27·3	0·66	85·8	0·91	118·1	0·91	118·1	0·79	102·7
Calculated absorption	1060		426·7		453·6		444·4		395·1		351·5		319·8

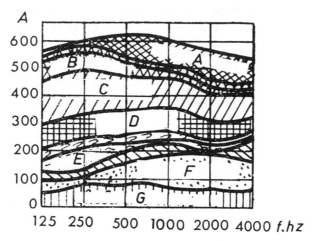

Fig. 7.24. The role of the sound absorption of various materials in the total absorption of the studio.

should be greater than A_{supp} given in Table 7 by the value of that area of the inner surfaces which will be covered by additional materials. If we take that into account, then from the calculated sound absorption taken from the last line of the table we must subtract the absorption of 1060 m² of plaster, and then the real additional absorption will be that given in the third line of Table 10. This additional absorption is close to that corresponding to optimum reverberation time, as can be seen from a comparison of the values of A_{supp} in Tables 7 and 10.

TABLE 10

f Hz	125	250	500	1000	2000	4000
$A_{\text{calculated}}$	426·7	453·6	444·4	395·1	351·5	319·8
A_{plaster}	12·7	13·8	18·0	21·1	24·5	26·5
A'_{supp}	414·0	439·8	426·4	374·0	327·0	293·3
A_0	124·0	149·8	186·8	229·8	218·0	233·7
A'	538·0	589·6	613·2	603·8	545·0	527·0
mean coefficient of absorption α_{mean}	0·212	0·231	0·240	0·236	0·214	0·206
$-\ln(1-\alpha_{\text{mean}})$ and	0·24	0·26	0·275	0·27	0·24	0·23
$-S\ln(1-\alpha_{\text{mean}})$	612	663	700	689	613	585
$+4m_B V$	—	—	—	29	63	203
$-S\ln(1-\alpha_{\text{mean}})$ $+4m_B V$	612	663	700	718	676	788

The calculation of total absorption in the studio (A') given in Table 10 allows us to determine the mean coefficient of absorption and the denominator of the fraction in expression 2.40. If we use the data in the last line of

Table 10 and formula 2.40, we can find the calculated reverberation time. This time for different frequencies is shown in Table 11.

TABLE 11

f Hz	125	250	500	1000	2000	4000
$T_{\text{calculated}}$	1·95	1·80	1·70	1·68	1·76	1·53
T	2·0	1·8	1·7	1·7	1·7	1·7
$\Delta T = T_{\text{calc}} - T$	−0·05	—	—	−0·02	+0·06	−0·17
$\dfrac{\Delta T}{T} \cdot 100\%$	−2·5	—	—	−1·7	+3·5	−10·0

As can be seen from Table 11 and Fig. 7.25 the frequency characteristic of reverberation time falls within the permissible limits.

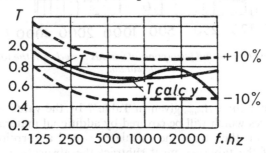

Fig. 7.25. Frequency characteristic of reverberation time of the studio.

3. *The arrangement of the sound absorbent materials.* The sound absorbent materials and constructions are positioned evenly round the enclosure in such a way that the arrangement of individual materials on parallel surfaces

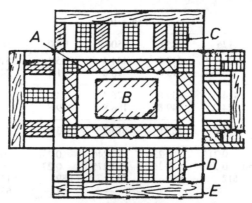

Fig. 7.26. Diagram of the arrangement of sound absorbent materials on the ceiling and walls of the studio.

is asymmetrical. As is shown in Fig. 7.26, plywood panels are placed along the lower part of the walls. Semi-cylindrical panels of two types and acoustic tiles are placed mainly on the walls. The ceiling houses a large part of the

266

porous vibratory systems and the Bekesy screens. That part of the floor farthest away from the window is covered with carpet, and the remaining part is covered with polyvinyl acetate.

4. *Calculation of the sound insulation of the boundaries.* The calculation of sound insulation from noise penetrating into the studio through the boundaries, is based on the determination of total noise level according to formula 6.44. Taking the plan of the enclosures shown in Fig. 7.23, the areas of those boundaries for which the level of exterior noise or sound insulation are dissimilar are inserted in column 2 of Table 12. From Appendixes 2 and 3 values of N and τ_{db} corresponding to the characteristics of the boundaries are copied into the table, and the values for insertion in formula 6.44 are calculated.

It follows from Table 12 that

$$\sum S \cdot 10^{0 \cdot 1(N - \tau_{db})} = 37624$$

and it can be seen from Table 10 that for a frequency of 500 Hz $A' = 613$ and consequently, the level of noise penetrating into the studio will be, on the basis of formula 6.44:

$$N_{\text{wall}} = 10 \log 37624 - 10 \log 613 = 45 \cdot 7 - 27 \cdot 9 = 17 \cdot 8 \text{ dB}$$

This level of noise is below the permissible level, and so the sound insulation of the studio against exterior noise is considered adequate.

TABLE 12

Description of boundary	S m²	N db	τ db	$\dfrac{N - \tau db}{10}$	$10^{0 \cdot 1(N - \tau_{db})}$	$S \times 10^{0 \cdot 1(N - \tau_{db})}$
Outer walls, double brick	380	75	75	—	—	—
Door from foyer	3·0	50	25	2·5	316	948
Wall bet. stud. and control	30·0	85	75	1·0	10	300
Window from control. Spec. 3-layer	3·0	85	55	3·0	1000	3000
Door to store	3·0	50	25	2·5	316	948
Wall div. stud. from store and foyer	49	50	75	—	—	
Wall bet. stud. and 2nd fl. rooms	132	85	75	1·0	10	1320
Upper floor (i.e. ceiling ferro-concrete with slag filler	726	75	60	1·5	31·6	22900
Lower floor ferro-concrete slabs with floor on beams	726	75	65	1·0	10	7260
Safety exit to yard (double, special, via lobby)	3·0	50	25	2·5	316	948
$\Sigma S \cdot 10^{0 \cdot 1(N - \tau_{db})}$						37624

5. *Calculation of the ventilation system and its damping.* Before selecting the ventilation system, we should first establish the necessary rate of air change, for which we need to know the amount of heat emitted in the enclosure. This heat originates from the lighting system and from the performers. The amount of heat emitted by a lighting system is

$$Q_{\text{light}} = \frac{\theta w_\lambda F}{F_\lambda}$$

where F is the total light flux, F_λ and w_λ are the light output and power of the selected type of lamp and $\theta = 860$ kilocalories per hour is the amount of heat per 1 kilowatt of electrical power expended on lighting.

As we know that the minimum permitted lighting is $E = 200\ lk$, and choosing a coefficient of reserve $k = 1\cdot3$ and a coefficient of light flow $\eta = 0\cdot4$, we calculate the full light flow necessary to illuminate the entire area of the studio $S_n = 726$ m according to the formula

$$F = \frac{KES_ni}{\eta} = \frac{1\cdot3 \,.\, 200 \,.\, 726 \,.\, 2\cdot2}{0\cdot4} = 1\cdot04 \,.\, 10^6 \text{ lumens}$$

In this formula i is the index of the enclosure defined as

$$i = \frac{l\,.\,b}{h_1(l+b)} = \frac{33\,.\,22}{6(33+22)} = 2\cdot2$$

where h_1 is the distance from the light source to the illuminated surface.

From Appendix 7 we find that with lamps NG-28, with $\omega_\lambda = 50$ watt and $F_\eta = 9100$ lumens the quantity of heat emitted by the light sources will be

$$Q_{\text{light}} = \frac{860\,.\,0\cdot5\,.\,1\cdot04\,.\,10^6}{9100} = 49700 \text{ kilocalories/hr}$$

As one person emits 100 kilocalories per hour, the total quantity of heat emitted in the studio will be

$$Q = Q_{\text{light}} + Q_{\text{perf}} = 49700 + 110\,.\,120 = 62900 \text{ kilocalories/hr}$$

In this case the necessary rate of air exchange is:

$$V_{\text{exch}} = \frac{Q}{0\cdot284\Delta t} = \frac{62900}{0\cdot284\,.\,4} = 57500 \text{ m}^3/\text{hr}$$

As can be seen from Appendix 7 such a quantity of air can be moved by an air conditioning installation KD 6061.

The total area of the grids of the input and extraction air ducts at a speed of the air passing through them of $v_p = 2$ m/sec will be:

$$q_p = \frac{2V_{\text{total}}}{3600\,.\,v_p} = \frac{2\,.\,57500}{3600\,.\,2} = 16 \text{ m}^2$$

If we reckon that the speed of the air in the trunking is $v_k = 5$ m/sec we can define the cross-sectional area of the ventilation ducts. It will be:

$$q_k = \frac{V_{\text{total}}}{3600\, v_k} = \frac{57500}{3600\,.\,5} = 3\cdot 2 \text{ m}^2$$

The decay of sound in ventilation trunking is found from the formula

$$\tau_{db_k} = N_a - N_B + 10 \log \left[\frac{4(1-\alpha)}{A} + \frac{1}{\Omega r^2} \right]$$

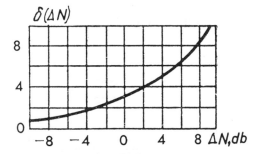

Fig. 7.27. For the calculation of level when non-coherent signals are combined.

To find this value, let us first calculate the level of aerodynamic noise generated by the ventilator:

$$N_a = 44 + 25 \log H + 10 \log V_{\text{total. sec}} + \delta$$

where H is the pressure of air at a flow equal to 50 kg/m², $V_{\text{total.sec}}$ is the quantity of air moved in m³/sec, δ is a value which has a value of 1 dB for inlet and 5 dB for extract.

For the studio being planned

$$N_a = 44 + 25 \log 50 + 10 \log \frac{60000}{3600} + 1 = 99 \text{ dB}$$

Now let us define the permitted value of noise level penetrating into the studio via the ventilation system, according to the formula:

$$N_B = N_{\text{wall}} + \Delta N$$

Defining from the formula the value

$$\delta(\Delta N) = N_{\text{supp}} - N_{\text{wall}} = 25 - 18 = 7 \text{ dB}$$

from Fig. 7.27 we find the value $\Delta N = 6$ dB corresponding to this value of $\delta(\Delta N)$, and calculate the permissible value of noise penetrating into the studio:

$$N_B = 18 + 6 = 24 \text{ dB}$$

269

Let us determine the required decay of sound in the ventilation trunking according to the formula quoted above, assuming that the ventilation grids radiate sound energy within a solid angle $\Omega = 2\pi$ and this decay is found at a distance from the grid of $r = 1$ m to be:

$$\tau_{db_k} = 99 - 24 + 10 \log \left[\frac{4(1 - 0\cdot2)}{613} + \frac{1}{6\cdot28} \right] = 68 \text{ dB}$$

Let us find the actual value of sound decay in the ventilation trunking of the system according to formula 6.51. To do this we must first consider the placing of the air pipes.

Let the main duct be of rectangular section with the sides in the ratio of $1 : 2$ and length $l_{total} = 20$ m. Further, let this duct branch into two with

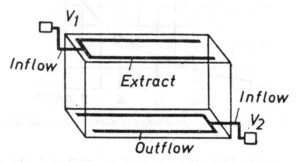

Fig. 7.28. Diagram of the arrangement of air ducts. B_1 and B_2 are the ventilation plants.

the same ratio of side lengths, the length of each branch being $l_{branch} = 44$ m (Fig. 7.28). Finally let the air ducts have three bends and be treated with sound absorbent materials with $\alpha = 0\cdot1$. Under these conditions, taking $h_k = 1\cdot25$ and $b_k = 2\cdot56$, the perimeter of the cross section of the channel before it branches will be

$$P_{total} = 2 \cdot 1\cdot25 + 2 \cdot 2\cdot56 \simeq 7\cdot6 \text{ m}$$

The perimeter and area of cross section of each channel after branching will be:

$$P_k = 2 \cdot 0\cdot9 + 2 \cdot 1\cdot8 = 5\cdot4 \text{ m} \quad \text{and} \quad q_k = 0\cdot9 \cdot 1\cdot8 \simeq 1\cdot6 \text{ m}^2$$

If we insert the values of P, l and q which we have found earlier, and defining $\phi(\alpha)$ by Fig. 6.38, we can calculate the attenuation of sound in the main channel as

$$\Delta\tau_{db} = 1\cdot1 \frac{P \cdot l}{q} \phi(\alpha) = 1\cdot1 \frac{7\cdot6 \cdot 20}{3\cdot2} 0\cdot1 = 5\cdot2 \text{ dB}$$

The decay in each channel after branching will be:

$$\Delta\tau_{db} = 1\cdot1 - \frac{5\cdot4 \cdot 44}{1\cdot6} \cdot 0\cdot1 = 16\cdot3 \text{ dB}$$

270

As the change in the transverse section of the duct creates additional absorption, let us define the ratio of the sections of the ducts before and after branching as $m_1 = q/q_k$, and the ratio of the sections of the channel after branching to that of the ventilation grids as $m_2 = q_k/q_p$. These ratios are expressed by equations:

$$m_1 = \frac{3\cdot2}{1\cdot6} = 2 \quad \text{and} \quad m_2 = \frac{1\cdot5}{16} = 0\cdot1$$

As can be seen from Fig. 7.29, the additional attenuation caused by the change in transverse section of the air duct for the two cases under consideration will be respectively:

$$\Delta\tau_{db_3} = 0\cdot5 \text{ dB}, \qquad \Delta\tau_{db_4} = 4\cdot5 \text{ dB}$$

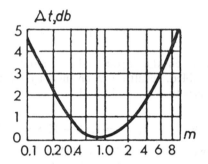

Fig. 7.29. Curve of the dependence of additional attenuation on the ratio of the sections of an air channel m.

Taking into account that at every bend in a duct additional absorption is introduced to the value of 2 dB, attenuation resulting from this factor will be

$$\Delta\tau_{db_5} = 3\cdot2 = 6 \text{ dB}$$

Thus the total attenuation of sound in the ventilating system will be

$$\Delta\tau_{db_B} = \Delta\tau_{db_1} + 2\Delta\tau_{db_2} + \Delta\tau_{db_3} + \Delta\tau_{db_4} + \Delta\tau_{db_5}$$
$$= 5\cdot2 + 2\cdot16\cdot3 + 0\cdot5 + 4\cdot5 + 6 = 46\cdot8 \text{ dB}$$

A comparison of $\Delta\tau_{db_k}$ and $\Delta\tau_{db_B}$ reveals that the duct cannot provide the necessary attenuation of sound, and so an acoustic filter has to be used to provide the additional attenuation. This additional attenuation is:

$$\Delta\tau_{db_f} = \Delta\tau_{db_k} - \Delta\tau_{db_B} = 68 - 47 = 21 \text{ dB}$$

By using formula 6.52 we can define the area of sound absorbent material needed for the treatment of the filter (where $\alpha = 0\cdot3$):

$$S = \frac{q \cdot \Delta\tau_{db_f}}{\alpha} = \frac{1\cdot6 \cdot 21}{0\cdot3} = 112 \text{ m}^2$$

271

Leaving the area of the section of the duct unaltered and reckoning that the area of absorbent material for each cell of the filter is twice the area of its section, the number of cells is found to be:

$$n = -\frac{112}{2 \cdot 1 \cdot 6} = 35$$

If the distances between the ribs of the cell are set at 40 cm, then the length of the filter will be:

$$l = 0 \cdot 4 \cdot 35 = 14 \text{ m}$$

The level of diffuse noise in the studio caused by the ventilation system is found from the equation:

$$N_{\text{diff}} = N_a - \Delta \tau_{db_B} - \Delta \tau_{db_k} + 10 \log \frac{4}{A}$$

$$= 99 - 46 \cdot 8 - 21 + 10 \log \frac{4}{613} = 9 \cdot 3 \text{ dB}$$

The noise power when the input and extractor systems are working simultaneously is doubled, resulting in an increase of its level by 3 dB, so that the total level rises to $N_{\text{diff}} = 12 \cdot 3$ dB.

We find that

$$\Delta N = N'_{\text{diff}} - N_{\text{wall}} = 12 \cdot 3 - 17 \cdot 8 = -5 \cdot 5 \text{ dB}$$

and from Fig. 7.27 we find that $\delta(\Delta N) = 1$ dB where $\Delta N = -5 \cdot 5$. Consequently the total level of noise in the studio when the ventilation system is in operation will be:

$$N = N_{\text{wall}} + \delta(\Delta N) = 17 \cdot 8 + 1 = 18 \cdot 8 \text{ dB}$$

Such a level of noise has no dangers for the recording or transmission of music, as it is considerably below the permissible level.

8 The Amplification of Sound in Enclosures and in Open Space

8.1 Systems for sound amplification and the demands made on them

In this chapter we shall consider the amplification of sound. The amplification of sound via an electro-acoustic system, resulting in its being adequately audible and evenly distributed to listeners spread out over a large area in an auditorium or in the open air, is one system. Amplification systems in which the sound fields of the primary and secondary sound sources are connected only by an electro-acoustic system are distinguished from those where, in addition to an electro-acoustic link, there is also a direct acoustic link.

In the first case the performer and the loudspeakers of the system are located in different enclosures or at a distance from one another; in the second case the performer and the microphone are situated close to the loudspeakers, as a result of which the sound field created by the latter is fed back to the microphone.

Systems of sound amplification without an acoustic connection between the receiver of sound (the microphone) and the radiator are normally used in public address systems in streets, parks, stadia, etc., when the hearer does not need to see the source. If, however, the amplified sounds are connected with the action of performers or with a speaker, for example in open air theatres or cinemas, in large enclosures, foyers, etc., i.e. where sight and sound should be inseparable, this makes an acoustic connection unavoidable, and dictates the use of this system of sound amplification.

The question as to whether a sound amplification system should be used can be settled in every actual case on the basis of a few simple calculations to evaluate the attenuation of the sound in the area which it is serving. For such calculations this formula, for instance, can be recommended:

$$\Delta N = N_0 + N_r - N_s$$

where N_0 is the mean level of intensity of sound from the performer when he is at a distance of one metre from the microphone, N_r is the level at the farthest point of the area covered, N_s is the mean noise level. Amplification is used where $\Delta N \leqslant 10$ dB.

Depending on how the loudspeakers are placed, there are four methods of radiating sound via an electro-acoustic system to a large area. These methods are the centralized method, the distributed method, the common zone method and the combined method.

In the centralized method one radiator or one group of radiators located together is used. If a group is used their acoustic axes are at an angle relative to each other.

In the distributed method the radiators are evenly distributed over the area to be served in such a way that each covers its own area.

In the common zone method the radiators are distributed at various points or concentrated in groups and provide a specific level of loudness as a result of mutual coverage of separate parts of the area served.

Finally the combined method involves a combination of two of the previous methods, e.g. the centralized plus the distributed, which allows a higher quality of sound to be achieved than do any of the others on their own.

Let us list the specific requirements of the electro-acoustic system of sound amplification.

1. The lowest level of transmission should exceed noise level by 10 dB, and should consequently be close to 50 dB for special open spaces and in enclosures, and 60–65 dB to cover streets or squares in a city.

2. The peak signal level should be 85–105 dB (the lower limit for speech transmission and for public address systems in streets, the upper for music) which combined with the first specification determines the dynamic range— 50–55 dB for sound amplification in halls and quiet areas, and 20–25 dB in street public address systems.

3. A variation in loudness over the whole area covered of no more than 5–10 dB. The first tolerance applies to systems where acoustic feedback exists and the second to systems where it does not.

4. The frequency range depends on the use to which the system is being put and on the transmission conditions. For systems intended mainly for speech (street PA systems) it is necessary to select a narrower range than would be the case for musical or dramatic transmission. The upper limit of the range is further restricted because of considerable losses in air (see 2.5) which is particularly important in the work of outdoor PA systems when the distance between the radiator and the listeners is up to as much as 200–300 m.

274

On the basis of these remarks, a frequency range of 1000 to 4000 Hz may be considered satisfactory for outdoor PA work, while for sound amplification out of doors or indoors the range can be from 50 to 8000 or 10,000 Hz.

5. When using a distributed or a common zone method of arranging loudspeakers in the open air, it is essential that the difference in levels ΔN caused by the speakers nearest to the listener should not exceed a certain value which depends on the delay time τ of the signal from the loudspeaker furthest removed from the listener. This limiting value can be found from the graph in Fig. 8.1. Where the 'centralized' method of siting loudspeakers is used they should be so arranged relative to the performer that the sound and visual images correspond.

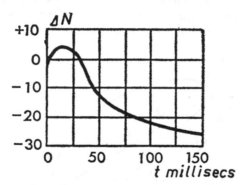

Fig. 8.1. Curve of the maximum relative level ΔN of a delayed signal, in relation to the time of delay, for masking by the original signal (of standard level).

6. A sound amplification system in which acoustic feedback is present should work under stable conditions. If this is not done, this usually results in the unpleasant distortions typical of these systems.

7. Frequency- and non-linear distortions should not significantly exceed the standards for transmissions radiated via an electro-acoustic system.

It should be particularly stressed that the conditions of irradiating an open space with sound correspond in almost every particular to the conditions of sound reproduction in open air cinemas, and therefore all the conclusions which may be reached about the former may be related also to open air cinemas. A further condition to be fulfilled in the latter case is that the image and sound should be synchronous. This is considered in 9.5.

8.2 Covering an open space with one loudspeaker

The task which is faced when attempting to cover an open space by means of one loudspeaker is to create over a specific area a level of loudness which is adequate to do so while keeping variations in the level of loudness within the permissible limits. In order to solve this problem, it is necessary to find the angle or height at which the loudspeaker is mounted above so that it covers the area to be irradiated, within the permitted limits of level, and also to calculate the loudspeaker power needed to ensure the necessary loudness.

275

The horn loudspeaker of circular section

Let a horn loudspeaker with this shape of section (Fig. 8.2) be suspended at a height h over the area to be irradiated with sound. The length of the area is l. The acoustic axis of the loudspeaker is at an angle γ to the vertical.

The sound pressure created by the loudspeaker at any point B lying at an angle θ to the acoustic axis will be very different from pressure at point C, where the acoustic axis impinges on the edge of the space to be irradiated. This difference is caused by the difference in distances r and r_0 to these points and by the varying sensitivity of the loudspeaker which changes with the angle from the axis.

By the directional characteristic is understood a curve $\Phi(\theta)$ representing a line of equal pressure, measured as the ratio of the pressure at a point D, at a distance from the loudspeaker as the point where the acoustic axis

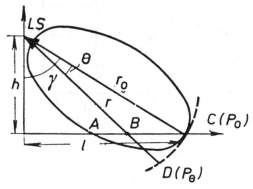

Fig. 8.2. Diagram of the positioning of a horn loudspeaker hung above an area to be irradiated with sound.

intersects the edge of the area point C, to the pressure at that point p_0. We can then write:

$$p_\theta = p_0 \Phi(\theta) \qquad 8.1$$

If we assume that sound pressure varies in inverse proportion to the distance, the pressure at point B may be expressed by the equation:

$$p = p_\theta \cdot \frac{r_0}{r} \qquad 8.2$$

It follows from equations 8.1 and 8.2 that

$$p = p_0 \Phi(\theta) \cdot \frac{r_0}{r} \qquad 8.3$$

On the basis of Fig. 8.2 we may write that

$$r_0 = \frac{h}{\cos \gamma} \quad \text{and} \quad r = \frac{h}{\cos (\gamma - \theta)}$$

and consequently the pressure at point B will be

$$p = p_0 \Phi(\theta) \frac{\cos (\gamma - \theta)}{\cos \gamma} \qquad 8.4$$

As Y. M. Sukharevsky showed, the directional characteristic of the horn can be expressed with accuracy sufficient for practical purposes, and for a sufficiently wide range of frequencies by the equation of an ellipse of eccentricity e' close to 0·9, and so we can write that

$$\Phi(\theta) = \frac{(1 - e'^2) \cos \theta}{1 - e'^2 \cos^2 \theta} = \frac{0 \cdot 2 \cos \theta}{1 - 0 \cdot 8 \cos^2 \theta} \qquad 8.5$$

The dependence of directionality on the angle of reception θ for this, and cases which approximate to it, is shown in Fig. 8.3.

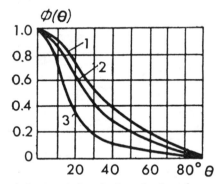

Fig. 8.3. The directional characteristics of a horn loudspeaker, represented as an ellipse with an eccentricity 0·85 (1), 0·9 (2) and 0·95 (3).

The distribution of the sound pressure along the length of the area to be irradiated with sound can be described by the ratio p/p_0, which, taking expressions 8.4 and 8.5 into account, is represented in the form:

$$R(\theta) = \frac{p}{p_0} = \frac{0 \cdot 2 \cos \theta}{1 - 0 \cdot 8 \cos^2 \theta} \cdot \frac{\cos (\gamma - \theta)}{\cos \gamma}$$

After division by $\cos \gamma \cdot \cos^2 \theta$ and a simple substitution of the trigonometric functions by tangents, this last equation can be written as

$$R(\theta) = \frac{1 + \tan \gamma \cdot \tan \theta}{1 - 5 \tan^2 \theta} \qquad 8.6$$

The level of sound pressure within the limits of this length will change in accordance with the formula

$$N = 20 \log R(\theta) = 20 \log \frac{1 + \tan \gamma \cdot \tan \theta}{1 - 5 \tan^2 \theta} \qquad 8.7$$

277

If curves of the level of sound pressure are constructed according to this formula as functions of the relative distance x/l between the base of the loudspeaker and any point B on the line AC (see Fig. 8.2) then these curves for equal values of relative height of suspension h/l will be represented as shown in Fig. 8.4.

As can be seen from a consideration of the curves in Fig. 8.4, the level of sound pressure rises quickly at first as the listener moves from point A to point C (see Fig. 8.2) and then, beginning from a certain value of x/l, begins to fall comparatively slowly. The maximum divergence of the level of sound pressure ΔN from the level at the axis of the loudspeaker, known as irregularity of received sound, changes its value according to the value

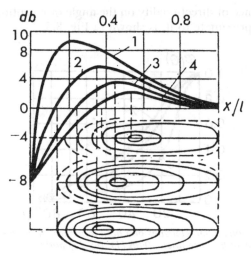

Fig. 8.4. Curves of the distribution of sound energy along the projection on the ground of the axis of the loudspeaker; (1) where $h/l = 0\cdot1$, (2) $h/l = 0\cdot2$, (3) $h/l = 0\cdot3$ and (4) $h/l = 0\cdot4$ ($e = 0\cdot9$).

h/l or, which amounts to the same thing, according to the angle of suspension γ. If we consider that irregularity of sound radiation must always be positive, it will fall with a reduction in height h/l.

At the same time there will be a reduction in the area of the ellipse on the ground within which the variation in the sound field changes in the way which has been indicated (see lower part of Fig. 8.4). Alternatively, if the level in the irradiated area is allowed to fall below that on the acoustic axis, then without exceeding the limits of the permissible variation, we can arrive at a state of sound radiation, almost without any reduction of the area to be served. This can be seen from the lower part of Fig. 8.4 (in dotted lines).

If we take the first course, bearing in mind possible changes in eccentricity of the ellipse which describes the directional characteristic of the loudspeaker, we can (for example by the method of constructing curves as in Fig. 8.4 and finding maximum points on them) draw curves relating the

278

variation of sound pressure ΔN to the height or angle of suspension of the loudspeaker. In Fig. 8.5 curves $a/l = f(\gamma)$ show the corresponding changes in length of the areas irradiated, which can easily be determined from Fig. 8.4 by measuring the length of the section along the axis x/l within which the curve of sound pressure lies below this axis.

The width of the area covered by the sound radiated from the loud-speaker can be determined if we write the eccentricity of the ellipse in the form:

$$l' = \sqrt{1 - (b'/a')^2}$$

where a' and b' are its longer and shorter axes, and that the eccentricity of the ellipse on the ground is defined by the equation:

$$e = e' \sin \gamma$$

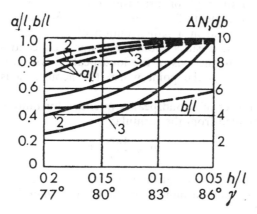

Fig. 8.5. Curves to determine the axial parameters of the area receiving the sound: solid lines are $N = f/(h/l)$; broken lines are $a/l = f (h/l)$ and $b/l = f (h/l)$, for $e = 0.85$ (1), $e = 0.9$ (2), $e = 0.95$ (3).

From these two equations, the short axis of the ellipse on the ground, or in other words the width of the area, will be:

$$b = a\sqrt{1 - e^2 \sin^2 \gamma}$$

Curves $b/l = f(h/l)$ calculated from this formula are shown in Fig. 8.5.

If the second method of solving the problem is selected, in which the variation is regarded as a fall from the maximum value of energy to the minimum, while allowing the latter to lie below the level of pressure on the acoustic axis of the loudspeaker (i.e. to have a negative value), then in this case the unevenness of sound distribution and the length of the ellipse on the ground becomes almost independent of the angle or the height of suspension.

Returning to Fig. 8.4 we can see that the variation in the sound distribution can be determined from formula 8.7 if the angle θ is replaced by angle

θ_{opt}, which corresponds to the value of x/l at which the curve has a bend. Consequently, the variability of sound distribution is

$$\Delta N = 20 \log \frac{1 + \tan \gamma \tan \theta_{opt}}{1 - 5 \tan \theta_{opt}} \qquad 8.8$$

where θ_{opt} is defined if the derivative $dN/d\theta$ is made equal to zero.
  In this case

$$\Delta N = 20 \log (0.5 + \sqrt{0.25 + 0.05 \tan^2 \gamma}) \qquad 8.9$$

In the lower part of Fig. 8.4 are shown distribution areas which have similar variations. Within these areas are drawn lines of equal loudness (at 1 dB intervals). The continuous lines correspond to levels lying above the level at the point of intersection of the axis of the loudspeaker point C, broken lines to levels below this nominal level. This part of the figure shows that a reduction in the angle of suspension (an increase in the relative height h/l) results in a movement of maximum sound pressure away from the axis of the loudspeaker and in a redistribution of sound energy. For a comparatively large angle of suspension, the power is concentrated in that part of the area closest to the axis of the loudspeaker; with a small angle it is concentrated nearer to the point where the acoustic axis intersects the ground.
  The acoustic power of a loudspeaker intended for outdoor amplification of sound is calculated from the formula:

$$P = \frac{4\pi r_0^2}{\Omega} I \qquad 8.10$$

where I is the sound intensity at the point where the acoustic axis intersects the ground. If we have in mind that the level of intensity is given by the equation

$$N = 10 \log \frac{I}{I_{threshold}}$$

where $I_{threshold}$ is the intensity at the threshold of hearing, equal to 10^{-12} watt/m^2, then the value of intensity I can be written as:

$$I = 10^{0.1(N-120)}$$

A substitution of this expression in equation 8.10 leads to the formula:

$$P = \frac{4\pi r_0^2}{\Omega} 10^{0.1(N-120)} \text{ watts} \qquad 8.11$$

It follows from Fig. 8.2 that the distance from the loudspeaker to the point where its acoustic axis strikes the ground is given by the equation

$$r_0 = l/\sin \gamma$$

By substituting for r_0 in formula 8.11 its value from the last equation, we can write this formula as:

$$P = \frac{4\pi l^2}{\Omega \sin^2 \gamma} 10^{0.1(N-120)} \qquad 8.12$$

280

Finally, if, on the basis of the second requirement in 8.1, peak signal level is taken as 100 dB, and the value of Ω is selected in an area of small variation corresponding to the horizontal part of the curve $\Omega = \phi(f/f_{\text{кт}})$ and equal to 20 (see Fig. 8.6), then the last formula can be rewritten as:

$$P = \frac{0 \cdot 5 l^2}{\sin^3 \gamma}\, 10^{-2} \text{ watts} \qquad\qquad 8.13$$

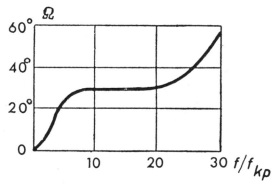

Fig. 8.6. The frequency dependence of the coefficient of axial concentration of a horn loudspeaker.

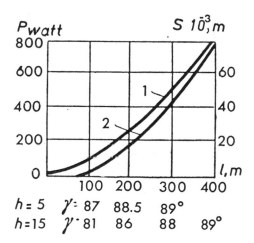

| $h = 5$ | $\gamma = 87$ | 88.5 | 89° | |
| $h = 15$ | $\gamma = 81$ | 86 | 88 | 89° |

Fig. 8.7 Graphs to determine the acoustic power (1) and the area radiated (2) for known values of h or γ.

The graph drawn from this formula in Fig. 8.7 enables us to determine the acoustic power of the loudspeaker needed to irradiate the given area of length l for the selected height of suspension of the loudspeaker. This graph also allows us to find the area S irradiated by the given loudspeaker for all other parameters of the system.

The sound column

The great defect of a horn speaker with a circular or square cross section is that its directional characteristic is the same breadth vertically as it is horizontally, which results in the small width of the area served by such a loudspeaker (see Fig. 8.2). To broaden the area that can be served by sound, it is preferable to use radiators which have a directional characteristic which is compressed vertically and broadened horizontally. This shape of directional characteristic is found in horn loudspeakers where the cross section is a rectangle of exaggerated height, and also in the so-called 'sound columns' which consist of a group of cone loudspeakers mounted one above each other.

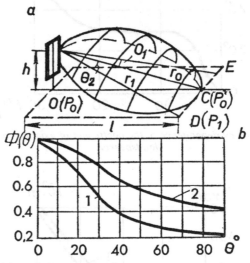

Fig. 8.8. The position of the sound column on an area to be radiated with sound (a) and its directional characteristics (b); (1) in the vertical and (2) in the horizontal plane.

Fig. 8.8b shows directional characteristics of the sound column 10 KZ-1 in the vertical (1) and the horizontal (2) planes. The behaviour of the directional characteristic in the vertical section is caused by the fact that as angle θ_1 increases (Fig. 8.8a), the phases of the waves from each of the radiators of the vertical column change, causing mutual cancellation. The broadening of the characteristic in the plane in which the listeners are seated is connected with the fact that the two vertical rows of loudspeakers are turned at an angle of 60° to each other.

The nature of the changes in the levels of sound pressure along the length of the area irradiated with sound can be found from Fig. 8.8a. From this sketch it follows that pressures p and p_0 at any point situated at a distance x from the support of the column and at the point of support 0 will be:

$$p = \frac{\Phi_1(\theta_1)}{r} \quad \text{and} \quad p_0 = \frac{\Phi_1(\pi/2)}{h}$$

282

where $\Phi_1(\theta_1)$ is the directional characteristic in the vertical plane. The distribution of sound energy is defined by the ratio:

$$\frac{p}{p_0} = \frac{h\Phi_1(\theta_1)}{r\Phi_1(\pi/2)}$$

8.14

It follows from Fig. 8.8 that $r = x/\cos\theta_1$ and so, substituting this value for r in equation 8.14, the relative change in level along a line OC can be expressed by the equation:

$$N = 20\log\frac{p}{p_0} = 20\log\frac{h}{x}\cdot\frac{\Phi_1(\theta_1)\cos\theta_1}{\Phi_1(\pi/2)}$$

8.15

The change of level as a function of the distance x/h from the support to any point on the line, expressed by the equation 8.15 is shown in the form

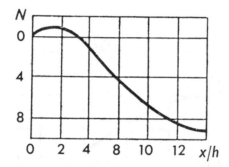

Fig. 8.9. Curve of change in level along the axis of the sound column projected along the ground.

of a graph in Fig. 8.9. In calculating from this curve, the values of $\Phi_1(\theta_1)$ and $\Phi_1(\pi/2)$ correspond to the characteristics in Fig. 8.8.

From the curve constructed from formula 8.15 it follows that for every change of the distance from the loudspeaker support to the listener by a value equal to the height of suspension of the loudspeaker, the level of sound pressure changes by approximately 1 dB. Moreover if a variation of 10 dB is permissible in direction OC, then the length of the area to be irradiated may be selected as twelve times the height of suspension of the loudspeaker.

The width of the area served in this case can be defined from the condition that where it is of rectangular shape it may have a level 3 dB lower at its farthest corners (points D and E) than at point C. In accordance with Fig. 8.8 the ratio of the pressures at these points will be:

$$\frac{p_1}{p'_0} = \frac{r_0}{r_1}\Phi_2(\theta_2)$$

where $\Phi_2(\theta_2)$ is the directional characteristic of the column in the horizontal plane.

283

The reduction of level in moving from point C to point D, reckoning that $r_0/r_1 = \cos \theta_2$ is expressed by the equation:

$$N_1 = 20 \log p_1/p'_0 = 20 \log \Phi_2(\theta_2) \cos \theta_2 \qquad 8.16$$

The dependence presented by equation 8.16 for a column 10 KZ-1, is given in the form of a curve in Fig. 8.10. If from this curve we define the angle θ_2 corresponding to the reduction stipulated above, and substituting it into equation

$$\frac{d}{2} = r_0 \tan \theta_2$$

found from Fig. 8.8, we can calculate the width of the zone to be irradiated with sound. It will be

$$d = 2r_0 \tan \theta_2 = 2 \tan \theta_2 \cdot \sqrt{h^2 + x^2}$$

or, as the permissible variation close to 10 dB takes place where $x = 12h$, then:

$$d = 2 \cdot 0 \cdot 5 \sqrt{145h^2} \approx 12h$$

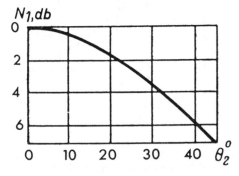

Fig. 8.10. Curve of the change in signal level as the listener moves perpendicularly to the projection of the axis of the column along the ground.

The length and breadth of the zone of sound coverage (Fig. 8.11) also depend on the angle of inclination of the sound column ϕ and on its length nd, where d is the distance between adjacent loudspeakers and n is the number in the column. A reduction in the angle of inclination or the length of the column leads to an increase in the length of the zone of sound coverage for the same height of suspension of the speaker h. So for a column suspended at a height of $h = 6$ m, a reduction in angle ϕ from 20° to 18°, or of the length of the column from 1·5 to 1·3 m, causes the length of the zone of sound coverage to double—from 30 to 60 m.

The acoustic power of the loudspeaker in the given case can be determined from formula 8.11, in which the distance r_0 is replaced for convenience of calculation by the height of suspension of the loudspeaker h.

284

With such a substitution

$$P = \frac{4\pi h^2}{\Omega \sin^2 \theta} 10^{0.1(N-120)}$$ 8.17

where N is the level of sound pressure near the sound column.

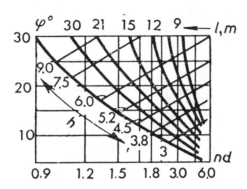

Fig. 8.11. Graph for the determination of the basic data of a sound radiation system using a single column.

Radial loudspeakers

Whereas when a horn loudspeaker or a sound column is used to irradiate a large area the acoustic axis remains almost parallel to the surface of the zone of sound coverage, making an angle with it of 10–15°, the use of a radial loudspeaker demands that the axis should be perpendicular to the zone to be covered. This is connected with the fact that a section of the polar characteristic of a radial loudspeaker in the plane perpendicular to its axis is in the form of a circle. Such loudspeakers are obtained by the use of a horn with an acoustic lens, by a ring form arrangement of a group of loudspeakers in a single housing, (loudspeaker 10 GDN-1) or with the use of both an acoustic lens and a ring arrangement (DGR-25).

When the loudspeaker is placed vertically relative to the zone of sound coverage it is omnidirectional, and so the ratio of the pressures at any point at a distance x from the support, and at the support itself (Fig. 8.12) will be:

$$\frac{p_x}{p_0} = \frac{h}{r_0} = \frac{1}{\sqrt{1 + \left(\frac{x}{h}\right)^2}}$$

and the relative change in the level of sound pressure is expressed as

$$N = 20 \log \frac{p_x}{p_0} = 20 \log \frac{1}{\sqrt{1 + \left(\frac{x}{h}\right)^2}}$$ 8.18

285

As curve 1, calculated according to this formula, shows, movement away from the support of $x = 3h$, causes the level to fall by 10 dB, which is the permissible limit of variation. It follows from this that the zone of sound coverage in the given case is a circle whose radius is equal to three times the height of suspension of the loudspeaker, or a square inscribed in this circle with sides $a = 4·2h$.

By changing the shape of the polar directional characteristic in the vertical plane, we can increase the dimensions of the zone of sound coverage while retaining the same degree of variation. If the loudspeaker is designed so that

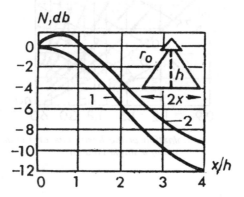

Fig. 8.12. Curves of the change of signal level with movement away from a radial loudspeaker for an elliptical (1) and a pear shaped (2) shape of the directional characteristic.

this characteristic is pear shaped, then as can be seen from curve 2 constructed for a radiator 10 GDN-1, the change in level becomes less, and the zone of sound coverage increases to $x = 4·5h$. This more than doubles the area of the zone of sound coverage.

The acoustic power of an individual loudspeaker can be calculated by formula 8.17, bearing in mind that in this case angle $\theta = 90°$.

From what has been said, we conclude:

1. Small areas can be covered by means of a horn or a radial loudspeaker, or by a sound column.

2. In working with a horn loudspeaker, the zone of sound coverage is elliptical in shape; with the other two types of loudspeakers the zone is much larger and has either a circular or a rectangular shape.

3. Maximum sound pressure from a sound column or a radial speaker occurs respectively at the edge or in the centre of the zone (see Figs. 8.9 and 8.12). With a horn speaker the point of maximum pressure moves away from the edge of the zone nearest to the loudspeaker as the angle of suspension is decreased (see Fig. 8.4).

4. In the first two cases the levels of sound pressure decrease as the distance from the loudspeaker increases. This decrease is of the order of 1 dB for a distance equal respectively to the height or to half the height of

286

suspension of the loudspeaker. In the third case (horn loudspeaker) the fall in level takes place on both sides of the point of maximum pressure, and with a comparatively large angle of suspension this decline in the direction of the loudspeaker is very rapid (see the lower part of Fig. 8.4).

5. An increase in the height of suspension for all types of loudspeakers, and an increase in the angle of suspension for horn loudspeakers result in an increase in the zone of sound coverage, but entail an increase in the power of the loudspeakers.

6. The power of a horn loudspeaker is most conveniently calculated from the range at which it is working (see Fig. 8.2); for a radial speaker and a sound column, the basis of the calculation is the height at which the speaker is placed.

8.3 The sound coverage of an open space by a group of loudspeakers

If the area is large, a group of loudspeakers is used to provide sound coverage, and they are usually arranged so that the axes of adjacent loudspeakers are either parallel (the common zone method) or intersecting (the centralized method). In both cases there are zones in the space in front of the loudspeakers where the action of two neighbouring loudspeakers is combined.

The overlapping zones of coverage of neighbouring loudspeakers leads to a reduction in the variation of irradiation of these zones. Moreover, incorrect selection of the width of the overlapping zone may result in distortion in the form of echo, which appears where there is little difference in the levels of the signals received from a pair of loudspeakers and where there is a noticeable difference in the time that these signals arrive at the listener.

The width of the zone of overlapping depends primarily on the distance between the loudspeakers. This zone becomes wider as the power of the radiators is increased, with a consequent increase in their range of action, and also with a decrease in the acuteness of their directional characteristics.

As a characteristic of the connection between the zones of operation of neighbouring loudspeakers, I. G. Dreisen[23] introduced the concept of the coefficient of the chain of nominal areas, by which is understood the ratio

$$k = \frac{u}{u + 1}$$

where u is the ratio of the width of the zone of overlapping c to the distance $2d$ between the loudspeakers (see Fig. 8.13b).

A group of horn loudspeakers

Let us consider the action of two loudspeakers arranged side by side and forming part of a common zone system of sound radiation. If these loudspeakers, which are separated by a distance $2d$, have identical powers and angles of suspension ϕ, and their axes are parallel, then on the basis of

Fig. 8.13a we can calculate the sound pressures p_1 and p_2 created by each of the loudspeakers at point M, which is in the overlapping zone. On the pattern of equation 8.3, we can write that

$$p_1 = p_0 \Phi(\theta_1) \frac{r_0}{r_1} \quad \text{and} \quad p_2 = p_0 \Phi(\theta_2) \frac{r_0}{r_2} \qquad 8.19$$

As the experiments of Mayer and Schodder show, the addition of the energies of two signals where one of them is slightly delayed relative to the other gives a slight error (1–2 dB). As such an error can have no real

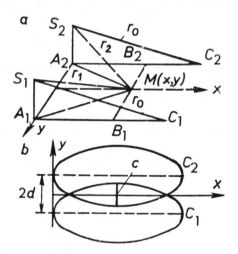

Fig. 8.13. Graphs for evaluating the variation in sound energy level from two loud-speakers by means of formula 8.20.

influence on the overall result, the rule of energy combination of signals may be used to give the resultant pressure at point M in the form of an equation:

$$p^2 = p_1{}^2 + p_2{}^2$$

The distribution of sound energy in the overlapping zone can in this case be shown as:

$$R^2(\theta) = \left(\frac{p}{p_0}\right)^2 = \left(\frac{p_1}{p_0}\right)^2 + \left(\frac{p_2}{p_0}\right)^2$$

or, on the basis of equations 8.19:

$$R^2(\theta) = \Phi^2(\theta_1) \cdot \left(\frac{r_0}{r_1}\right)^2 + \Phi^2(\theta_2) \cdot \left(\frac{r_0}{r_2}\right)^2 \qquad 8.20$$

If we find from the triangles $A_1 S_1 M$ and $A_1 M B_1$ the value of r_1 in the form $r_1{}^2 = h^2 + x^2 + (d + y)^2$ and from triangles $A_2 S_2 M$ and $A_2 M B_2$ the value of $r_2{}^2 = h^2 + x^2 + (d - y)^2$ and insert these values into equation 8.20, we obtain:

$$R^2(\theta) = \Phi^2(\theta_1) \frac{r_0{}^2}{h^2 + x^2 + (d + y)^2} + \Phi^2(\theta_2) \frac{r_0{}^2}{h^2 + x^2 + (d - y)^2}$$

where x and y are the coordinates of point M.

The change in the level of sound pressure in the area between the projections of the acoustic axes of the loudspeakers will be:

$$N = 20 \log R(\theta) = 10 \log [\Phi^2(\theta_1)] \frac{r_0{}^2}{h^2 + x^2 + (d + y)^2} +$$

$$+ \Phi^2(\theta_2) \frac{r_0{}^2}{h^2 + x^2 + (d - y)^2} \qquad 8.21$$

Formula 8.21 allows us to determine the variation inside the zone indicated above, and to resolve the question whether it is necessary to decrease or increase the distance between the two loudspeakers in order to keep this variation within the permitted values.

If a cluster of horn loudspeakers (a centralized system) is used to provide sound coverage of an open space, the danger of distortion from interference is avoided, as the length of the paths of sound from the series of loudspeakers to a common point in the overlapping zone will be identical. In this case the problem to be solved is to decide on an adequate value for the chain coefficient of the zones, which may be replaced by specification for the angle between the axes of the neighbouring loudspeakers (Fig. 8.14b), ensuring that the variation in the overlapping zone does not exceed the standards laid down.

As follows from Fig. 8.14a, the pressure at point N, the distance to which is equal to the distance between the loudspeaker and the point where its acoustic axis intersects the ground, will be:

$$P_N = p_0\Phi(\theta_N)$$

The connection between angle θ_N and its projection in the plane of the zone of sound coverage, which is represented by angle $\delta/2$, is expressed in the following equation

$$\theta_N = \delta/2 \sin \gamma$$

Taking into account that for angles of suspension of loudspeakers which are employed in practice $\sin \gamma$ is very close to unity, we can reckon with sufficient accuracy that $\theta_N = \delta/2$. Thus, the reduction in sound level at point N compared with the level at point C_2 may be written in accordance with equation 8.5 in the form:

$$\Delta N = 20 \log \frac{p_N}{p_0} = \Phi(\delta/2) = \frac{0 \cdot 2 \cos \delta/2}{1 - 0 \cdot 8 \cos^2 \delta/2} \qquad 8.22$$

289

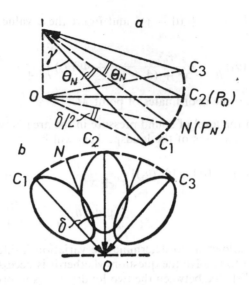

Fig. 8.14. Graphs to determine the angle between the projections of the axes of adjacent speakers in a centralized system.

In Fig. 8.15 curve 2 depicts the law which is analytically represented by equation 8.22. It follows from the curve that if we allow a 5 dB drop in level at point N, angle δ may be given a value of from 45° to 50°.

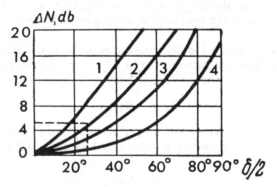

Fig. 8.15. Curves of the loss of signal level for changes in the angle between the projections of the axes.

The length of the zone of sound coverage as for one loudspeaker, allows the question of the angle of suspension of the radiator and its power to be decided, and the breadth of the area is the basis for the required number of such radiators, bearing in mind that each of them serves an area within an angle δ.

290

A group of sound columns

Sound columns are most conveniently used by arranging them in a straight line chain along a broad section of the zone to be covered. In this case it is essential, as the columns are placed at equal distances across the breadth of the area, to specify the distance between neighbouring pairs which will avoid variations greater than can be permitted in the overlapping areas.

With the aim of increasing the length of the zone of sound coverage, as was done in the case of a single column, the far boundary of the zone along the line of the acoustic axis (point C in Fig. 8.16) should be chosen from consideration that the maximum variation permitted is 10 dB. There should be no point along this boundary between the zones of operation of neighbouring columns where the level is less than the level at point C.

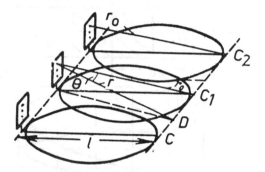

Fig. 8.16. A group of sound columns.

Thus the zones of action of neighbouring columns should overlap each other in such a way that along line CDC_1 levels are constant. This can be done if it is allowed that the level at D may be 3 dB lower than at point C. If this is allowed, (see 8.2) the width of the zone of a single column, and consequently the distance between columns, will be $12h$.

A group of radial loudspeakers

As a cross section of the directional characteristic of radial speakers in the plane of the zone of sound coverage is circular, we should consider two variants of the way in which they can be used. The first variant may be applied to areas of elongated form, and the second to square areas which would be given better sound coverage by a number of loudspeakers distributed over the area in a square array.

In the first case the distance between the speakers may vary according to the width of the zone which the whole chain has to cover. Having established that at the extreme points D_1 and D_2 (Fig. 8.17) variation should be maximum (within the defined limits), i.e. equal to 10 dB, we can establish a variation for point C which will be less than at points D_1 and D_2.

291

If there is a reduction in variation at this point, the width of the zone served by the chain decreases, and the distance between loudspeakers increases. This conclusion is reached by the following considerations.

As follows from Fig. 8.17, a reduction in level at point D_2 compared with point O is expressed by the equation:

$$N = -10 \log \frac{h^2 + d^2 + b^2}{h^2} + 3 \qquad 8.23$$

where $2b$ and $2d$ are the width of the zone of sound coverage and the distance between the loudspeakers.

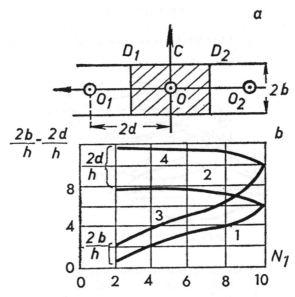

Fig. 8.17. A group of radial loudspeakers (a) and curves for the determination of the width of the area served (1, 3) and the distance between loudspeakers (2, 4) for the given drop in level (b).

The level at point D_2 is increased by 3 dB, having in view the fact that at this point two signals from equidistant loudspeakers sited at O and O_2 are combined.

Taking N to be 10 dB, as was stipulated above, this last equation can be rewritten in the form:

$$\log \left[1 + \left(\frac{d}{h} \right)^2 + \left(\frac{b}{h} \right)^2 \right] = 1 \cdot 3 \quad \text{or} \quad \left(\frac{d}{h} \right)^2 + \left(\frac{b}{h} \right)^2 = 19 \qquad 8.24$$

The reduction in level at point C, on which the neighbouring loudspeakers in the chain have negligible influence, can be expressed as:

$$N_1 = -10 \log \frac{h^2 + b^2}{h^2} \qquad 8.25$$

It follows from this equation that:

$$\left(\frac{b}{h}\right)^2 = 10^{-1 \cdot 0 N_1} - 1 \qquad\qquad 8.26$$

If we insert the value of b/h from 8.26 into 8.24, this last can be shown as:

$$\left(\frac{d}{h}\right)^2 = 18 - 10^{-0 \cdot 1 N_1} \qquad\qquad 8.27$$

Formulae 8.26 and 8.27, as was indicated earlier, enable us to make a reasonable choice of the distance between the loudspeakers of the chain, depending on the width of the zone of sound coverage. If half the width of this belt is inserted into formula 8.26 related to the height of suspension, the value of the reduction in level at C is determined. Then the established value N_1 is inserted into formula 8.27, from which the distance $2d$ between loudspeaker supports is found. This distance corresponds to the condition that variation at point D_2 shall not exceed 10 dB.

As can be seen from curves 1 and 2 in Fig. 8.17b, which are calculated from formulae 8.26 and 8.27 an increase of 20–30 per cent in the distance between loudspeakers $2d/h$ can be achieved, if the width of the zone of sound coverage $2b/h$ is reduced.

When the zone is so wide that several chains disposed side by side are needed to cover it, the choice of the distance between the loudspeaker both longitudinally and laterally, and the length of the area, are determined in a rather different way.

Line $D_1 D_2$ may be moved away from the sources O_1 and O_2 in such a way that the level at point D_2, caused by sources O and O_2 is reduced to 13 dB, having in mind that when a further adjacent chain of loudspeakers O' and O_2'' is in operation, the intensity of sound at this point will be doubled and the permissible level of 10 dB will be ensured at this point. Such a movement entails the need to increase the right hand sides of equations 8.23 and 8.25 by 3 dB, after which formulae 8.26 and 8.27 take the form:

$$\frac{b^2}{h^2} = 10^{-0 \cdot 1 (N_1 - 3)} - 1 \quad \text{and} \quad \frac{d^2}{h^2} = 38 - 10^{0 \cdot 1 N_1} \qquad\qquad 8.28$$

Curves 3 and 4 in Fig. 8.17b, calculated from formula 8.28, show that, while observing the limits of permissible variation, movement of the loudspeakers can take place in various ways. If the distance between neighbouring speakers across the breadth of the area is reduced, a certain increase may be made in the distance separating the loudspeakers along the length of the area. The most profitable distribution is the one in which $2b/h = 2d/h = 10$, as the area served by each of the loudspeakers will be as large as it can be, and the arrangement will be the simplest.

A calculation of the common zone method with radial loudspeakers is best begun with the selection of the type of loudspeakers to be used, which immediately determines their power. From the power, using formula 8.17,

293

we find the height of suspension which will allow the maximum level at the base of the loudspeaker. Then the graphs in Fig. 8.17b are used to determine the distance between neighbouring loudspeakers and the area which will be served by each of them. This last value enables us to determine the number of loudspeakers necessary to cover an area of the given dimensions.

To sum up, here are some observations:

1. Sound coverage for large areas by means of horn loudspeakers may be arranged according to the centralized or to the common zone system, while if sound columns or radial speakers are used for this purpose, only the common zone system is used.

2. The selection of the centralized system using horn speakers makes it necessary to decide first on the angle between the acoustic axes of neighbouring speakers, and to find an angle which will avoid variations in sound energy which exceed the permitted values in the area covered.

3. To solve the problem of providing sound coverage by means of the common zone system, and also to determine parameters connected with the operation of a single radiator, it is necessary to calculate the longitudinal and lateral distances between the loudspeakers in the area. The object of these calculations is to achieve adequate uniformity in the sound field over the area of sound coverage.

8.4 Acoustic feedback in sound amplification

The simplest system of sound amplification for the open air is a microphone, an amplifier and a loudspeaker (Fig. 8.18). Let there be a sound source placed in front of the microphone and emitting sinusoidal signals.

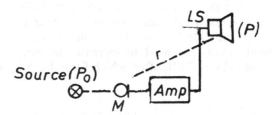

Fig. 8.18. Diagram of a sound amplification system

Then, apart from the action of the sound source on the microphone, acting at pressure p_0, pressure p, caused by the operation of the loudspeaker in the system, will also be acting on the microphone. The action of a signal from the output of the system on its input is defined by the coefficient of acoustic feedback, which is expressed by the equation

$$\beta = \frac{p}{p_0 + p} \qquad\qquad 8.29$$

It is clear that for any ratio of pressure p and p_0, this coefficient can never be greater than unity.

294

Signals from the primary source and from the loudspeaker will reach the microphone with a phase difference. The phase difference arising in this case will change depending on the frequency of the signal and on time t necessary for the sound wave to travel from the loudspeaker to the microphone, i.e.:

$$\phi = 2\pi ft = \frac{2\pi}{c_0} fr$$

Thus the coefficient of feedback in a general form can be expressed by the equation:

$$\beta = \frac{p}{p_0 + p} (\cos \phi + j \sin \phi)$$

From this we find that

$$\frac{p}{p_0} = \frac{\beta}{(\cos \phi - \beta) + j \sin \phi}$$

or, being interested only in the absolute value:

$$\left| \frac{p}{p_0} \right| = \frac{\beta}{\sqrt{(\cos \phi - \beta)^2 + \sin^2 \phi}} = \frac{\beta}{\sqrt{1 + \beta^2 - 2\beta \cos \phi}} \qquad 8.30$$

The level of sound pressure at any point x before the loudspeaker is defined from the equation

$$N_x = 20 \log \frac{p_x}{p_0} = 20 \log \frac{p_x}{p} \cdot \frac{p}{p_0} = |\Delta N_x| + 20 \log \left| \frac{p}{p_0} \right|$$

where ΔN_x is the level of the amplified signal at point x without reckoning the action of feedback.

After the substitution for p/p_0 of its value from 8.30, we can write that:

$$|N_x| = |\Delta N_x| + 20 \log \frac{\beta}{\sqrt{1 + \beta^2 - 2\beta \cos \phi}} \qquad 8.31$$

When, as a result of a change in signal frequency, ϕ takes the value $2n\pi$ or $(2n + 1)\pi$, the signal level at point x will be expressed by equations:

$$|N'_x| = |\Delta N_x| + 20 \log \frac{\beta}{1 - \beta} \qquad 8.32$$

or

$$|N''_x| = |\Delta N_x| + 20 \log \frac{\beta}{1 + \beta} \qquad 8.33$$

The second part of equations 8.32 and 8.33, for a specific value of β will have, respectively, positive and negative values which lead to an increase or to a reduction in level at point x. The change of sign in the second part indicates that in sound amplification both positive and negative feedback can take place changing from one to another as the signal frequency changes. Thus feedback in a sound amplification system results inevitably in an irregularity in the frequency characteristic, which, as can be seen from equations 8.32 and 8.33 can be expressed in the form:

$$N'_x - N''_x = 20 \log \frac{\beta}{1 - \beta} - 20 \log \frac{\beta}{1 + \beta} = 20 \log \frac{1 + \beta}{1 - \beta}$$

The curve in Fig. 8.19, constructed from the last formula shows that the irregularity of the sound field decreases rapidly as coefficient β is decreased.

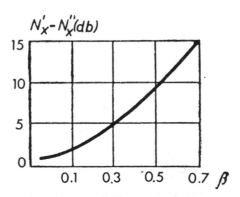

Fig. 8.19. The influence of the coefficient of feedback β on the irregularity of the frequency characteristic of a sound system.

A frequency characteristic for $\beta = 0.3$ is illustrated in Fig. 8.20. The peaks in the characteristic coincide with frequencies at which positive feedback has its largest values, and dips are observed at frequencies where negative feedback is at its maximum.

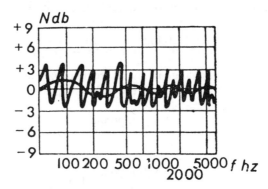

Fig. 8.20. The frequency characteristic of a sound amplification system with feedback ($\beta = 0.3$).

An increase in the coefficient of feedback results not only in a rapid increase of frequency distortions, but also creates (as it draws near to unity) conditions under which, as a result of a significant increase of positive

296

connection at some frequencies, the installation goes into a state of self-oscillation and begins to howl. The danger of the self-oscillation of a system makes it necessary to reduce considerably the feedback coefficient, and this is also useful in preventing frequency distortions.

This kind of distortion is directly connected with another form which is peculiar to sound amplification systems.

When the prime source is switched off, the last impulse produced by the loudspeaker, after time $t = r/c_0$ again acts on the microphone, as a result of feedback, albeit with reduced intensity. This impulse, again reproduced by the loudspeaker, acts once more on the microphone after a similar delay and with still more attenuation is again emitted by the loudspeaker.

Thus the repeated signal gradually decays and ceases to be audible. This phenomenon, which resembles the process of the decay of sound in an enclosure, and is caused by the circulation of the signal in the sound distribution system, is called regenerative reverberation.

The sound of regenerative reverberation is very different from that of the effect of natural reverberation in an enclosure, as the unevenness of the frequency characteristic of the system has a noticeable effect on the decaying signal. As a result of the multi-toothed form of the frequency characteristic the signal reproduced by the loudspeaker loses more and more of individual frequency components after each reflection, and the sound acquires a special ringing coloration. In order to avoid these distortions, feedback has to be reduced still further than is demanded by the avoidance of howl-round in the installation.

The moment when self oscillation appears and the level to which it grows depends not only on the coefficient of feedback but also on such parameters as the amplitude and phase characteristic of amplification and its time-constant. For every installation it is possible to establish a method of operation such that the possibility of the appearance of howl-round is removed while allowing for sufficient amplification of the signal.

The methods usually employed to avoid the danger of self-oscillation, and to overcome other forms of distortion caused by acoustic feedback, make use of the directional properties of loudspeakers and microphones and their correct relative positioning. Where a sound amplification system is used in an enclosure, the overall absorption of the enclosure and the distribution of sound absorbent materials relative to its end boundaries is also of great importance.

To reduce feedback it is necessary to prevent as far as possible the direct radiation from the loudspeaker from reaching the microphone. This can be accomplished if directional microphones and loudspeakers are used for the sound amplification, and if they are arranged so that the sound from the loudspeaker falls into the least sensitive zone of the microphone (Fig. 8.21), or if the microphone itself faces into the zone of lowest sensitivity of the loudspeaker (Fig. 8.21b). In this last case the reduction of the sensitivity to the loudspeakers in the area in which the microphone is located is achieved by turning the axes of the microphones through angle π, and by switching their sound coils into anti-phase.

Acoustic feedback can also be reduced by the use of two unidirectional microphones, and feeding them in antiphase to the input of the amplifier (see Fig. 8.21c). The second microphone, with its back turned to the performer, receives the same amount of direct energy from the loudspeaker as does the first, and as it were, balances it out because of its own reverse connection to the input.

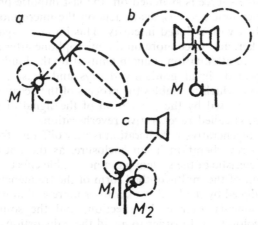

Fig. 8.21. Various methods of reducing feedback in sound amplification.

A more complex method is the use of a chain of similar, let us say, unidirectional microphones, which in the plane corresponding to the line in which the microphones are placed has a very highly directional characteristic. It is clear that to reduce feedback, the loudspeaker should be in the same plane as this characteristic.

Thus, to summarize:

1. When a sound amplification system is in operation, acoustic feedback appears which may be positive or negative depending on the frequency of the signal produced.

2. As a result of this feedback a number of specific distortions arise, which show themselves by the equipment going into self-oscillation, as considerable non-linearity of the frequency characteristic, and as regenerative reverberation.

3. The appearance and magnitude of these distortions depends on the degree of feedback, defined by the coefficient of acoustic feedback β, the maximum value of which is equal to unity.

4. To protect the system from self-excitation and to reduce the other two forms of distortion β should be no greater than 0·2–0·3.

5. Such a low coefficient of feedback with an adequate power of the amplified sound can be obtained by using a number of measures which by the use of directional microphones and loudspeakers and by their placing can ensure a minimum action of the direct radiation of the loudspeaker on the microphone.

8.5 Sound amplification in an enclosure

The sound processes which go on when a sound amplification system operates in an enclosure are much more complicated. Apart from the direct signals from the source and the loudspeaker, as in open air conditions, the microphone also receives signals from many delayed and gradually decaying reflected signals generated by both the primary and secondary sound sources.

The effect of each such signal on the microphone is characterized by its coefficient of feedback, the strength of which is determined by the pressure developed by the signal. The frequencies at which this coefficient takes on its maximum and minimum values for each signal is particular to that signal, as the path lengths for signals are different one from another, and depend on the dimensions and shape of the enclosure, the position of the microphone and the number of reflections association with each signal.

Thus, sound processes in an enclosure in which sound amplification apparatus is working are determined by a whole system of feedbacks operating with time shifts.

This last condition means that signals reach the microphone with very varied phases, which together with that of the evenness of the sound field they create is a sufficient reason for considering sound processes in the enclosure from the standpoint of the statistical theory. In a consideration of this kind we should remember that the formation of the sound field in the enclosure in this case is very different from the normal. In fact, when a sound amplification system is operating in an enclosure there are two sound sources of different powers. The action of each of them on the enclosure can be described by the equation of energy balance which is found in the statistical theory.

The first equation, stipulated by the presence in the enclosure of the primary sound source, essentially repeats equation 2.18 which refers to a normal enclosure, namely:

$$V \frac{dE_1}{dt} + V\delta E_1 = P_{a_1}$$

where P_{a_1} is the power of the primary source, and E_1 is the density of energy created by it in the enclosure, and δ is $c_0 \alpha_{mean} S / 4V$, the index of sound decay.

The second sound source, in the form of a loudspeaker of a sound amplification system, creates its own sound field, the differential balance equation for which has the form:

$$V \frac{dE_2}{dt} + V\delta E_2 = P_{a_2}$$

In this equation the power of the secondary sound source is determined by the power of the first and by the conditions of sound amplification characterized by the coefficient of acoustic feedback. Due to the acoustic feedback, the power radiated by the loudspeaker depends on the reflected energy acting on the input to the system. Consequently the right hand part of the last equation can be presented as the sum of the energies acting on the microphone. The first part defines the power of the primary sound source, the second, that part of the power which has circulated as a result of acoustic feedback, and finally, the third part is connected with the reflected energy. Thus:

$$P_{a_2} = k^2[P_{a_1} + P_{a_2} \cdot q^2 + (E_1 + E_2)\delta q^2] \qquad 8.34$$

where k is the amplification of the system, q^2 is a value describing the relative amount returning from the loudspeaker to the microphone compared with the direct energy. The value of q is given in equation 8.30.

The two differential equations shown above, which together express the sound processes in the enclosure when a sound amplification system is in operation, can be a basis for the determination of reverberation time in such an enclosure.

For this they must be rewritten, keeping in mind that when the primary source ceases to operate $P_{a_1} = 0$. Under this condition, the equations of balance are

$$\frac{dE_1}{dt} + \delta E_1 = 0 \quad \text{and} \quad \frac{dE_2}{dt} + \delta E_2 = \frac{k^2 q^2}{V}(E_1 + E_2) \qquad 8.35$$

As the processes described by equations 8.35 take place in the enclosure simultaneously, it is adequate to add them in order to arrive at the overall characteristic of the process. As a result, we get:

$$\frac{dE}{dt} + \delta E + \frac{k^2}{V} q^2 E$$

where $E = E_1 + E_2$. Rearranging the terms, the equation we obtain can be presented in the form:

$$\frac{dE}{E} = - \left(\delta - \frac{k^2}{V} q^2 \right) dt$$

Its solution, describing the process of energy decay, is, by analogy with 2.32,

$$E = E_0 e^{-[\delta - (k^2/V)q^2]t} \qquad 8.36$$

If we compare expressions 8.36 and 2.23, it can easily be seen that the exponent of e in the first has a smaller absolute value than in the second. This means that the sound in the enclosure with the amplification system in operation decays more slowly than when the system is not working.

In 2.5 it was demonstrated that reverberation time and the index of decay δ are connected by the following equation:

$$T = \frac{6}{\delta . \log e}$$

As the index of decay in expression 8.36 occurs in the factor included in brackets as an exponent of e, the reverberation time when the sound amplification system is in operation takes the form:

$$T_r = \frac{6}{\log e \cdot \left(\delta - \dfrac{k^2}{V} q^2\right)}$$

or, after the insertion of the value of δ from 2.23 and transferring $c_0 s/4V$ outside the brackets:

$$T_r = \frac{6.4}{c_0 \log e} \cdot \frac{V}{S\left(\alpha - \dfrac{4}{c_0 s} k^2 q^2\right)} = \frac{0 \cdot 164 V}{S\left(\alpha - \dfrac{\tau}{V} k^2 q^2\right)} \qquad 8.37$$

It follows from formula 8.37 that the sound absorption capacity of the enclosure in the given case is expressed by the difference term in brackets. The mean coefficient of absorption, when the sound amplification system is working, decreases as it were by a value proportional to the mean time of one reflection τ and to the ratio $k^2 q^2/V$. This is equivalent to the enclosure having acquired negative absorption due to the operation of the sound amplification system.

If in formula 8.37 q^2 is replaced by its value from equation 8.30, then the reverberation time is:

$$T_r = \frac{0 \cdot 164 V}{S\left(\alpha - \tau \dfrac{k^2}{V} \dfrac{\beta}{(1 + \beta^2 - 2\beta \cos \phi)}\right)} \qquad 8.38$$

i.e. it depends not only on τ, but also on the coefficient of feedback β and on the phase shift ϕ between the direct and repeated signals acting on the microphone. For a comparatively large value of β and for a phase shift such that feedback becomes positive ($\cos \phi = 1$) the 'negative' coefficient of absorption expressed by the second part of the denominator of the fraction of equation 8.38, can approach the value of the mean coefficient of absorption α, and then reverberation time T_r can become very large.

This corresponds to the case where, when there is considerable positive feedback, the system continues to generate even after the sound source has been switched off. With a gradual reduction in the value of β, 'negative' absorption, and consequent regenerative reverberation, becomes progressively less. This results in a gradual reduction in effective reverberation time to the value obtaining when the sound amplification system is not working. This can only occur when acoustic feedback has been reduced to a minimum by special measures.

Thus, as the coefficient of feedback is increased, the reverberation time characteristic of the given enclosure, gradually increases because of regenerative reverberation. The reverberant sound begins to acquire a characteristic tone colour—a ringing effect—as β is further increased. Finally, at a certain value, which depends on the value of the mean coefficient of absorption, the system goes into a self oscillatory state and reverberation time, in

301

accordance with formula 8.38, becomes infinite. The change in reverberation time we have described is illustrated by the curves in Fig. 8.22, drawn by means of formula 8.38.

Fig. 8.22 allows us to make some additional conclusions.

An increase in the total absorption by the introduction of additional absorbent materials creates conditions in which regenerative reverberation has little influence on the quality of sound reception at increased values of the coefficient of feedback. If we say that the ratio T_r/T should not exceed 1·7 (see Fig. 8.22), then for a mean value of $\alpha = 0·7$ we can without any special risk increase the level of amplification of the system to a point where the coefficient of feedback becomes 0·4–0·5.

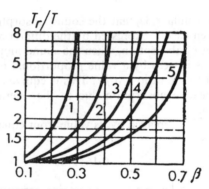

Fig. 8.22. The dependence of the relative change in regenerative reverberation on the feedback coefficient for different values of the absorption coefficient. 1, $\alpha = 0·1$ 2, $\alpha = 0·2$ 3, $\alpha = 0·3$ 4, $\alpha = 0·5$ 5, $\alpha = 0·7$

A good method of fighting regenerative reverberation is to introduce highly effective absorbent materials specially into the enclosure. If these are concentrated in the zones of high sensitivity of the microphone and the loudspeaker (assumed directional), the reflected energy returning to the microphone is considerably reduced, and so is regenerative reverberation.

Returning to equation 8.34, it is important to obtain the necessary power of the loudspeaker by the first term of the equation. Increasing the second and third terms, which are related to the feedback, might bring the system into an unstable condition which could easily develop into a howl. These terms may be reduced so that the system will work in a stable regime by reducing q. This can be done by increasing the directional properties of the microphone and loudspeaker, by increasing the distance between them and by reducing the coefficient of sound absorption (as $\delta = c_0 \alpha S/4V$).

With the same aim in mind it is desirable to reduce the power of the primary source P_{a_1} which can easily be done by increasing the distance between the microphone and the performer. With this in mind, the necessary increase of the first term, and consequently of the energy radiated by the loudspeaker, can be arranged only by altering the coefficient of sound amplification of the system k.

302

Let us state our conclusions in brief:

1. The feedback coefficient of a sound amplification system in an enclosure should be less, and measures to protect the microphone from repeatedly acting signals should be more effective, than when the system is in the open air.

2. Regenerative reverberation, added to the reverberation of the enclosure, results in an impairment of the quality of sound transmission even for comparatively small values of the feedback coefficient. The influence of regenerative reverberation is not detectable in practice if the amplification is reduced by 6 dB below the level at which regeneration begins.

3. Additional increase in reverberation time when a sound amplification system is working depends on the dimensions and shape of the enclosure and also on the mean absorption coefficient and on the directional characteristics of the microphone and loudspeaker.

4. To increase the level of loudness in the enclosure it is necessary to try to get the maximum increase in the amplification coefficient with the greatest possible distances between the microphone on the one hand and the performer and the loudspeaker on the other. It is also necessary to select more directional microphones and loudspeakers, to damp the enclosure more, and to place the sound absorbers so that maximum absorption is on those areas which coincide with the directions of greatest sensitivity of the microphone and the loudspeaker.

9 The Acoustic Characteristics of Enclosures for Sound Transmission

9.1 General observations

As the human voice and most musical instruments have poor directional qualities and relatively small power, the level of loudness in enclosures for direct listening falls as the distance between the performer and the listener increases, thus impairing the clarity of speech and the quality of music for listeners at the back of the hall. Unlike halls for direct listening, halls intended for sound transmission through an electro-acoustic system contain secondary sound sources whose power can be considerably changed. In this connection, the question of increasing loudness in halls for sound transmission by the use of reflected energy is of no significance. The use of secondary sources of sound introduces another peculiarity: because loudspeakers have fairly narrow directional characteristics, the signal received by the listener has a relatively higher content of direct energy, improving the clarity of speech.

We should not forget a further peculiarity of halls for sound transmission, arising from the possibility of using several loudspeakers with differing directional characteristics placed at various points in the hall with their acoustic axes running in the necessary directions. In these conditions it is easier to distribute the sound energy uniformly in time, because the sound sources are fixed and do not move, as frequently happens in halls for direct listening.

Even when a stereophonic system is used for sound transmission, a system in which the energy is constantly redistributed between channels, the permanent position of the loudspeakers has a significant influence on the character of the sound field. This influence finds expression, for example, in the fact that the sound field and the distortions introduced by the stereophonic system are defined by the rates of energies in the channels.

These peculiarities of enclosures for sound transmission require acoustic conditions which differ from those in enclosures for direct listening, or to primary enclosures used for transmission via broadcasting channels or for sound recording.

Enclosures of this type include cinema auditoria, halls with sound amplification equipment and multi-purpose type halls.

9.2 Acoustic conditions in cinema auditoria

When considering the question of acoustic conditions in enclosures of this type, we have to take into account some further peculiarities characteristic of them alone. In particular it must be remembered that acoustic conditions in a cinema auditorium depend on the acoustics of the primary enclosures where the sound was recorded. In considering this relationship account must be taken of the following two contradictory circumstances:

1. It would seem that the enclosure in which the sound recording was made should be acoustically neutral, as in recording or in the process of recording which is customarily used to introduce artificial reverberation into the recording attempts are made to ensure that the sound recorded should have optimum acoustic properties. This seems even more correct because the listener who perceives the sound from the secondary source cannot dissociate himself from the reverberation inherent in this sound as he can in direct listening conditions.

2. Where there is an electro-acoustic connection (see 2.8) the secondary enclosure, if it is more damped than the primary, has a very insignificant influence on the resultant reverberation. Moreover the reproduction in a heavily damped enclosure of sound recorded in optimum conditions does not produce optimum results. Music sounds dead, and listeners in a heavily deadened enclosure quickly tire.

These two circumstances together suggest the necessity of supplying in the hall some reverberation which would overcome short-comings in the perception of sound.

In analysing the peculiarities of the acoustics of cinema auditoria, one cannot overlook the claims of versatility, by which we mean that the reproduction of speech and music are of equal importance. If it is easy to satisfy optimum conditions for speech transmission, ensuring clarity resulting from a short reverberation time in the hall, it is extremely difficult to meet, at the same time, the conditions connected with the good reception of music, as for this purpose a comparatively long reverberation time is necessary.

Thus in a cinema auditorium attempts must be made, while ensuring an optimum reverberation time for music, to ensure good clarity of speech.

If we place absorbent material with a high absorption coefficient on the areas of a few first-order reflections, the decay of sound energy in the enclosure can be divided into two stages (see 3.3). The first stage, associated with the areas of first-order reflections which are placed on the part of the ceiling nearest to the source, and on the lower part of the walls, proceeds more quickly than the second, from the later reflections from lightly damped surfaces. Calculations show[41] that for large enclosures the function between the first and second sections of the sound decay curve is at the point $T_1 \approx 150$ msec.

It is also known (see 2.12) that the energy borne by those few first-order reflections arriving in the interval before time $t_0 = 50$ msec, combines with

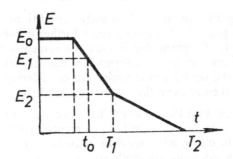

Fig. 9.1. The decay curve of sound energy in an enclosure with an uneven distribution of sound absorbing material.

the direct energy, increasing the loudness level and is therefore the useful energy. If this is borne in mind, then on the basis of Fig. 9.1 and from the definition of the clarity factor, this latter may be expressed in the following form:

$$Q = \frac{\displaystyle\int_0^{t_0} E(t)\mathrm{d}t}{\displaystyle\int_{t_0}^{\infty} E(t)\mathrm{d}t + \int_{T_1}^{\infty} E_1(t)\mathrm{d}t} \qquad 9.1$$

where $E(t)$ and $E_1(t)$ are functions representing the decay of energy in the first and second stages.

If we assume that in each stage the decay is exponential, we can write

$$E(t) = I_0\, e^{-\delta_1(t-t_0)} \quad \text{and} \quad E_1(t) = I_1\, e^{-\delta_2(t-T)} \qquad 9.2$$

If we insert these values into equation 9.1 and integrate, as for equation 2.98, we obtain

$$Q = \frac{I_0/\delta_1(1 - e^{-\delta_1 t_0})}{I_0/\delta_1[1 - e^{-\delta_1(T_1-t_0)}] + I_1/\delta_2} \qquad 9.3$$

306

or, dividing the numerator and denominator of the right-hand part of the equation by I_0/δ_1, we have

$$Q = \frac{1 - e^{-\delta_1 t_0}}{[1 - e^{-\delta_1(T_1 - t_0)}] + \dfrac{I_1 \delta_1}{I_0 \delta_2}} \qquad 9.4$$

The values of δ_1 and δ_2 are easily determined by the method used to obtain formula 2.60. Thus, if we assume that energy decay starts from a level of 60 dB, we obtain

$$\delta = \frac{6}{T \log e}$$

In the given case the decay amounts to ΔL in the first stage and $60 - \Delta L$ in the second stage, so

$$\delta_1 = \frac{\Delta L}{\log e \,.\, T_1} = \frac{\Delta L}{0 \cdot 43 T_1} \quad \text{and} \quad \delta_2 = \frac{60 - \Delta L}{0 \cdot 43(T_2 - T_1)} \qquad 9.5$$

If we rewrite formula 9.4, taking account of the values of δ_1 and δ_2, in equation 9.5 we obtain

$$Q = \frac{1 - e^{2 \cdot 3 \Delta L(t_0/T_1)}}{1 - e^{\{2 \cdot 3 \Delta L[1 - (t_0/T_1)]\}} + \dfrac{I_1 \Delta L}{I_2(60 - \Delta L)} \cdot \left(\dfrac{T_2}{T_1} - 1\right)} \qquad 9.6$$

This formula enables us to determine the value of the clarity factor as a function of the total reverberation time T_2 in the enclosure for a given slope of the decay curve in the first stage (from value ΔL).

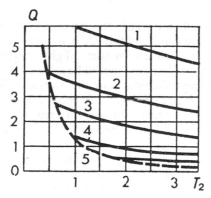

Fig. 9.2. Graph of the dependence of the clarity factor on reverberation time for various steepnesses of the decay curves.

If into this formula we insert the values of t_0 and T_1 indicated above, and carry out the calculations, then we can obtain data from which the curves in Fig. 9.2 were drawn. It follows from the curves that the clarity (Q) of the transmission does not decrease significantly with increase in total

307

reverberation time, for any value of the steepness of the initial section. The curves in Fig. 9.3, from the same data, show that the steepness of the decay curve in the first stage has the greatest influence on the clarity of speech. It is enough to increase the steepness of the decay curve by 5–6 dB in order to obtain the same clarity for a total reverberation of $T_2 = 3$ sec as for $T_2 = 1$ sec.

Thus, in the conditions of a cinema auditorium, it is useless to try to achieve a very short reverberation time with an even sound field, because this would not only entail great expense for absorbing materials and complication of the treatment of the inner surfaces of the hall, but also because if they ensure the necessary clarity of speech these conditions cannot produce optimum conditions for music transmission.

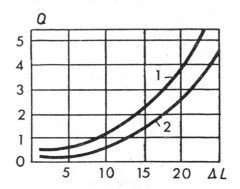

Fig. 9.3. Curves of the dependence of the clarity factor on the slope of the energy decay curve in the first stage of reverberation.

It is clear that in halls for which the necessary steepness in the initial stage of the decay curve can be obtained, it is wiser to have a sufficiently large overall reverberation time with some slight unevenness of the sound field. The overall reverberation time should be selected so as to give optimum conditions for music reproduction, and the steepness of the initial stage so as to ensure a high degree of clarity of speech transmissions.

In practice, there is no great difficulty in creating such conditions, as they can be obtained by treating the ceiling and the lower parts of the walls with efficient absorbent materials.

After a consideration of some of the peculiarities of a cinema auditorium and the evident possibilities of improving the acoustic conditions by means of special treatment of the inner surfaces of the enclosure, it is important to determine in what conditions these peculiarities have so great an effect that they should be used in practical acoustic design. To answer this question we have first to clarify what basic acoustic requirements have to be met at the present time for cinema auditoria.

In the building norms SN 30–58, it is indicated that acoustic conditions in an auditorium should be such that the optimum reverberation time calculated for the given auditorium should be related to its volume (Fig. 9.4).

The curves in the figure are based on the assumption that calculations are carried out allowing for a 75 per cent capacity audience. Moreover, unlike the out-of-date norms GOST 2691–44, optimum reverberation time is calculated not at six, but only at two frequencies—128 and 512 Hz.

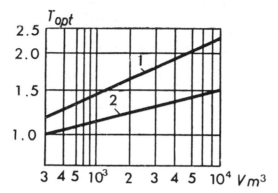

Fig. 9.4. The dependence of optimum reverberation time on the volume of a cinema auditorium from the norms SN 30–58, for frequencies of 128 (1) and 512 Hz (2).

If we base our calculations on the established norms, whose terms are the result of many years' experience in the use of auditoria for sound films, it should be observed that they can hardly be considered adequate. To specify the acoustic properties of an enclosure more strictly, it must be recom-

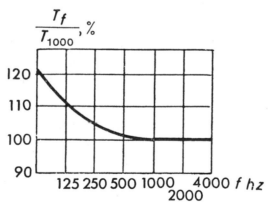

Fig. 9.5. The frequency characteristic of optimum reverberation for a cinema auditorium.

mended that optimum reverberation time is obtained for several lower and higher frequencies, for example, 50, 1000, 2000 and 4000 Hz. In order to find the values of optimum reverberation time at these frequencies, the frequency characteristic of reverberation time shown in Fig. 9.5 should be used.

309

In calculations carried out in accordance with the norms, significant differences may be observed in a number of cases particularly where the hall is of fairly large volume between calculated and actual acoustic parameters of the hall. Moreover these divergences increase as the number of the audience increases. Researches into a large number of auditoria in cinemas in Moscow and Leningrad[40,45] have convincingly shown two reasons for these divergences.

The first reason is that in the calculation of total absorption of audience, it is not always correct to define it as the product of the sound absorption of one person multiplied by the number of spectators. The error in such calculations, although small for a small number of spectators, grows considerably as the number of spectators increases and as they are more

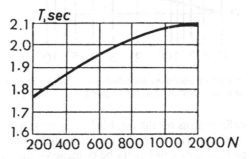

Fig. 9.6. *The data obtained from this curve and from Fig. 9.5 are the basis of further calculations of the acoustic parameters of the hall without taking account of the sound absorbing capacity of the audience.*

densely distributed in the hall. The second reason results from the fact that filling the hall with spectators produces a relatively high absorption on one of its surfaces which, in accordance with Rozenberg's conclusions (see 3.2) makes the field in the auditorium uneven and the formulae of the statistical theory used for calculations inaccurate in consequence.

It is not difficult to exclude the influence of these two reasons on the results of calculations if the calculations are carried out for empty halls. For this it is necessary to know the dependence of optimum reverberation on the volume of an empty hall. This dependence, or the corresponding dependence of optimum reverberation time on the number of seats in the hall, was obtained experimentally in the form of the curve in Fig. 9.6. The data obtained from this curve and from Fig. 9.5 are the basis of further calculations of the acoustic parameters of the hall without taking account of the sound absorbing capacity of the audience.

We have considered above three various approaches to the question of establishing optimum acoustic conditions in auditoria. It has been pointed out that this can be done on the basis of the recommendations of SN 30–58 for a hall 75 per cent full, from experimental data for an empty auditorium, and finally on the basis of the possibility of creating these optimum conditions simultaneously for speech and music. The first two approaches to

310

the establishment of optimum conditions envisage the use of the statistical theory, and consequently an even distribution of both absorbent and good reflecting materials on the bounding surfaces; the third is based to a large extent on the geometrical method and presupposes an uneven distribution of sound absorbers.

The existence of a number of ways of solving one and the same problem obliges us to make a choice of the best of them, or to define the conditions under which one or another of these methods should be preferred.

As was observed in 3.3, the last of the methods we have considered of creating optimum conditions in an auditorium should be used only when an initial drop of energy in the first stage of decay of no less than 5–6 dB can be achieved by the effective damping of the ceiling and the lower part of the walls. Apart from this, this method is of interest when optimum conditions for speech and music transmissions are very different from one another. The necessary damping in the first stage and the significant difference in reverberation time for optimum reception of speech and music (see Fig. 3.21) occur only for halls of large volume. So this method of establishing optimum conditions could correctly be recommended for halls designed for a large audience, and particularly for the multi-purpose type halls where the creation of conditions enabling good reception of speech and music is particularly important.

If the necessary drop of energy in the first stage of decay cannot be achieved, and the number of spectators in the given hall is sufficiently large, we can try to obtain the necessary reverberation in an empty hall. Finally, in a comparatively small auditorium where the difference in the optimum reverberation time for speech and music transmission is not very great, and where the sound absorbing capacity of the spectators does not have a real influence on achieving the necessary reverberation, optimum conditions should be created on the basis of the recommendations of the norms SN 30–58.[63]

9.3 Acoustic conditions in a hall for stereophonic sound reproduction

The special nature of stereophonic sound is to locate the apparent sound source correctly and to give naturalness and 'spaciousness' to this sound.

In an enclosure for stereophonic reproduction, conditions should be observed which would exclude the negative and stress the positive influence of its acoustic properties on the character of stereophonic transmissions.

In order to determine these conditions, it is necessary to consider the process of the formation of the sound field in multi-channel sound reproduction.

In conditions of single-channel sound reproduction, each point of the enclosure is reached, after the direct energy, by the energy of first-, second-order, etc. reflections which create a system of image sources (see Fig. 3.9), variously orientated around this point. In stereophonic reproduction the array of these sources becomes more complex, primarily because the number of real sources increases and each of them is placed at a certain distance from its neighbour, and secondly because the signals themselves produced by the loudspeakers have specific time-shifts which change with changes in the position of the performer relative to the microphone system during recording.

The intensity of each of the real sources is also constantly changing, and it may happen that the intensity of one or several of them will become significantly less than that of some image sources caused by first- and even second-order reflections of radiators which are operating more loudly. Thus in stereophonic reproduction the sound field is formed in a different way from that in single-channel sound reproduction.

We can assume that the presence of a large number of image sources, noticeably and systematically changing the overall ratio of sound intensity,

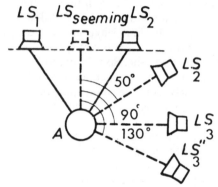

Fig. 9.7. Diagram of the distribution of loudspeakers for a specific influence of reflections on the location of the apparent sound source.

will have a definite influence both on the location of the apparent sound source created by the stereophonic system and also on the naturalness and 'spaciousness' of the sound transmissions.

These considerations contradict the results of experiments conducted by Haas with real sound sources, one of which created the direct wave, and the other the 'reflected'. It became clear from the experiments that where the intensities of the direct and the 'reflected' sounds were equal, and even where the latter had a distinctly higher level, if it was delayed relative to the first by a time of $\Delta t = 10$ msec, only the source of the direct wave was subjectively significant.

In order to clarify to what extent this law is valid for stereophonic transmission, i.e. when the position of an apparent source of sound, and not the actual source, is considered, new experiments were conducted by Figver. The conditions under which these experiments were conducted will be understood from Fig. 9.7.

The intensities of the sounds from the first and second loudspeakers were made equal, as a result of which the apparent source appeared to the listeners to be in Zone A, in the centre. The third loudspeaker, emitting the 'reflected' sound, was placed at various angles in relation to the apparent

312

source, and was fed with signals which could be varied in intensity and delay time relative to the basic signal. Some results of the experiments are shown in Fig. 9.8.

As follows from the curves in Fig. 9.8, the apparent sound source noticeably changes its angular position γ according to the delay time of the 'reflected' signal Δt and on the angle β direction. When the signal is delayed, up to 15 msec, the apparent sound source moves towards this signal: with a delay of more than 40 msec the movement takes place in the opposite direction. The magnitude of the departure of the apparent source from its normal position also depends on the difference of levels of the direct and 'reflected' signals. The less the difference, the greater the movement of the apparent sound source.

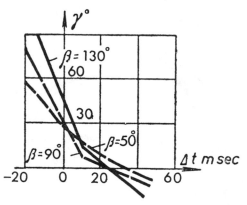

Fig. 9.8. The dependence of angular movement γ^0 of the apparent sound source on the time difference in the arrival of direct and reflected signals Δt for various angular situations of the source of the latter β and a difference in their intensities of 2·5 dB.

Thus the first- and even second-order reflections which, particularly in weakly damped enclosures, carry energy comparable with the direct energy, can cause a further movement of the apparent sound source. This distorting movement is most noticeable to listeners situated near the reflecting surfaces.

A large number of image sources and their individual dispositions can result not only in movement of the seeming sound source, but also in an inaccurate definition of its situation (sharpness).

Experiments to determine the influence of the acoustics of the enclosure on the accuracy of location of the seeming sound source were conducted with a two-channel stereophonic system for various reverberation times in the enclosure[57]. The listeners were placed at five different points in the enclosure; the first and second points were situated on the axis of the stereophonic system $x/l = 0$ at distances from the line of the loudspeakers y/l, equal to 2 and 4. The three other points were moved to the right: two of them relative to the first point at a distance $x/l = 0.5$ and one relative to the second at a distance $x/l = 1$. With the aim of achieving

313

greater generality, the position of the listeners, as also of the apparent source was expressed in relative values (by its relation to half the distance l between the loudspeakers). Between thirty and forty listeners took part in the experiments, and each of them made one observation at each in turn of the five indicated points in the enclosure.

The listeners determined the position of the apparent sound source, which changed when the difference of signal levels between the channels of the system was changed. It was found that the position of this source

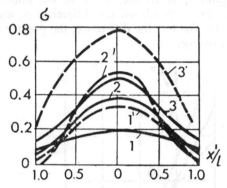

Fig. 9.9. *The dependence of the standard deviation σ on the position of the apparent source x'/l for central (solid) and side seats (broken). Curves 1 and 1', for T = 0·3 sec. 2 and 2', for T = 1·1 sec. 3 and 3', for T 3·8 sec.*

determined by each of the listeners showed a marked variation from the mean value calculated from all the results. If, from these experimental results, we construct curves of the dependence of the standard deviation σ on the position of the apparent source x/l (Fig. 9.9) then it is easily established that the value of this variance, which characterizes the accuracy of location of the apparent source, changes with the position of this source and, which is very important, increases with increase in reverberation time

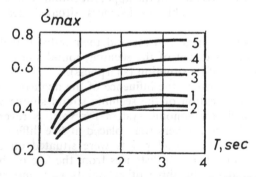

Fig. 9.10. *The dependence of the maximum value of the standard deviation on reverberation time: curves 1 and 2 for central seats at a distance from the system of y/l = 2 and 4; 3 for x/l = 0·5 and y/l = 2; 4 for x/l = 1 and y/l = 2; 5 for x/l = 1 and y/l = 4.*

314

and with changes in the position of the listener in relation to the stereo-phonic system and the reflecting surfaces.

It should be noted that the maximum value of the standard deviation σ grows with reverberation time T, at first very quickly (Fig. 9.10) and then, as the reverberation time increases from 1·2 to 4 sec, slowly and independently of the point in the hall occupied by the listener. (The numbers of the curves in the figure correspond to the number of the observation points.) From this follows the conclusion that it is necessary to reduce the reverberation time of cinema auditoria to 1·2 sec or less, in order to increase accuracy of location.

Returning to the dependence of the accuracy of location of the apparent source on the position of the listener, let us examine the curves in Fig. 9.11 from which it can be seen that as the listener approaches the side wall (goes away from the axis of the stereophonic system) the error in location

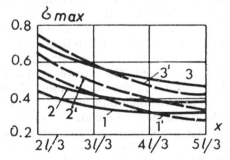

Fig. 9.11. Curves of the dependence of the standard deviation on the distance x between the listener and the wall where T = 0·5 sec. (1 and 1') T = 0·8 sec. (2 and 2,) and T = 3·8 sec. (3 and 3'). Curves 1, 2 and 3 for distances to the system of y/l = 2; 1', 2', 3', for y/l = 4.

grows even more, and particularly when reverberation time in the enclosure is long. This can be explained by the disturbing effect of the first, most intense reflections, coming from the side walls and the ceiling.

Thus the impression is created that the accuracy of location of the apparent source falls away not so much because of a large value of reverberation time in the enclosure as of the influence of the first-order reflections which carry comparatively large sound energy.

This conclusion is confirmed by experiments conducted by A. N. Kacharovich.[43] First, speech and music were reproduced from a recording through one channel in such a way that each signal could be reproduced twice or three times with different time delays and different amplitudes. Then delayed signals simulating reflections were fed through three different channels, and the loudspeakers at the output of the second and third channels were moved horizontally and vertically relative to the listener.

Repeated listening to the sounds reproduced with different time delays and at different angles of arrival of the delayed signals led to a number of conclusions. It became clear that only for certain optimum conditions,

315

when the delay times $\Delta t_1 = \Delta t_2$ were both equal to 25 msec, the amplitudes of the second and third signals were less than the basic signal by 3 and 6 dB respectively, and the angle of displacement of the loudspeakers is 90°, does the quality of sound significantly improve. The tone colour becomes richer, the sound is more natural and closely linked with the enclosure, and the clarity of speech is preserved.

Any departure from these optimum conditions, particularly as regards delay time, results in a worsening of sound quality. This becomes particularly noticeable where return signals have a delay time longer than 30 msec. Thus it is clear that the correct position of the apparent sound source and the accuracy of its location depend to a great degree on the structure of the first-order reflections connected with the initial stage of the reverberation process.

In considering the second peculiarity of stereophonic transmission, its naturalness, it should be noted that this qualitative property can be intensified if the enclosure is sufficiently reverberant. Experiments show that in stereophonic transmission, reflections of sound from the boundary surfaces of the enclosure cause additional fullness and a broadening of the apparent sound source, i.e. they can provide greater naturalness and spaciousness in the perceived sound. The perception of a three-dimensional quality is particularly important in the transmission of music, and so it is correct, for this form of stereophonic transmission, to choose a reverberation time sufficiently long and close to the musical optimum.

Such a solution to the problem as regards music transmissions contradict the conditions for correct location in speech transmission, when reverberation time should be as short as possible.

The contradictory requirements for stereophonic transmission and the possibility of achieving high clarity of speech and correct location of the apparent source by correct attention to the initial stage of the decay process lead to the following recommendations.

The reverberation time in halls for stereophonic reproduction should be selected on the basis of the conditions for good transmission of music, i.e. reverberation time should be sufficiently long. However, bearing in mind the increased effectiveness of reverberation in the stereophonic sound recording process, this time should be approximately 20 per cent less than in halls for single channel sound reproduction. To create good conditions for the reception of stereophonic transmission of speech and to improve the sharpness of the image, steps have to be taken to obtain a sharp fall on the initial portion of the decay curve. This can easily be achieved by placing absorbing materials on the areas of first order reflections on the ceiling and walls. By means of a correct combination of sound absorbent and reflecting surfaces, optimum conditions for speech reception may therefore be created.

In some cases stereo-reverberation installations are used to amplify the sound of stereophonic transmissions for the listeners. These installations enhance the naturalness of the sound by giving it the appropriate reverberation colour. They use magnetic reverberators. The sound signals from a

316

number of tracks of the stereophonic recording (Fig. 9.12), after mixing, are fed to a sound recording head which records the signals on a tape loop. A group of mutually spaced heads reproduce the signals which, after amplification, are fed to loudspeakers specially arranged in the hall. Feedback from one of the heads and the varied time shifts caused by the successive reproducing heads create the necessary effect of reverberation.

Devices known as 'reverbophones' which have mechanical delay circuits are used for the same purpose. In these devices part of the pressure of the sound signal is fed to driving coils acting on several ferrite rotors. The

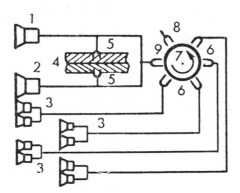

Fig. 9.12. Plan of a stereo-reverberation installation: (1), (2) loudspeakers of the stereophonic system; (3) loudspeakers of the reverberation installation; (4) magnetic stereo recording; (5), (6) sound reproducing heads; (7) magnetic tape loop; (8) erase head; (9) sound recording heads.

rotation of these latter results in a deformation of spiral spring delay lines which, in turn, produces delayed signals in receiving rotors which are placed at the far ends of the delay lines.

An analysis of the acoustic condition in enclosures for stereophonic sound reproduction brings us to the following conclusions:

1. Reverberation in an enclosure has an adverse effect on the accuracy of location of the image, which results in additional movement and in a reduction in the accuracy of its spatial placing. But the reverberation process produces sound which is more natural and more three-dimensional, particularly in the transmission of music.

2. To try to avoid the contradictions inherent in the approach to the specification of the acoustic parameters of a hall designed for stereophonic transmission, a reverberation time appropriate to its volume should be selected, with sound absorbing materials being so arranged as to allow the steepness of the decay curve to be significantly increased and for reflections with a delay greater than 25 msec to be eliminated.

3. Bearing in mind that in stereophonic recording the effective reverberation time is increased, reverberation time in halls for stereophonic

reproduction should be reduced by 20 per cent compared with the reverberation time recommended with the use of a single-channel sound reproducing system.

9.4 Acoustic conditions in halls with sound amplification systems and in halls of general application

Halls in which a sound amplification system is used are, from the acoustic point of view, a combination of enclosures for direct listening and enclosures intended for sound reproduction. In halls of this type, simultaneous use is made of both primary and secondary sound sources. These halls differ from cinema auditoria firstly in that the sound field in them is created by the performers at the time of hearing, and simultaneously, the loudspeakers and reflections. Secondly, in these halls the role of the primary enclosure in which the sound, has been recorded for subsequent reproduction has no influence. Halls with a sound amplifying installation are themselves both the primary and the secondary enclosures.

In determining acoustic conditions for halls of this type a simplified approach based on the division of the sound energy into two parts is possible. One part is associated with the effect of the primary sound source, and the second with the loudspeakers of the amplification system. In this approach the reverberation times of the 'primary' and the 'secondary' enclosures are identical and consequently the acoustic conditions in the hall when a sound amplification system is being used can be defined on the basis of Sapozhkov's formula (2.15) for coupled enclosures.

On the basis of this formula we can consider that reverberation time in halls with sound amplification is increased by approximately 20 per cent. This is further justified by the fact that the increased loudness level caused by the system, just as in a cinema auditorium, gives the impression of a greater reverberation time than in fact exists.

Thus, in halls with a sound amplification system, the calculated reverberation time should be 20–30 per cent less than in halls for direct listening.

Such a choice of reverberation time can be justified if the loudspeakers of the system are of comparatively low power, are not highly directional, and are situated centrally. But if directional radiators of fairly high power are used in sound amplification, or if a distributed system of radiators is in operation, the influence of reverberation on the sound quality is not only not increased, but on the contrary can be significantly weakened, and the sound amplification system can even be used to correct the acoustic characteristics of the hall.

As a result of sound amplification it is possible to improve the clarity of speech in enclosures with clearly unsatisfactory acoustic conditions and reverberation times of from 3 to 5 sec. This is a result of considerable amplification of loudness level in halls in which natural sound sources are not capable of producing the loudness necessary for normal hearing.

Moreover, similar results can be achieved by the use of a distributed

318

system of radiators in which each of a large number of low powered loud-speakers serves a single listener or a small group of listeners. Enclosures of this type include indoor stadia, factory floors, railway stations and so on.

It must be borne in mind in considering sound amplification in theatre auditoria and concert halls, that this amplification should ensure acoustic conditions similar to those of direct listening in the best enclosures of similar application. Such results can be obtained only in halls which do not have major acoustic deficiencies by creating in them a sufficient and evenly distributed loudness and a reverberation time which will allow high clarity of speech transmission to be combined with the essential reverberation for the transmission of music.

Success in meeting the first condition depends on the degree of diffusion of direct energy, as in halls of this type diffuse energy is normally distributed evenly enough. The dispersal of direct energy depends on a number of factors. Apart from the volume, shape and acoustic treatment of the hall, with which we shall deal later, it is also affected by the shape of the directional characteristics and by the disposition of the radiators (including their positions relative to each other and to the boundary surfaces of the enclosure, the height of suspension, and the horizontal and vertical angles of suspension).

For a correct combination of these factors, it is necessary to ensure:

1. For all listeners a comparatively small value of the acoustic ratio R, allowing the equivalent reverberation to be kept close to the optimum.

2. The most accurate coincidence between the auditory images created by the performer and the sound amplification system.

In most cases, these demands can be met by using a centralized system consisting of narrowly directional loudspeakers placed close to the performer, allowing the aural images to fit. In this case directional characteristics, and correctly chosen height and angles of suspension of the radiators will ensure the essential equivalent reverberation and loudness for listeners at the back of the hall.

If the hall is so large that a centralized system cannot ensure the correct distribution of direct energy and create the necessary equivalent reverberation in a remote part of the hall, this can be corrected by bringing in additional loudspeakers to serve areas where the loudness is insufficient.

Often, particularly in large halls in which even the combination of the centralized and a distributed system does not give good results, use is made of a widely distributed system which serves the listeners by means of individual low power loudspeakers.

The second condition, that of ensuring good transmission of speech and music, can be met in two ways: the first is essentially a compromise between the optimum values of reverberation time for speech and music transmission, and the second involves the use of supplementary acoustic systems which enable reverberation time to be controlled within the limits necessary to ensure optimum conditions in either case. These supplementary systems are normally either ambiophonic installations or artificial reverberation systems.

319

If these above conditions are met, a situation develops in which the influence of the loudness level of the performer and the natural reverberation time of the hall on the quality of speech transmission is comparatively small. Thus, correct use of sound amplification systems, and systems for the control of acoustic conditions can avoid the interfering action of large volume and long reverberation time, as can be judged from Fig. 9.13.

Fig. 9.13a shows curves corresponding to direct hearing, and Fig. 9.13b curves obtained with a system of sound amplification. A comparison of these figures shows that in the latter case the zones of improved hearing

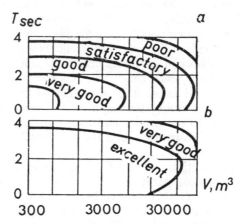

Fig. 9.13. Curves defining the quality of hearing (a) for direct listening and (b) for sound amplification.

are considerably widened and indicates the possibility of a significant improvement in acoustic conditions by the use of a sound amplification system.

Multi-purpose halls which are used for film projection, for staging plays and concerts, and also for speech-only functions such as conferences or meetings, have to meet more severe demands than cinema auditoria or halls with a sound amplification system. Despite the fact that in both these groups of halls a sufficiently high level of loudness in all parts of the auditorium and an optimum reverberation time for both speech and music are of importance from the point of view of acoustic conditions, the quantitative values of these parameters differ slightly in the different groups.

This is mainly so for optimum reverberation time, which, if the influence of regenerative reverberation in the amplification of speech is taken into account, should be 10–15 per cent lower than in the reproduction of recorded sound. For direct listening to large music ensembles a reverberation time longer than for the reproduction of recorded sound should be chosen.

Thus in multi-purpose halls it is essential to be able to obtain an optimum reverberation time for direct listening, for the optimum reception of sound from the sound track of a film (i.e. reduced by 20–25 per cent) and for the clear reception of speech via a sound amplification system (i.e. a further

320

reduction of 10–15 per cent). It is hardly possible to arrive at such a variety of acoustic conditions by purely acoustic means, and so, as in a hall for sound amplification, here too electro-acoustic means are necessary.

Two methods of establishing in an enclosure acoustic conditions which will meet the demands of the various types of sound transmission can be distinguished. The first method stems from the fact that acoustic means (acoustic treatment of surfaces, the installation of sound absorbers, etc.) allow the creation of optimum conditions for that type of transmission which demands the longest reverberation time. It is clear that for a multi-purpose hall this type of transmission is the direct transmission of music ensembles. A reduction in reverberation time to its optimum value for the reproduction of recorded sound can be achieved by the use of electro-acoustic amplification.

So, for example, the use of distributed and centralized systems can ensure the necessary acoustic power by providing the listeners with the correct equivalent reverberation. A still further reduction in reverberation time to a level appropriate to optimum conditions for speech can be attained by the use of a distributed system of sound amplification alone, in which the importance of the direct energy grows, in comparison with the previous case, which results in a further reduction in equivalent reverberation.

In order to make a correct choice of the sound amplification system, to determine the type and position of radiators in this method of creating the necessary acoustic conditions, we must first find the value of the drop in direct energy when the sound source is switched off; it is this value which results in the necessary equivalent reverberation. It can be calculated from Fig. 2.33. As, according to this figure:

$$10 \log (1 + 1/R) = 60 - \Delta L \qquad\qquad 9.7$$

then equation 2.92 can be rewritten as

$$60 \, T_{equiv} = (60 - \Delta L)^2 \frac{T}{60}$$

whence

$$\frac{60 - \Delta L}{60} = \sqrt{\frac{T_{equiv}}{T}}$$

and the value of the drop in energy when the sound source is switched off will be

$$\Delta L = 60 \left(1 - \sqrt{\frac{T_{equiv}}{T}} \right) \qquad\qquad 9.8$$

Once we know this value it is easy to determine the acoustic ratio and thus to find to what extent direct energy should exceed reflected energy in order to create the necessary equivalent reverberation. The next step is to choose a sound amplification system, the type of loudspeakers with appropriate directional characteristics, and the positions in which they are to be placed, which will allow the necessary acoustic ratio to be obtained.

321

21

The second method enables optimum reverberation time for the transmission of speech via a sound amplification system to be achieved by acoustic methods. An increase in reverberation time to values ensuring optimum conditions for the reproduction of recorded sound and the direct transmission of music is carried out by means of an artificial reverberation system or by an ambiophonic installation. As this method can be used only where there is no need for sound amplification in speech transmission, its use is recommended only for halls of comparatively small volume. For large volume multi-purpose halls the first method should be used.

For very big halls the question of the acoustic equipment is often decided by means of a compromise, i.e. by the simultaneous use of electro-acoustic apparatus as indicated for both the first and the second methods.

It is precisely thus that the problem of the acoustic installation of the large hall of the Palace of Congresses in the Kremlin, built in 1961 (acoustic design by V. Furduev and A. Kacherovich), was successfully solved.

This hall, with a volume of 52000 cubic metres and a capacity of 6150, has a reverberation time of 1·5 sec, which is close to the optimum for speech. To achieve this reverberation time, sound absorbing constructions were used which have a coefficient of sound absorption of almost 0·8 and this changes comparatively little over a wide frequency range. The fact that the constructions are predominantly placed on the surface of the ceiling has resulted in the fact that filling the hall to capacity has almost no effect on its acoustic characteristics.

The hall is equipped with sound amplification installations and with an artificial reverberation system. The sound amplification system consists of a group of loudspeakers in the proscenium arch and of individual loudspeakers fitted into the backs of the seats distributed throughout the entire hall. The separate or joint operation of these systems enable the equivalent reverberation time to be altered and the reverberation perceived by the listeners to be significantly reduced.

A multi-channel ambiophonic system, and reverberation installations with magnetic reverberators and an echo chamber, working on a large group of loudspeakers distributed along the walls of the hall, allow reverberation time to be increased within wide limits, and can even create an echo effect.

9.5 The volume and dimensions of an auditorium

The volume of an auditorium is determined by the purpose to which it is to be put and by the size of the audience for which it is designed. In deciding on the volume the basis of the calculations is to achieve a rational agreement between the requirements for good acoustic, lighting, and hygienic conditions, while bearing in mind the economics of the project.

The development of wide screen and panoramic cinematography and of the technique of stereophonic sound reproduction, together with attempts to achieve maximum economy in the construction and use of cinemas, have given rise to a tendency to build auditoria for large audiences. According

322

to the norms and the technical conditions for the design of cinemas (SN 30–58) it is recommended that cinemas in towns should be large, with 600, 800, 1200 or 1600 seats.

Apart from this, multi-purpose halls of 2500, 4000 or 6000 seats are being built with a view to their use not only for film projection, but also for theatrical presentations and concerts. Halls of from 200 to 300 seats with normal film projection facilities are, according to the norms, recommended only for small settlements.

The overall norm value for all auditoria is the unit area laid down by hygienic conditions. This is calculated at 0.8 m² per person in the audience. Unit volume depends on the purpose of the auditorium and on the necessary acoustic and lighting conditions. Taking in the latter, the height of the auditorium must be determined by the height of the screen and by the permissible vertical viewing angle. However, this is frequently in contradiction with acoustic requirements for which excessively high enclosures are undesirable, because in this kind of hall, clarity of speech is worsened and the establishment of optimum acoustic conditions becomes much more complex.

The combination of acoustic, lighting and hygienic demands has resulted in the recommendation of a unit volume not less than 3.5, 4.0 and 4.5 m³ per spectator respectively for cinemas of 300, 600 and 1600 capacity. The unit volume for halls intended for both cinema and concert use has not been the subject of a norm, and in plans which have already been passed is anything up to 9 or 10 m³ per person. Thus the overall volume of the auditorium can be determined by one of the formulae:

$$V = NS_N h \qquad \text{or} \qquad V = NV_N \qquad\qquad 9.9$$

where S_N and V_N are respectively the unit area and unit volume, and h is the height of the hall.

The dimensions of the hall are also decided from the need to satisfy simultaneously lighting and acoustic conditions.

As far as acoustic conditions are concerned, in choosing the dimensions of a hall it must be remembered that these dimensions must not be equal to, nor multiples of, each other, nor yet very different from one another. Where dimensions are identical or multiples, some of the normal modes of vibration of the enclosure disappear while a significant difference between the dimensions results in a great delay in the decay of oblique modes compared with tangential and particularly with axial modes. In either case, the sound field becomes less uniform. The choice of the length of the hall is of great importance. The length of throw of modern projection apparatus allows the length of the hall to be as long as 60 m or more. However, such a length results in the asynchronism between the image projected on the screen and the accompanying sound being more than the permissible amount. As this asynchronism may not exceed 1/24 sec, the norms SN 30–58 set the maximum length of the hall at 40 m.

In placing the screen, from the standpoint of good visibility in the presence of a raked floor which allows the correct transmission of various frequencies, it is recommended that the height of the auditorium should be calculated by the formula:

$$h = 1\cdot6 + 0\cdot2l$$

which predetermines the following ratio of the height and the length of the hall: for the longest, 1:4·3 and for short halls, 1:3.

To make the most sensible use of the floor area, the breadth of the auditorium should be 70 per cent of the length. However, in a normal cinema, to reduce any great difference between the arrival times of direct and reflected signals at the listener's ear, this value may be reduced to 60 per cent.

In large auditoria with wide screen or panoramic projection, the great width of the screen may make it desirable for the width of the auditorium

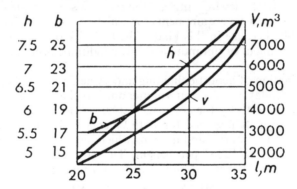

Fig. 9.14. Curves to determine the length, breadth and height of the auditorium for a wide screen cinema according to its volume.

to be increased to 80 per cent of the length. So, for normal cinemas with a capacity of up to 300 seats, the ratio of dimensions is represented as 1:2:3, and for wide screen, 1:3·3:4·3.

If the auditorium does not have a balcony, then, on the basis of these calculations, we can construct curves of the dependence of breadth b and height h on length l (Fig. 9.14) which allow the necessary dimensions to be found easily.

As the designs for auditoria with a capacity of more than 1600 usually include balconies, the dimensions and position of the balconies have to be determined. And in so doing it is essential to ensure a sufficiently high loudness level in the back rows of the balcony and of the space under the balcony, and to allow as large as possible an area for the aperture linking the auditorium with the under-balcony space. On this basis the height of the hall above and below the balcony should be greater than 2·5 m and the corresponding depth of space should be related to the height in the ratio 2–2·5:1.

Halls for film projection and concerts with capacities of 2500, 4000 and 6000 spectators are designed with balconies and have lengths respectively of

40, 50 and 60 m. In a hall of 4000 seats, in order as far as possible to exclude divergence between image and sound, almost the same asynchronism (1/24 sec) is allowed for members of the audience in the front and back rows.

In a hall of 6000 seats, in view of the fact that a departure from synchronism of sound and vision is harder for spectators at the back of the auditorium to detect than for spectators at the front, asynchronism of up to 1/12 sec (with sound delay) is permissible for those at the back, while for those at the front, asynchronism of up to 1/24 sec (with sound in advance) is allowed.

For an auditorium of 2500 capacity and trapezoidal form the width of the front and rear walls should be 0·69 and 1·01 of the length of the hall; for trapezoidal halls of 4000 and 6000 capacity, respectively, 0·7 and 1·03, and 0·71 and 1·04.

Because of the disadvantage of comparatively great height of the auditorium, namely the need to increase the quantity of sound absorbing materials in its internal treatment, the heights of very large multi-purpose auditoria are made as small as possible, just enough to allow screens of the necessary dimensions to be installed. The heights of these halls is made only 1 m higher than the panoramic screens installed in them, which, for auditoria of the dimensions indicated above are 12·9, 16·5 and 20 m. Thus the heights of these theatres are 35 per cent of the length of the hall.

The conclusions are as follows:

1. The volumes and dimensions of auditoria are determined by the need to create in these auditoria conditions for the optimum reception of sound and vision while meeting the demands of hygiene and economy. These values are determined according to the technical conditions in SN 30–58 which give the recommended values of unit area and unit volume.

2. The ratio of length to breadth and height of auditoria changes slightly with an increase in volume. The change tends towards an increase of the latter two parameters.

3. For multi-purpose cinema auditoria it is permissible to exceed the normally allowed length, and one also has a rather freer choice of the other dimensions. However, these divergences are not very large and proceed from a compromise between the demands of acoustics, lighting, and hygiene.

9.6 The shape and acoustic treatment of a cinema auditorium

The selection of the shape of an enclosure for direct listening is linked with the need to ensure within the enclosure a high degree of diffuseness of the sound field, an equal loudness level throughout its area, and an absence of conditions which would lead to a concentration of sound and to the creation of an echo. These requirements apply to a large extent also to a cinema auditorium.

The first condition, which means essentially that the enclosure should have a sufficiently dense spectrum of characteristic frequencies, decaying at roughly the same speed, is of great importance for a cinema auditorium. It can be fulfilled, firstly, as follows from the wave theory, by ensuring that the dimensions of the hall are not multiples of one another, and secondly by avoiding the possibility of setting up a stable system of standing waves such as would result from the presence of parallel surfaces in the enclosure. Consequently, the boundaries of the hall should not be parallel and its dimensions should not be multiples of one another.

As was pointed out in 9.1, the second requirement for a cinema auditorium can be fulfilled by using sufficiently powerful sound sources so arranged as to ensure that an equal loudness level is created at both the front and the

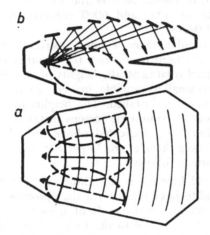

Fig. 9.15. A good shape for a cinema auditorium (a) plan (b) section.

back of the hall by direct energy alone. However, it should be borne in mind that such a distribution of energy can be achieved if the shape of the hall and the ratio of its dimensions are correctly chosen.

This conclusion is reached from an analysis of formula 2.84, from which follows that for a given loudspeaker power, equal energy density throughout the whole area of the hall will be achieved when the ratio $\Phi(\theta)/r$ is constant. The invariability of this ratio can be achieved only by attaining the necessary agreement between the shape of the hall, the directional characteristic of the loudspeaker, and its placing relative to the hall. Thus as far as the second demand made of a cinema auditorium—that of ensuring an equal loudness level throughout the area—is concerned, the shape and the ratio of dimensions have some importance, although this is certainly less than in enclosures for direct listening.

This requirement can be satisfied by a reduction in the depth of the hall, by making it wider, particularly at the end, and by using several loud-speakers, the outer ones of which are turned away from one another by an angle depending on their directional qualities and the angle of the side walls (see Fig. 9.15).

326

Apart from this the creation of an even field throughout the area of the hall is helped by giving to the proscenium, to the walls near the sound source and to the ceiling a shape which will allow part of the reflected energy with a small delay time to be directed to the first rows of seats (shown by arrows).

This is particularly important for cinema/concert auditoria which are used for direct listening or for sound reproduction, where the loudspeakers are placed at a great height. This is often done in large auditoria in order to supply the farthest listeners with direct energy.

While we are considering the uniform distribution of sound energy throughout the auditorium, we must not forget that identical conditions

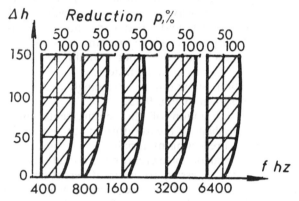

Fig. 9.16. *The reduction in sound pressure in percentages as a result of the absorption of sound by listeners when they are placed at various heights (Δh) above those sitting in front of them.*

must be present for the transmission of various frequencies, which, as follows from the experiments of Bekesy[5], is also associated with the shape of the hall. The data of these experiments (Fig. 9.16) show that the loss of energy (reduction in sound pressure) resulting from the sound absorption of listeners sitting in front depends on the height to which they are raised above the row in front. Changes in the timbre of sound resulting from dissimilar reduction in sound pressure are significantly decreased if the floor is given a slope such that listeners in one row are from 20 to 30 cm higher than those of the row in front. This leads to the necessity of having a floor raked at an angle which, in combination with the necessary shape of the ceiling, results in a vertical cross section of the hall as shown in Fig. 9.15*b* being most acceptable.

The third requirement for a cinema auditorium is also connected with the relationship between its shape and the shape of the directional characteristics and the disposition of the loudspeakers. In reality, an excessive concentration of sound in any given part of the hall can result not only from the presence of concave surfaces, but also from an incorrect choice and orientation of the loudspeakers relative to one another.

The presence of concave surfaces can result in the redistribution of reflected energy in the hall. This depends on the frequency of the signal inasmuch as the degree of concentration depends on the ratio of wavelength to the radius of curvature of the surface. For a given curvature of the surface, low frequencies, the wavelengths of which are greater than that of the radius of curvature, will not be focused so much as middle and high frequencies, which will be concentrated close to a particular point in the hall.

A choice of loudspeakers which is not correct from the point of view of directional characteristics, or an incorrect choice of the angle of suspension and the angle of azimuth, cause concentrations of energy as the result of unnecessary overlapping of their zones of action. Thus, to avoid concentration of sound, the use of concave surfaces in the ceiling and walls of the enclosure should be avoided, and great care should be taken over the choice and siting of loudspeakers. The appearance of flutter echo is avoided by ensuring that the enclosure does not contain large curvilinear, and comparatively non-absorbent parallel surfaces.

All these considerations show quite convincingly that the best shape for an auditorium is that shown in Fig. 9.15.

But the complete satisfaction of the requirements considered above depends not only on the shape of the auditorium but also on the distribution of sound absorbing and reflecting materials.

It follows from 9.2 that in order to create good conditions for the reception of speech and music it is necessary to ensure different degrees of steepness for the initial and final stages of the decay curve, which suggests that we should aim at an almost total damping of the ceiling of the auditorium. This, however, results in a very uneven sound field (see 3.2), as when the floor is simultaneously heavily damped (by the presence of an audience) the sound field is transformed from a three-dimensional field into a plane field.

If we consider these mutually contradictory conditions, we come to the conclusion that it would be more satisfactory to deaden only the front part of the ceiling, particularly in large halls. This in itself would create the necessary difference in steepness between the initial and final slopes of the decay curve. In this case the front part of the ceiling could be a plane surface, while the rear part can be made to scatter by dividing it into small areas at varying angles (see Fig. 9.15b), so that it evens out the loudness level in the farthest parts of the auditorium and in the balcony.

If there is a balcony, the ceiling of the under-balcony area should be made up of similar elements angled in such a way as to ensure that well dispersed reflections can be directed into the under-balcony space.

Varying treatments of the wall surfaces can also have an influence on a change in shape of the auditorium.

The desire to avoid the harmful influence of long-delay signals, which have been reflected from the ceiling, and then from the rear and front walls and back to the listeners in the front rows, often leads to the rear wall being completely damped. In this case its shape is of no great significance, and it can be either plane or concave.

328

The interrelation between the side walls, which gives the auditorium its trapezoidal form, can be changed if sound absorbent materials are alternated with reflecting panels or if their surfaces are made up of separate elements which create a dispersed reflection. If this method of acoustic treatment is correctly employed, the side walls may be made parallel.

While bearing in mind the beneficial influence on the acoustic conditions of the auditorium of surfaces made up of scattering elements, it is important to determine the limits of this influence. As can be seen from Fig. 9.17a, a dispersed reflection, which can overcome a number of acoustic defects in

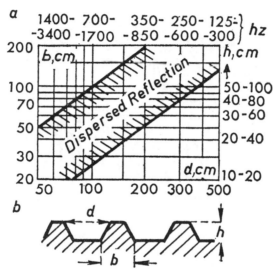

Fig. 9.17. The zone of dispersed reflection (a) where the surfaces of the hall are divided into elements of various dimensions. (b) Method of division.

the hall, can be achieved within wide limits by varying the distance between elements d, and the breadth b and height h of the individual elements. Moreover, the shape of the elements can be plane or curved. The region of greatest dispersal is limited to a specific frequency band, depending on the dimensions of the elements—the profile of the surface. As the profile becomes more uneven, the frequency range of the dispersed reflection becomes narrower and moves towards the low frequency region. In order to broaden the frequency range of the dispersed reflection the use is recommended of elements of varying dimensions, as shown in Fig. 9.18.

An analysis of the influence of the shape of the hall, the choice and siting of loudspeakers and sound absorbent materials, and the profile treatment of the interior surfaces on the acoustic conditions in a cinema auditorium shows that if all these factors are wisely combined they can ensure a high quality of sound in the hall. Moreover, use of the three latter factors alone can go a long way towards neutralizing the influence of the shape of the hall. In particular, the correct choice and siting of loudspeakers can eliminate the need to give the floor of the auditorium a steep rake; treatment of the ceiling and the rear wall with effective absorbent materials and an appropriate combination of sound absorbers with individual elements of the wall surfaces results in its being possible for the shape of the hall to approximate to a rectangular parallelepiped.

Thus, because the shape of the auditorium is far from being the only parameter defining acoustic conditions, it can in many ways depart from the recommendations given above (see Fig. 9.15). If this is the case, the requirements set out at the start of this section must be met by use of the other

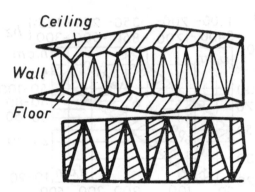

Fig. 9.18. Diagram of the division of a wall into elements of various dimensions.

factors (the disposition of loudspeakers and sound absorbent materials, the profiling of surfaces, etc.).

The following considerations can be the basis for choosing a particular shape for a hall. If the hall has large dimensions, or is frequently used as a multi-purpose hall for direct listening, the best shape is trapezoidal, because of the ease with which it meets the various requirements of the hall, and because it simplifies the solution of the problem of acoustic treatment of the interior surfaces. The shape of small-volume halls, intended mainly for showing films, can be nearer to a parallelepiped, provided that the acoustic treatment and the selection and siting of loudspeakers are carefully considered.

These conclusions are in close accord with the practice of cinema construction, as can be seen from Fig. 9.19, which shows plans of cinema and concert auditoria with 800, 1600 and 2500 seats.

The selection of the form of an enclosure for sound reproduction is simplified if we agree with the point of view of A. N. Kacherovich,[41] who states that the attempt to create a diffuse sound field in a cinema auditorium is not justified. Kacherovich bases his arguments on the need to ensure a significant difference between the rates at which sound energy decays at the initial and final stages of the decay process, which ensures suitable conditions for good reception of speech and music, and on the possibility of obtaining sufficient loudness and clarity at all points in the hall by using powerful loudspeakers. From these considerations he recommends the choice of the simplest possible shapes for enclosures—rectangular or trapezoidal with a flat ceiling. In this case the treatment of the ceiling and the lower parts of the walls with efficient sound absorbers, because of the fact that these surfaces are areas of primary reflections, results in a significant increase in the steepness of the initial slope of the decay curve while the final stage is more gently inclined. This meets the conditions necessary for the good reception of speech and music reproduced from a recording.

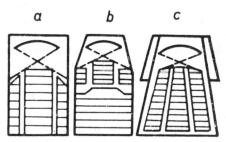

Fig. 9.19. Outline plans of typical cinema auditoria for (a) 800, (b) 1600 and (c) 2500 seats (to different scales).

In brief, the conclusions on the choice of the shape for an enclosure for sound reproduction can be set out as follows:

1. The shape of a hall, the nature of its acoustic treatment, and the type and position of loudspeakers have an influence on the quality of sound reproduced.

2. The shape of the hall has the greatest influence in this respect when the hall is small in volume and where it is used for direct listening. In these cases a trapezoidal plan is recommended for the hall, with a prescribed angle of rake in the floor and angling of the ceiling, together with an appropriate form of the proscenium (panoramic cinemas, cinema/concert halls).

3. For medium size cinemas, we can recommend a simplified shape for the hall with a flat horizontal ceiling (Fig. 9.19b), while for small cinemas, the shape can be that of a rectangular parallelepiped.

4. Sound absorbing materials and loudspeakers should be placed so as to favour the even distribution of direct sound energy and to allow a sufficient dispersal of reflected energy along the surfaces where listeners are placed.

5. To ensure the necessary dispersal of sound energy, use should be made of elements in the surfaces of the ceiling and the walls, with such a distance between elements as to ensure, as far as possible, identical conditions for the transmission of different frequencies.

6. In large auditoria, in order to create adequate acoustic conditions for the reproduction of speech and music, and particularly in order to obtain better locations in stereophonic reproduction, an uneven distribution

331

of sound absorbent materials is recommended with the most effective being positioned primarily on the areas of the first few orders of reflections.

7. In small-volume halls, in which the optimum reverberation time is close to that necessary for speech transmission, sound absorbent materials may be placed evenly on all the interior surfaces of the hall.

9.7 The method of acoustic planning of cinema auditoria

The purpose of the acoustic planning of cinema auditoria is to calculate a number of acoustic parameters which determine the choice of materials for the treatment of the interior surfaces of the enclosures.

In carrying out the acoustic planning of small or medium sized cinema auditoria, in which absorbent materials can be distributed quite evenly on the interior surfaces, the formulae of the statistical theory may be used for calculation. In this case the order of calculations of a hall which is being designed from scratch is as follows.

1. The volume of the auditorium and its useful area S_{useful} are determined with reference to the number of audience seats according to formula 9.9.

2. The height of the hall is determined from the ratio V/S_{useful}, and from the recommended ratio or, from Fig. 9.14, the other two dimensions of the hall l and b are found.

3. The unit volume and the unit area are checked to see that they do not exceed the existing norms, and if they do not, the volume, the area of individual surfaces, and the overall area are calculated accurately.

4. By using the curves in Figs. 9.4 and 9.5, according either to the volume of the hall or the number of spectators, the optimum reverberation time for a frequency of 1000 Hz is determined, and then it is calculated for the following frequencies: 125, 250, 500, 2000 and 4000 Hz.

5. The total sound absorption which will give the optimum reverberation time is calculated from formula 2.42, from which, using the values of T_{opt} for the indicated series of frequencies, S and V, the necessary values of α_{mean} are found. The total sound absorption for each of the frequencies is found as the product $\alpha_{mean} S$.

6. Sound absorbing materials are selected with such coefficients of absorption and with such a ratio of their areas that the total absorption afforded by them at all frequencies corresponds as closely as possible to the calculated values of total absorption. The results of the calculations should be drawn up in a table, with the names of the absorbent materials listed vertically and, in horizontal columns, their areas, their coefficients of absorption, and the total sound absorption at each frequency and for each of the materials.

For the acoustic calculation of an existing hall, we begin by calculating the total sound absorption of the materials which are already in position on its interior surfaces, and then select supplementary materials whose introduction into the hall will create a total absorption close to that demanded by the calculations.

332

7.   By dividing the total absorption of all the materials by the area of all the surfaces, the calculated coefficient of sound absorption for each frequency is found, and according to formula 2.33 the calculated value of T_{opt} is determined. The difference between the calculated value of T_{opt} and the required value should not be greater than $\pm$ 10 per cent.

8.   A plan is drawn up for the positioning of the sound absorbent materials in the hall, using the recommendations given in 9.6.

The calculation of the sound insulation of the auditorium is carried out in the same sequence as for a film studio (points 12–15 in 7.7).

In the acoustic planning of large halls, for which the creation of the best conditions for the reception of speech and music requires an uneven distribution of sound absorbent materials, the sequence of calculations must be rather different.

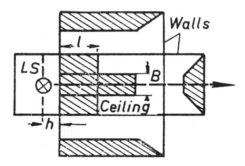

Fig. 9.20. The enclosure laid out flat. The areas where effective sound absorbing materials are placed are indicated by cross hatching.

The purpose of the uneven distribution of materials to create a diffuse sound field, which makes the use of the formulae of the statistical theory for the calculations impossible. In this case the calculations can be carried out in accordance with the theory contained in 3.3 and 9.2. The sequence of these calculations can be as follows:

1.   After the volume, the dimensions of the enclosure and T_{opt} have been determined in accordance with points 1–4 of this section, the position of the loudspeaker is chosen, and by this is determined the height of its suspension h and the height of the areas of first order reflections on the walls of the enclosure. (This height can be considered approximately equal to the height at which the loudspeaker is mounted). The area of the lower part of the walls which will be occupied by efficiently absorbing materials is then calculated.

2.   The breadth and length of the areas of first- and second-order reflections on the surface of the ceiling is found by the use of formulae 3.12 and 3.13, and the total of these areas is calculated. The plan of the distribution of the areas occupied by effective materials is shown in Fig. 9.20.

3.   From formulae 3.15, 3.14 and 3.16, a calculation is made of the delay time of the first and second order reflections from all the bounding

surfaces of the enclosure for a certain mean position of the listener. The longest delay time, as follows from 3.3 will be the time of completion of the first stage of decay—i.e. one of the items (T_1) which define reverberation time in the given enclosure.

Time T_1 can be immediately determined if, for the selected point of situation of the listener, the longest ray from all the rays of second order reflections is determined by means of graphic constructions. Most probably this longest ray will be a ray reflected from the ceiling and the rear wall of the hall. Then T_1 can easily be determined from formulae 3.14 and 3.15. Finally, for preparatory calculations, we can, as was pointed out in 9.2, assume that T_1 is equal to 0·15 sec.

4. For the treatment of the areas of first-order reflections, choice is made of a material with a high sound absorption coefficient and with a frequency characteristic similar to that of the frequency characteristic of optimum reverberation time. From the tables is found the coefficient of absorption of the undamped surfaces of the enclosure.

5. The value of the difference in levels N_1 at which the initial stage of the decay of sound energy ends is calculated as the difference $60 - \Delta L$, where, in accordance with formula 2.30:

$$\Delta L = 60(1 - \alpha)^2$$

6. The mean coefficients of sound absorption of the ceiling and walls are calculated, bearing in mind that they are treated with materials having differing absorption of sound energy. The mean coefficient of absorption of the floor is found, taking into account the absorption by the audience.

7. From formula 3.23 the coefficient of absorption α is found, which characterizes the speed of decay of sound energy over the final stage, and from formula 3.20 the time of this decay T_2 is found. The reverberation time of the whole enclosure is defined as $T_1 + T_2$.

All calculations under points 5, 6 and 7 are carried out for a series of frequencies (125, 250, 500, 1000, 2000 and 4000 Hz), and the frequency characteristic of reverberation time obtained as a result of calculations is compared with the characteristic of optimum reverberation time.

9.8 An example of the acoustic planning of a large cinema auditorium

Task

To carry out the acoustic planning of an auditorium for a wide screen cinema for 2500 audience, in a building which overlooks a tree-lined street with a noise level of 70 dB.

Plan of the enclosures

The auditorium is placed within the building so that its front wall abuts on to the street; on one side wall is a cloakroom with a noise level of 85 dB, and on the other a quiet corridor with a noise level of 70 dB. Behind the

334

rear wall is a foyer with a small stage. (*Translator's note:* Some Soviet cinemas provide live, light music or occasional personal appearances of actors and directors for members of the audience waiting for the next showing to start). The noise level in the foyer is 90 dB.

Geometrical calculation of the hall

In accordance with the norms SN 30–58 the auditorium of a 2500-seater cinema auditorium is planned with a balcony and should be 40 m long. Then, according to 4.9, where the hall is trapezoidal in shape, its width across the front wall will be:

$$B_F = 0·7 . L = 0·7 . 40 = 28 \text{ m}$$

and along the rear wall:

$$B_R = 1·01 . 40 = 40·4 \text{ m}$$

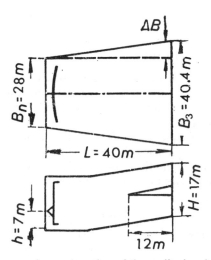

Fig. 9.21. Diagram plan and section of the auditorium being designed.

We determine the height of the hall on the basis of the equation:

$$H = H_{SC} + \Delta H_H + \Delta H_B$$

where $H_{SC} = 0·35 \, L$ is the height of the screen, ΔH_H and ΔH_B, are the distances from the floor or the ceiling to the edge of the screen, which are selected as respectively 2 and 1 m. Thus:

$$H = 0·35 . 40 + 2 + 1 = 17 \text{ m}$$

The area of the side walls (Fig. 9.21) comprises:

$$S_\delta = 2L_\delta . H = 2\sqrt{L^2 + \Delta B^2} . H = 2\sqrt{40^2 + 6·2^2} . 17 = 1380 \text{ m}^2$$

335

The area of the front and rear walls is:

$$S_{F,R} = B_F . H + B_R . H = 28 . 17 + 40·4 . 17 = 1153 \text{ m}^2$$

The area of the floor and the ceiling is:

$$S_{\text{floor,ceiling}} = 2L \frac{B_F + B_R}{2} = 2 . 40 . 34·2 = 2740 \text{ m}^2$$

If the depth of the balcony is taken to be 0·3L, the area of the balcony floor and the ceiling of the underbalcony space is:

$$S_{\text{balcony}} = 2 . 0·3L . B_R = 2 . 0·3 . 40 . 40·4 = 970 \text{ m}^2$$

The total area of the bounding surfaces is:

$$S = 1380 + 1163 + 2740 + 970 = 6353 \text{ m}^2$$

and the volume of the hall:

$$V = L . \frac{B_F + B_R}{2} = 40 . 34·2 . 17 = 23,300 \text{ m}^3$$

Optimum reverberation time

According to Fig. 9.4 we determine that at a frequency of 500 Hz with single channel sound reproduction, reverberation time should be 1·7 sec. Remembering that the auditorium will be used for the reproduction of stereophonic sound tracks, reverberation time should be reduced by 20 per cent, i.e. at a frequency of 500 Hz it should be taken as equal to 1·4 sec. At all other frequencies this time should be altered in accordance with the frequency characteristic shown in Fig. 9.5. Consequently reverberation time at the frequencies at which it is calculated should be as shown in Table 13.

TABLE 13

Freq. (Hz)	125	250	500	1000	2000	4000
$T_{\text{necessary}}$ (sec)	1·6	1·5	1·4	1·4	1·4	1·4

Determination of the dimensions of the areas of first-order reflections

If the height of suspension of the loudspeaker is $h_{LS} = 7$ m, then the area of the lower part of the walls which should be treated with good absorbent materials is:

$$S_{\text{lower}} = 2L_{\text{side}} . h_{LS} + B_R h_{LS} + B_F h_{LS}$$
$$= 2 . 40·5 . 7 + 40·4 . 7 + 28·2 = 906 \text{ m}^2$$

and

$$S_{\text{higher}} = 2L_{\text{side}}(H - h_{LS}) + B_R(H - h_{LS}) + B_F(H - h_{LS})$$
$$= 2 . 40·5 . 10 + 40·4 . 1·0 + 28 . 15 = 1634 \text{ m}^2$$

336

The dimension of the areas or the surface of the ceiling which provide later reflections will be determined from equations 3.12 and 3.13 as:

$$b'_{area} = \frac{H - h_{LS}}{2H - h_{LS}} = \frac{17 - 7}{34 - 7} \cdot 28 = 10{\cdot}4 \text{ m};$$

$$b''_{area} = \frac{17 - 7}{34 - 7} \cdot 40{\cdot}4 = 15 \text{ m};$$

$$l_{area} = \frac{H - h_{LS}}{2H - h_{LS}} \cdot L = \frac{17 - 7}{34 - 7} \cdot 40 = 14{\cdot}6 \text{ m}$$

As the ceiling provides areas of no more than the first three orders of reflections, effectively absorbing material should be placed over all its surface (Fig. 9.22).

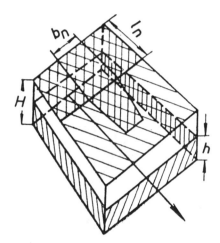

Fig. 9.22. The positioning of sound absorbent materials on the ceiling and walls of the hall in course of design. Double hatching shows the areas of first- and second-order reflections on the ceiling.

Calculations of the delay time of first-order reflections

To determine the delay time of the series of first order reflections it is necessary first to calculate the length of path of the direct and reflected rays. This can be done by use of formulae 3.16 and 3.14, if for the calculation of the length of rays reflected from the ceiling Fig. 3.16 is used, and for the length of those reflected from the walls use is made of Fig. 9.23.

For the calculations, let us choose a listening point on the axial line of the hall in the area of best visibility at a distance of $x = 25$ m from the central loudspeaker.

The length of the direct ray is:

$$l_0 = \sqrt{h^2 + x^2 + y^2} = \sqrt{49 + 625} = 26 \text{ m}$$

The length of a ray which has undergone a single reflection from the ceiling is:

$$l_1^I = \sqrt{(2H - h)^2 + x^2 + y^2} = \sqrt{(2 \cdot 17 - 7)^2 + 25^2} = 36 \cdot 5 \text{ m}$$

The length of rays which have undergone a single reflection from the walls is determined from the formulae:

$$l_1^{II} = \sqrt{h^2 + (2L - x)^2 + y^2} = \sqrt{7^2 + (2 \cdot 40 - 25)^2} = 55 \cdot 5 \text{ m}$$

$$l_1^{III} = \sqrt{h^2 + x^2 + (2B + y)^2} = \sqrt{7^2 + 25^2 + (2 \cdot 17 \cdot 1)^2} = 42 \cdot 9 \text{ m}$$

$$l_1^{IV} = \sqrt{h^2 + x^2 + (2B - y)^2} = \sqrt{7^2 + 25^2 + 34 \cdot 2^2} = 42 \cdot 9 \text{ m}$$

Results of the calculation of the length of rays of first- and second-order reflections are shown in column 4 of Table 14.

The delay time of each reflected ray compared with the direct ray is entered in column 5 of the same table.

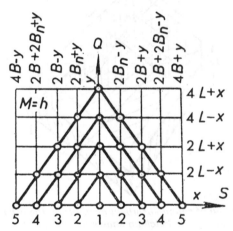

Fig. 9.23. Nomogram to determine the length of rays reflected from the walls of the enclosure.

Calculation of the time of energy decay in the initial stage

To determine this time it is necessary to calculate the difference in the levels of intensity of sound energy reaching the selected point after the nth reflection on the one hand and reaching the same point directly from the source on the other. This difference in levels is defined by the equation:

$$N_1 = N_n - N_0 = 20 \log l_0 - 20 \log l_n + \sum^n 10 \log (1 - \alpha_n)$$

The reduction in signal level resulting from reflections, depends on the absorbing capacities of the surfaces of the enclosure. Let us consider that for an untreated part of the walls, $\alpha_B = 0 \cdot 02$, for that part of the walls

338

treated with efficient sound absorbent materials (wood-fibre panels) $\alpha_H = 0.5$, and for the floor and ceiling $\alpha_{\text{floor}} = 0.5$ and $\alpha_{\text{ceiling}} = 0.5$. In this case, recalling that an effective absorber is placed along the lower part of the walls to a height of $h_{LS} = 7$ m, we can calculate the mean coefficient of absorption for the walls as:

$$\alpha_{\text{walls}} = \frac{(2B_{\text{side}} + B_{\text{rear}} + (H - h)\alpha_B + (2B_{\text{side}} + B_{\text{rear}})h\alpha_H}{S_{\text{side}} + S_{\text{rear}}}$$

$$= \frac{(2 \cdot 40.5 + 40.4) \cdot 10 \cdot 0.02 + (2 \cdot 40.5 + 40.4) \, 7 \cdot 0.5}{1380 + 686} = 0.218$$

Now let us determine the difference in levels for reflected rays which have undergone one reflection:

$$N_1{}^I = 20 \log 26 - 20 \log 36.5 + 10 \log (1 - 0.5) \quad = -6.0 \text{ dB}$$

$$N_1{}^{II} = 20 \log 26 - 20 \log 55.5 + 10 \log (1 - 0.218) = -7.3 \text{ dB}$$

$$N_1{}^{III} = 20 \log 26 - 20 \log 42.9 + 10 \log (1 - 0.218) = -6.8 \text{ dB}$$

These data, and also the results of the calculations of the difference in levels for rays which have gone through two reflections, are put into column 6 of Table 14.

TABLE 14

No.	No. of refs.	Surface reflected from	Length of reflected ray	Delay time (sec)	Level difference (dB)
1	1	From ceiling	36·5	0·031	− 6·0
2	1	From wall	55·5	0·087	− 7·3
3	1	From wall	42·9	0·050	− 6·8
4	1	From wall	42·9	0·050	− 6·8
5	2	From ceiling	61·5	0·105	−11·5
6	2	From ceiling	50·2	0·071	− 9·8
7	2	From ceiling	50·2	0·071	− 9·8
8	2	From floor	48·0	0·065	−11·3
9	2	From wall	105	0·232	−18·1
10	2	From wall	65·0	0·115	−13·5
11	2	From wall	72·8	0·138	−14·9
12	2	From wall	65·0	0·115	−13·5
13	2	From wall	72·8	0·138	−14·9

From the data in column 5 of Table 14 we construct the initial stage of the curve of the decay of sound energy in the auditorium. From this curve, shown in Fig. 9.24, and from Table 14, it follows that the time of the delay of energy within the limits of the initial sector $T_1 \simeq 0.4$ sec, and the difference in levels which corresponds to this is $N_1 \approx 15$ dB.

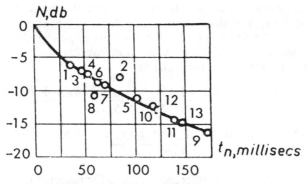

Fig. 9.24. Initial portion of the curve of energy decay.

The calculation of effective reverberation time

From Appendix 1 we find, and transfer into Table 15, the coefficients of sound absorption for the selected materials (dry plaster tiles and tiles ROSNIMS) and their areas, and then, from the formulae

$$\alpha'_{floor} = \frac{S_{gangway}\alpha_{gangway} + N\alpha_N}{S_{floor}}$$

and

$$\alpha'_{walls} = \frac{S_{upper}\alpha_{upper} + S_{lower}\alpha_{lower}}{S_{walls}}$$

where $S_{gangway}$ and $\alpha_{gangway}$ are the area and the coefficient of absorption of the gangways in the auditorium, we find the mean coefficients of the floor and the walls (α'_{floor} and α'_{walls}).

As the total breadth of the gangways in the hall is arrived at by allowing 0·6 m per 100 persons, the sought-for mean coefficients of absorption at a frequency of 125 Hz will be:

$$\alpha'_{floor} = \frac{0{\cdot}6\,.\,0{\cdot}01\,.\,2500\,.\,40\,.\,0{\cdot}02 + 2500\,.\,0{\cdot}37}{1856} = 0{\cdot}49$$

and

$$\alpha'_{walls} = \frac{1634\,.\,0{\cdot}02 + 906\,.\,0{\cdot}48}{2540} = 0{\cdot}18$$

Let us determine these coefficients for all frequencies and transfer them into Table 15 (lines 5 and 4).

From formulae 3.23 we calculate the value $\bar{\alpha}$ at all frequencies, and we insert the results of the calculation in the same table. For example, at a frequency of 125 Hz:

$$\bar{\alpha} = \frac{2{\cdot}2\,.\,0{\cdot}02\,.\,1634 + 0{\cdot}47\,.\,906 + (3{\cdot}3\,.\,0{\cdot}18 + 2\,.\,0{\cdot}47 + 0{\cdot}5)1860}{2{\cdot}2\,.\,1634 + 906 + 6{\cdot}3\,.\,1960} = 0{\cdot}26$$

340

and the time of the final stage of the decay of sound energy at this frequency, according to equation 3.20 is:

$$T_2 = 0.027(6 - 0.1N_1) \frac{V}{- S \ln (1 - \bar{\alpha})}$$

$$= 0.027(6 - 0.1 \times 1.5) \frac{23,600}{- 6353 \ln (1 - 0.26)} = 1.51 \text{ sec}$$

The time T_2 calculated in this way for all the other frequencies is written into Table 15 (line 8). Finally we find the effective reverberation time as the sum of T_1 and T_2.

Comparing the calculated value of effective reverberation time with the necessary reverberation time derived from Table 13, we note that the difference between them should not exceed the permissible limits of ± 10 per cent.

The calculation of sound insulation of a cinema auditorium, and also the calculation of a ventilation system, are carried out in the same way as for a music studio (see 7.8).

It should be pointed out that in the acoustic design of halls of small capacity (up to 800), it is usual to employ the method of planning based on the calculations of total sound absorption. An example of such a calculation is given in 7.8.

TABLE 15

No.	Value calculated	Frequencies in Hz					
		125	250	500	1000	2000	4000
1	α (plaster)	0·02	0·05	0·06	0·08	0·04	0·06
2	α (wood fibre)	0·47	0·52	0·5	0·55	0·58	0·63
3	α (ceiling)	0·47	0·52	0·5	0·55	0·58	0·63
4	α (walls)	0·18	0·218	0·22	0·25	0·23	0·28
5	α (floor)	0·49	0·53	0·59	0·64	0·65	0·65
6	ᾱ (mean)	0·26	0·286	0·28	0·32	0·31	0·33
7	ln (1 − ᾱ)	0·30	0·34	0·33	0·39	0·37	0·40
8	Final decay time	1·52	1·33	1·37	1·16	1·23	1·13
9	Initial decay time	0·14	0·14	0·14	0·14	0·14	0·14
10	Effective R.T.	1·65	1·47	1·51	1·30	1·37	1·27
11	Desired R.T.	1·6	1·5	1·4	1·4	1·4	1·4
12	Excess of R.T.	−0·05	+0·03	−0·11	−0·1	+0·03	+0·13
13	% Excess	−3·0	+2·0	−7·8	−7·2	+2·1	+9·8

10 Acoustic Measurements in Enclosures

10.1 Types of methods of measurements and measuring signals

Acoustic measurements in an enclosure should make it possible to evaluate quantitatively the significance of variables which influence the quality of the sound transmission in the enclosure. A comparison of the measured variables with the corresponding variables in acoustically faultless enclosures enables the basic defects of an enclosure to be discovered.

An exhaustive quantitative evaluation is complex because the values which define the qualitative side of the question are associated, as has been shown above, with extremely complex processes of excitation, formation and disappearance of the sound field in the enclosure. The task is made still more complicated by the fact that in evaluating the acoustic qualities of an enclosure, due notice must also be paid to the psycho-physiological laws of the perception of sound.

In evaluating the acoustic conditions in an enclosure we must bear in mind not only the peculiarities of aural perception, but also the accumulated ideas of audiences, who habitually listen to enclosed music rooms and with a certain distance between the performers in an orchestra, and who are able to connect the aural image with the visual.

The conditions can be quantitatively evaluated by the reverberation time, which in this respect must be regarded as the most important value. But knowing this value is far from sufficient to be able to give a complete evaluation of the acoustic character of an enclosure.

To describe the acoustic properties of an enclosure more completely, the frequency characteristic of reverberation time is first determined;

342

sometimes measurements are made of the fluctuation of sound during the decay process, of the acoustic ratio or of effective reverberation. All these measurements depend on the frequency of the vibrations radiated by the sound source, on the sound absorbent property of the materials in the enclosure, its geometrical parameters, and on the position in which the listeners are placed.

These measurements, which basically give the temporal characteristics of the sound processes, enable us to explain and quantify certain faults, but, as they are statistically based, they do not always allow us to indicate the source of these disturbances.

As the statistical theory is based on the assumption that the sound field in the enclosure is random, measurements of reverberation time and of other values which are based on the statistical theory, should be carried out in conditions which satisfy the requirements of a diffuse field. From this point of view, sinusoidal signals, which create a clear interferential picture, cannot be used in the measuring process.

The conditions of randomness are more fully satisfied by the use of narrow band signals which include a cluster of frequencies comparatively close to one another. Such signals include frequency-modulated signals (warble tone), obtained by the periodic modulation of a sinusoidal signal, multitones, which are a series of pure tones of equal amplitude at frequencies very close to each other, and bands of random noise, which are selected from the noise spectrum by the use of bandpass filters.

If measurements are being carried out in sound cinemas, good results can be obtained by using a recording of orchestral music combined with octave filters, enabling measurements to be carried out across the whole frequency range.

In using a sound source of the first type it is essential that the frequency of modulation should be from one-fifth to one-tenth the maximum divergence of the component frequencies from the mean frequency, so that the warble tone contains ten or more mutually close frequency components. Moreover, in order to get a clearer frequency discrimination it is desirable not to have an excessively wide band of component frequencies, so that divergences from the mean should be not greater than 20 per cent. The drawback of this type of signal is that the amplitudes of the component tones are significantly different one from another.

To obtain a signal of the second type (multitone) a special generator is used which gives ten or twelve pure tones of equal amplitude which are close to each other in frequency. The extreme frequencies of the band differ from the mean by 20 per cent. In using multitone it is important that the component signals should not have frequencies which are multiples of one another, to ensure a high degree of constancy of amplitude throughout the enclosure.

The advantage of a measuring signal of the third type is the constancy of the amplitude of all its many components, distributed in a frequency band of from half to a whole octave.

The influence of the type of signal on the decay curve is well illustrated in Fig. 10.1. A comparison of the graphs in Fig. 10.1 shows a preference

343

for sources of statistical noise, and to natural sources of an orchestral type.

Methods of measurement for obtaining the temporal characteristics of an enclosure cannot fully satisfy modern demands. The information about the acoustic properties of an enclosure obtained in this case is sometimes so incomplete, and even erroneous in that enclosures which have identical temporal characteristics are very different in acoustic properties. This happens because these characteristics do not take into account a large number of factors which have an influence on the formation of the sound field in the enclosure. Among the factors omitted are the shape of the enclosure, the disposition of absorbent and reflecting materials, the position of the sound source and its directional characteristic.

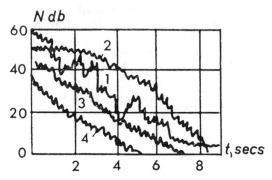

Fig. 10.1. Sound decay curves for different forms of signal: (1) pure tone, (2) warble tone, (3) statistical noise, (4) recording of an orchestra.

The role of these factors can be clarified if a method of measurement is used which allows the spatial acoustic properties of the enclosure to be found. This can be done by a study of the sound processes in a steady state.

We know that incorrect positioning of reflecting surfaces and the presence within the enclosure of focusing elements has an adverse effect on the distribution of sound energy and results in an unevenness of the sound field in the enclosure. Measurements of the level of sound pressure at various points in the enclosure while the source is working and with a signal of fixed frequency enable us to construct spatial characteristics of the unevenness of distribution of the level of sound pressure in the enclosure at different frequencies.

In steady state conditions, with the help of a narrowly directional receiver it is possible not only to determine the variations in the field but also to establish the direction of the surface creating the unevenness.

These measurements also enable us to determine what is known as the index of diffusion of the field. Finally, in a steady state condition we can determine the frequency characteristic of the sound transmission for each point in the enclosure.

344

It should be understood that in order to eliminate the harmful action of standing-wave effects a narrow band signal, e.g. warble tone, should be used. In measurements of the spatial variation of the field and of the index of diffuseness. The variations with frequency can be measured by use of a signal of sinusoidal form.

The reverberation and steady state measurement take into account mainly the effect of one group of factors on the acoustic properties of the enclosure. To get a complete characteristic of the enclosure it is necessary to use a method which would take into account all the factors which have an influence on the acoustic conditions. This is done best by the pulse method of investigation.

By analysing the oscillogram of a decaying sound pulse, it is possible to establish the sequence of arrival at the microphone of separate reflections from the various surfaces of the enclosure. The pulse oscillogram allows us to determine not only reverberation time but also the amplitudes of the strongest reflections which modify the sound of the original signal. Moreover, oscillograms enable us to find the surfaces or the individual elements of the interior of the enclosure which are responsible for the disturbing reflections, and the places at which these reflections are most clearly distinguishable.

For the quantitative evaluation of the acoustics of an enclosure by the pulse method, use is made of the coefficient of clarity (see 2.12). The pulse method can be used on models of enclosures. This is particularly important, because it means that similar experiments can be carried out before building starts, and all the acoustic defects discovered in the model can be eliminated either in the planning process or during construction.

The signal used for pulse measurements has a square time-variation, with a duration of 10–20 msec. The length of the pulse can usually be adjusted, allowing it to be set to exceed slightly the standard reverberation time in the enclosure under test. The pulse is filled either with sinusoidal vibrations, the frequency of which may change, or by a band of statistical noise. A short electrical discharge is often used as the pulse signal.

10.2 The reverberation method of research

The reverberation method of examining the acoustic properties of an enclosure is in practice carried out by a number of devices which operate according to differing principles. There is a large group of instruments, whose common feature is that they determine reverberation time by integrating a parameter which depends on the rate of decay of the signal.

For reverberation measurements we can use instruments which measure the relative change in sound pressure consequent upon transferring a stable sound source from an enclosure with a known reverberation time into the enclosure being examined. The reverberation time can also be found by the use of an oscillograph which records the oscillogram of the decay of sound energy in the enclosure. The instrument, known as a logarithmic automatic recorder, can be used for the same purpose.

Integrating instruments

Among the instruments of the integrating variety is the ballistic reverberometer which uses an RC chain as the integrating device.

If during the decay of sound pressure in the enclosure, a rectified voltage is fed from the output of a microphone to an RC circuit, the condenser charges, and the charging current is:

$$i \approx \frac{U_{input}}{R} \qquad\qquad 10.1$$

A voltage U_{input} at the input to the circuit decreases with time as the sound pressure acting on the microphone changes; i.e. it is shown as:

$$U_{input} = U_0\,e^{-\delta t}$$

where U_0 is the voltage at the input at the moment the source is switched off. Consequently, the charging current, taking equation 2.60 into account, is:

$$i = \frac{U_0}{R}\,e^{-\delta t} = \frac{U_0}{R}\,e^{-(13\cdot 8/T)t} \qquad\qquad 10.2$$

It is also known that voltage across the condenser can be found from the equation:

$$U_c = \frac{Q}{C} = \frac{1}{C}\int_0^\infty i\,dt \qquad\qquad 10.3$$

After the value of i found from equation 10.2 is inserted into this equation, it takes the form:

$$U_c = \frac{U_0}{RC}\int_0^\infty e^{-(13\cdot 8/T)t}\,dt = \frac{U_0}{RC}\cdot\frac{T}{13\cdot 8} = k\frac{U_0}{RC}\,T \qquad 10.4$$

or

$$T = \frac{RC}{kU_0}\,U_c \qquad\qquad 10.5$$

Setting a time constant for RC much larger than any measured reverberation time, and keeping the pressure U_0 constant in all measurements, the reverberation time in the enclosure, as expression 10.5 shows, can be measured with the help of a voltmeter connected in parallel with the condenser. The indication of this instrument, as can be seen from equation 10.4, is proportional to the integral of the decay function of the sound pressure, or, which amounts to the same thing, to an area bounded by the axes U_c and t and the decay curve of the enclosure.

Under the conditions set out above, the indications of the instrument depend not only on decay time T, but also on the fluctuations in sound pressure showing as unevenness of the decay curve. As the fluctuations decrease as the sound decays, the positive and negative fluctuations do not compensate each other, and this reduces the accuracy of the measurements.

346

The influence of the fluctuations decreases if their value is almost independent of the signal level. This can be achieved by introducing a logarithmic device before the integrating element, thus making the delay curve almost linear.

For practical application of this method, the measuring instrument should consist of a microphone M (Fig. 10.2a), a pre-amplifier A_1, a rectifier Re, a logarithmic amplifier of constant voltage A_2, an integrating device RC and instruments N and T, the first of which enables the output voltage U_0 to be kept constant, and the second measures reverberation time. In Fig. 10.2b, beneath the elements of the circuit of the ballistic reverberometer are shown the signals produced by the decay at these stages. In the upper

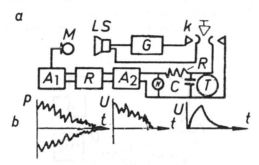

Fig. 10.2. Block diagram of a ballistic reverberometer (a) and the change in sound pressure and voltage in the various elements of the circuit.

part of Fig. 10.2a is shown the circuit of the sound-radiating apparatus, which consists of a loudspeaker LS, a multi-tone generator G and a key k, which switches on the measuring instrument T at the instant the loudspeaker is switched off.

An integrating device with an RC circuit is used at present to obtain automatically the frequency characteristic of reverberation time. For this purpose it is used in an apparatus with an automatic pen recorder. The complete instrument has been developed by the firm of Bruel and Kjaer (see end of this section).

Oscillographic instruments

Photographing the sound decay curve in an enclosure from the screen of an oscillograph tube, or direct observation of the curve if the screen has a long afterglow, enables us to make a visual judgment of the character of decay. If this curve is photographed together with a special grid, a quantitative evaluation of the reverberation process can be made.

A numerical value for reverberation time can be found with an oscillograph, between the oscillograph and the microphone is inserted an exponential amplifier (Fig. 10.3a) with an amplification of $k = B\,e^{\mu t}$. If the amplifier is switched on simultaneously with switching off the sound source, then the voltage at the Y plates of the CRT is:

$$U = kA\,e^{-\delta t} \qquad\qquad 10.6$$

or, by substituting the value of k:

$$U = AB\,e^{(\mu-\delta)t} \qquad\qquad 10.7$$

If amplification is selected so that $\mu = \delta$, this voltage is $U = A \times B$ i.e. with a linear display of the signal, instead of the decay curve appearing on the screen, there appears a comb-shaped curve, the mean value of which is constant (Fig. 10.3b). The decay is compensated by amplification, and from the value of this amplification (from the value of μ) reverberation time can be judged. It may happen that the choice of amplification may result in its being possible to keep the mean amplitude constant only for part

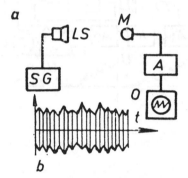

Fig. 10.3. Block diagram of an oscillograph reverberometer with an exponential amplifier A switched in to the horizontal plates of the tube.

of the oscillogram. This indicates that the sound decay curve is not exponential throughout its length, and represents, for example, the sum of exponential curves, as is observed for coupled enclosures. In this case, by selecting amplifications corresponding to each part of the oscillogram, the reverberation time for each of the component parts can be determined.

Apart from the oscillographic method using exponential amplification which has been considered above, there is a method of determining reverberation time which is known sometimes as the method of exponential display.

In this method the signal received by the microphone after the sound source has been switched off is amplified by a normal amplifier and fed to the Y plates of a cathode ray tube. The movement of the spot on the screen caused by this signal can be defined by the equation:

$$y = A\,e^{-\delta t}$$

The X plates of the tube, instead of being fed with voltage from the linear display, is fed from an exponential amplifier, as a result of which

348

the movement of the spot on the screen changes in accordance with equation

$$x = B\,e^{\mu t}$$

If we solve these equations jointly, we obtain,

$$y = \frac{A}{B} \cdot x\,e^{-(\delta-\mu)t} \qquad\qquad 10.8$$

or, where $\delta = \mu$

$$y = \frac{A}{B}x \qquad\qquad 10.9$$

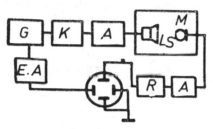

Fig. 10.4. Block diagram of an oscillograph reverberometer with exponential display: (G) square wave pulse generator, (K) electronic key, (A) amplifier, (EA) exponential amplifier, (R) rectifier.

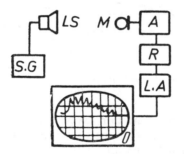

Fig. 10.5. Block diagram of an oscillograph reverberometer with a logarithmic amplifier in the Y-plate circuit of the tube.

Consequently, under these conditions the spot traces on the screen a straight line with fluctuations which depend on the evenness of the sound field (on A), from the inclination of which the reverberation time can be calculated.

Apparatus of this type is used in modern practice. In particular, Fig. 10.4 gives the block diagram of such an apparatus developed by the Bulgarian Cinema and Radio Institute[29,30] and used for measurements in enclosures and models.

An oscillogram of the decay of sound energy in an enclosure similar to that obtained by the method of exponential display is obtained if a logarithmic amplifier is inserted into the microphone chain, and the voltage

349

transformed by the amplifier is fed to the Y plates of the oscillograph tube. In this case the display should be linear. An apparatus of this type is shown schematically in Fig. 10.5.

Automatic pen recording instruments

The method used most frequently for recording the decay curve of sound pressure in an enclosure is a device which records rapidly changing levels mechanically. The amplified and rectified output of a microphone is fed to a pen recorder which records the level on waxed paper. The pen recorder is connected to the slider of a potentiometer at the output of the amplifier.

As the signal at the output decreases, the pen recorder moves, adjusting the potentiometer in such a way that the voltage at the input is slightly

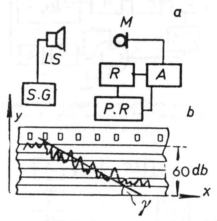

Fig. 10.6. Block diagram of a logarithmic pen recorder (a) and the record obtained by its use (b).

increased. This partial compensation takes place according to a logarithmic scale. Thus, the movement of the pen is proportional to the logarithm of the rectified voltage fed to the apparatus from the output of the microphone. When the sound source is switched off, the pen of the recorder traces on a moving sheet of paper a straight line with small fluctuations caused by the fluctuation of sound pressure in the enclosure.

From the decay curve of the enclosure shown in the lower part of Fig. 10.6, we have:

$$x = vt = y \cot \gamma \qquad \qquad 10.10$$

from whence

$$t = \frac{y}{v} \cot \gamma \qquad \qquad 10.11$$

Together with this, we can write by analogy with equation 2.90:

$$\frac{t}{T} = \frac{N}{60}, \quad \text{or} \quad T = \frac{60}{N} t \qquad \qquad 10.12$$

350

If we substitute for t in this last equation its value derived from equation 10.11, we can obtain a formula for the calculation of reverberation time:

$$T = \frac{60}{Nv} y \cot \gamma \qquad\qquad 10.13$$

where v is the speed of the paper; y is the movement of the pen when the sound source is operating; N is the signal level corresponding to this movement; γ is the angle made by the decay curve with the horizontal axis.

To find reverberation time quickly, a transparent rectangular protractor with a scale round its edge is used. The left hand side of the grid is aligned with the decay curve traced on the paper with its corner on one of the horizontal lines of the recording paper; the value of reverberation time is read from the scale where it is intersected by the line.

The circuit of a simplified apparatus for the determination of reverberation time on a pen recorder is shown in Fig. 10.6a. As the process of measuring reverberation time with an apparatus of this kind is a fairly lengthy

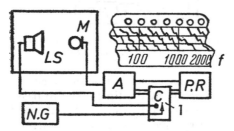

Fig. 10.7. Block diagram of apparatus for the automatic recording of decay curves. (NG) noise generator, (C) revolving commutator, (A) analyser, (PR) pen recorder.

operation, its circuit may with advantage be elaborated to enable the results of the measurements to be obtained automatically. Fig. 10.7, for instance, shows the circuit of an apparatus which allows decay curves to be recorded for different values of signal frequency.

This apparatus uses a revolving commutator C, which brings the motor of the pen recorder into motion and transmits this motion to the noise analyser A and the key 1. The key periodically switches on and off the loudspeaker which works from the noise generator NG. The revolving commutator moves a selector switch which switches the filters of the analyser in turn between the microphone and the pen recorder. Measurements are usually made using filters which pass a frequency band of one third of an octave. As the filters are changed, at the moment the loudspeaker is switched off, the pen recorder traces the decay curves as shown in the right hand upper corner of Fig. 10.7.

A slightly different apparatus is shown in Fig. 10.8 which automatically records the frequency characteristic of reverberation time.

The sound signal created by the loudspeaker is maintained at a constant level, usually by means of an amplitude limiter connected to the generator G. At the moment when the revolving commutator C switches out the loudspeaker, key 1 switches the battery to the *RC* unit until the output voltage of the microphone amplifier *A* has decreased by 40 dB.

At this moment key 2 disconnects the condenser *C* and switches the *RC* unit through a *DC* to *AC* converter to the level recorder.

The point of the pen recorder traces a curve of which the ordinates are proportional to the voltage at condenser *C*, and which, consequently, correspond to the reverberation time for a signal of the given frequency. Next, the selector switch changes the frequency of the signal, and the measurements are repeated.

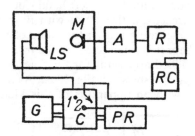

Fig. 10.8. Block diagram of apparatus for automatic recording of the frequency characteristic of reverberation time.

The scale of the frequency characteristic obtained on the paper chart is determined in the *x*-direction by the time taken to pass through the whole frequency range, and is governed by the speed of the cable drive which connects the motor of the pen recorder with the generator. The scale along the *y*-axis depends on the value of the electrical capacity and resistance of the *RC* circuit. These values are usually chosen so that a reverberation time of one second should correspond to a chosen number of divisions (2 to 4) on the recording paper.

Pen recorder instruments which are used for the measurement of reverberation time, are manufactured by the Vibrator factory in Leningrad, by the firm of Bruel and Kjaer in Copenhagen, and by others.

10.3   Studying a stationary sound field in an enclosure

The fluctuations which are observed in the decay of sound in an enclosure indicate that the sound field is not constant in time. There are a number of causes which lead to spatial and frequency irregularity in the sound field. Such irregularities prevent identical conditions of hearing for listeners at various points of the enclosure. As a result of this, the irregularity characteristic of the sound field in a steady state assumes some importance in the evaluation of the acoustic conditions.

Spatial irregularity of sound pressure level

If we have an instrument consisting of a generator *G*, a loudspeaker *LS*, a microphone *M* and a level measuring instrument *L* (Fig. 10.9), then by

352

moving the microphone from one part of the enclosure to another we can measure the levels of sound pressure and assess the spatial irregularity of the sound field. Such measurements should be made in the low, middle and high frequency ranges, and for each range measurements should be made at from five to eight frequencies differing from each other by 5–10 per cent. Averaging the results of these measurements, obtained with the use of a generator of complex waveforms (either multi-tone or noise) helps to eliminate the influence of standing waves created in the enclosure.

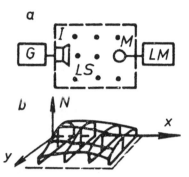

Fig. 10.9. Block diagram of equipment to display the spatial irregularity of the sound field (a) and the spatial characteristic of levels (b).

Spatial irregularity can be evaluated on the basis of the formula

$$\phi(x, y) = \sum_n N_{n_{\max}} - \sum_m N_{m_{\min}} \qquad 10.14$$

where $N_{\max}$ and $N_{\min}$ are the levels of sound pressure lying respectively above and below a certain mean level. A visual idea of the irregularly can be given by the spatial characteristics of the field constructed on the basis of measurements for each frequency. One such characteristic is shown in Fig. 10.9b.

Frequency-spatial characteristic of irregularity of the field

Each point within the enclosure can differ from each other point not merely in the level of sound pressure but also in the transmission of various frequencies. By changing the frequency of the radiated signal and recording the curve of the level of sound pressure by means of a pen recorder, we can determine any irregularity of frequency transmission for any point in the enclosure. Numerically this irregularity is defined by formula 10.14. It is clear that this irregularity is greater for enclosures which are less damped and acoustically less perfect, and this is confirmed by the curves shown in Fig. 10.10.

Establishing the frequency transmission irregularity for a number of different points in the enclosure makes it possible to locate the places where sound transmissions are best and worst received.

353

When the characteristics of frequency irregularity were being studied by means of an apparatus in which the radiated signal is in the form of a band of random noise of variable width, a variation of the irregularity with the width of the band was discovered. As the band was widened the irregularity at first was reduced, and then, after a certain width, became constant. As this limiting width is less for enclosures which are best from an acoustic point of view, it can serve as a measure of the diffuseness of the sound field.

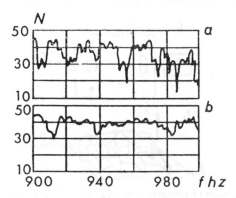

Fig. 10.10. Frequency characteristics of level in a boomy (a) and damped (b) enclosure.

Spatial distribution of the field of reflections

To obtain a measure of the diffuseness of the sound field it is important to know not only how sound energy is distributed in the enclosure but also what rays arriving at the given point cause this distribution. An even distribution of energy can be caused both by the arrival of sound energy mainly from one side, and by the equal probability of its arriving from various directions. A diffuse field in an enclosure should be uniform with respect to the directions of arrival of reflected signals at all points.

The distribution of reflections received at the given point from various directions can be characterized by a spatial diagram in which the point itself is represented by a small sphere with rays coinciding in direction with the direction of arrival of reflected signals, and having a length proportional to the sound energy arriving from the direction. Such a diagram (Fig. 10.11) enables us not only to assess the diffuseness of the field, but also to find which elements of the interior design of the enclosure are upsetting the diffuseness.

To obtain a quantitative evaluation of the diffuseness of the field, the method of statistical averaging is used. First, from the reflection distribution diagram, the mean value of the energy at the point of reception is determined:

$$\bar{A} = \frac{1}{n} \sum_i A$$

where $A_1, A_2 \ldots A_i$ are values of energy reaching the point, expressed by the length of the corresponding rays of the diagram, and n is the number of such rays. Then the absolute mean ΔA and the relative mean m of divergence are calculated:

$$\Delta A = \frac{1}{n} \sum_i |A_i - \bar{A}| \quad \text{and} \quad m = \frac{\Delta A}{\bar{A}}$$

Finally a value known as the index of diffuseness, which is defined as

$$id = \frac{m_0 - m}{m_0} \qquad 10.15$$

is found, where m_0 is the relative value of direct energy.

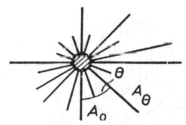

Fig. 10.11. Characteristic of the distribution of reflections received from various directions.

It is clear that in theory the index of diffuseness can range from zero, when the enclosure is so heavily damped that there are no reflections, and $m = m_0$ to one, when reflections bring in identical energy from all directions and the field is ideally diffuse.

The principle which underlies this method of investigation determines the method of solving the problem instrumentally. To construct a diagram of reflection distribution, it is necessary to have a generator of complex vibrations with a loudspeaker at its output, and a receiving instrument in the form of a highly directional microphone and a device for measuring sound pressure.

With the sound source in operation, the microphone is mounted on a turntable which has facilities for measuring the angle through which it is turned. It is first pointed towards the sound source, and then to the boundary surfaces of the enclosure. The microphone is evenly rotated and the levels of signals reaching the microphone from various angles are measured. These measurements are used to construct a diagram of the field of reflections, and to calculate the index of diffuseness.

The field may be considered sufficiently even if the index of diffuseness at various points in the enclosure is not smaller than 0·6.

The drawback of this method of investigating enclosures is that all reflections, irrespective of their delay time compared with the direct signal, are counted in the calculations, i.e. no differentiation is made between useful and harmful parts of the reflected energy.

355

Studying the diffuseness of the field by means of a directional microphone

The degree of diffuseness of the field can be measured by means of a single- or double-sided microphone and a sound source emitting a complex multi-frequency signal. Such a microphone is placed in the field of the sound source, and measurements are made of the pressure at the output of the microphone as it is revolved through various angles from 0 to 2π. The results of these measurements permit the construction of a polar directional characteristic $\Phi(\theta) = U_\theta/U_0$ which for an ideally diffuse field should be in the form of a circle. In the absence of any diffuse field (in open space) the directional characteristic from such measurements should coincide exactly with the directional characteristic of the microphone itself (Fig. 10.12).

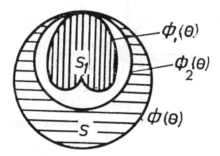

Fig. 10.12. Polar directional characteristics for a diffuse Φ (θ) and for a real Φ_2 (θ) field. Φ_1 (θ) is the directional characteristic of the measuring microphone.

Finally, in practice, in a closed enclosure, a directional characteristic so obtained is represented by a curve $\Phi_2(\theta)$ lying between the two indicated extreme characteristics. It is clear that the less diffuse the sound field, the more nearly will the actual directional characteristic coincide with the microphone's own directional characteristic, and the greater the area lying between a circle and the characteristic obtained in the given enclosure. Thus, this area can serve to represent the diffuseness of the sound field.

As a measure of the diffuseness of the field at the microphone, V. V. Furduev and Chen-Tun[85] who developed this method of measurement, proposed that the following value should be used:

$$d = \frac{S_2 - S}{S} \qquad\qquad 10.16$$

where S is the area contained between the circle and the characteristic obtained in the given enclosure (shaded in Fig. 10.12) and $S_2 = S_0 - S_1$. S_0 and S_1 are the areas of a circle of unit radius and of the directional characteristic of the microphone. As follows from formula 10.16 the degree of diffuseness of the field can vary from zero (where $S = S_2$) to one, (where $S = 0$).

356

Fig. 10.13 shows the results of the measurements carried out by the method under consideration. The basic directional characteristics (Fig. 10.13c) were made in the enclosure (Studio 2 of the State Radio) using a cardioid microphone. Then, using a planimeter, the areas which define the diffuseness of the field were measured and the measurements and calculations were entered on the graph shown in Fig. 10.13a. The continuous curve corresponds to a case of reduced total absorption, and the broken curve was obtained when additional sound absorption was produced. Fig. 10.13b shows the frequency characteristics of reverberation corresponding to these two cases.

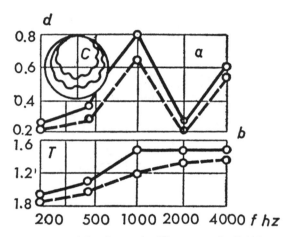

Fig. 10.13. Frequency curves of the change in diffuseness of a field in an enclosure (a) and of change in reverberation time (b) (c) diagrams of direction obtained for each case.

A consideration of the two latter pairs of curves makes it plain that the degree of diffuseness of the field depends on reverberation time. However, at a frequency of 2000 Hz the simple connection between these values breaks down and the diffuseness of the field falls as a result of the influence of other factors.

10.4 The pulse method of investigation

While the reverberation method of investigation allows us to establish the time sequence of the arrival of reflected signals at any given point, and investigation of a steady sound state makes it possible to discover the direction of the signals and the energy they are carrying, the pulse method allows both sets of information to be obtained simultaneously. This method reveals the dependence of the acoustic conditions in enclosures both on the volume of the enclosure and the extent to which it is damped, and on the shape of the enclosure, the form and position of the reflecting surfaces, the position of the sound source and of sound absorbing materials.

357

The measurement of the clarity coefficient

The diagram of apparatus used for pulse measurements is shown in Fig. 10.14. The transmitting part consists of a sound generator G, a modulator Mod, which forms square-wave pulses, reproduced after amplification via the loudspeaker LS. The receiving part of the apparatus includes an omni-directional microphone M, an amplifier A, an instrument for measuring sound pressure I, an octave filter F, and an electronic oscillograph EO. The transmitting and receiving parts are connected via a synchronizer S, which switches on the display of the oscillograph at the moment the sound pulse is emitted.

A supplementary device which is switched into the level recorder enables an automatic division to be made between the total energy of all reflections reaching the reception point with a delay of up to 50 msec and the energy

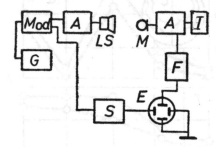

Fig. 10.14. Block diagram of a pulse measuring installation.

of reflections arriving at a delay of greater than 50 msec. This allows the clarity coefficient to be numerically defined in accordance with expression 2.98:

$$D = \frac{\int_0^{50} p^2 \, dt}{\int_0^{\infty} p^2 \, dt} \qquad\qquad 10.17$$

If we define the mean value of the clarity coefficient and the standard deviation from the formulae:

$$\bar{D} = \frac{1}{n} \sum_k D_k \quad \text{and} \quad \Delta D = \sqrt{\frac{1}{n} \sum_k (D_k - \bar{D})^2} \qquad 10.18$$

we can get an idea of the qualitative properties of the enclosure. It is clear that the smaller the scatter ΔD, the more even is the sound field. The permissible mean value of the clarity coefficient, as experiments have shown,[13] depends on the purpose for which the enclosure is intended. For good studios this value should be approximately 60 per cent; for concert halls it should have a mean value of from 50–55 per cent. As follows from

358

Fig. 10.15 which shows data of a number of measurements, the clarity coefficient does not in fact depend on the volume of the enclosure.

An impulse oscillogram obtained simultaneously with the data necessary for the determination of the clarity coefficient, gives much valuable information about the acoustics of the enclosure. From the distribution of various reflections of the impulse we can assess the delay of the reflections relative

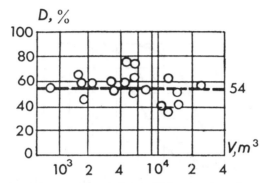

Fig. 10.15. The dependence of the clarity coefficient on the volume of the enclosure.

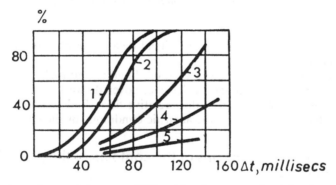

Fig. 10.16. The dependence of the perceptibility of interference on the delay time with differences in levels of the direct and delayed signals: 10dB (1); 0dB (2); − 3dB (3); − 6dB (4); − 10dB (5).

to the direct sound and establish whether they are useful or harmful. The magnitude of the reflected pulses gives an idea of their masking action, and helps in establishing the presence of an echo and in explaining the causes of its appearance.

As a criterion for deciding whether interference caused by reflections is noticeable, a critical delay time at which 50 per cent of the listeners perceive this interference is selected. If we consider the curves of the change in perceptibility of interference depending on the delay time of reflected signals, curves which were obtained by Haas[88] and shown in Fig. 10.16, we can establish that for the difference in levels between the delayed and direct signals of 10, 0, − 3 and − 6 dB, the critical delay time is respectively 60, 68, 108 and 175 msec.

These data relate to an enclosure with a reverberation time of 0·8 sec; enclosures with different reverberation times will have different critical delay times (Fig. 10.17). Critical delay time also depends on speech tempo. For slow (3·5 syllables per sec), normal (5·3 syllables per sec) and rapid

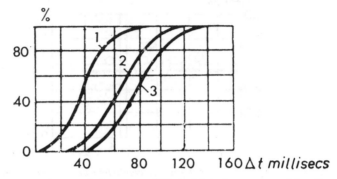

Fig. 10.17. The dependence of the perceptibility of interference on delay time for reverberation times of 0 seconds (1); 0·8 seconds (2); 1·6 seconds (3).

(7·4 syllables per sec) speech, critical delay time is respectively 90, 68 and 40 msec.

By examining the data of each oscillogram and including in our calculations the critical values we have discovered, we can find the interfering reflected signals and from their relative delay find the surfaces which have an adverse influence on the acoustic conditions of the enclosure.

Investigation by means of the contours of the perceptibility of interference

If we take into account the comparatively small influence of reverberation time on the magnitude of interference connected with the delay of reflected signals, we can use one more criterion for the evaluation of the acoustic properties of an enclosure. This criterion is a special network of curves which are known as the contours of perceptibility of interference (difficulty of perception). These contours (Fig. 10.18) define the limits at which respectively 10, 30, 50, 70 and 90 per cent of listeners detect interference at given level differences between the reflected and direct signals and at a given relative time lapse between them.

If we superimpose this grid of curves on the impulse oscillogram, provided that their scales are identical in both the horizontal and vertical axes, we can see how frequently the oscillogram goes beyond the limit of the contour of 10 per cent perceptibility of interference. If the oscillogram

360

fits below this contour then the quality of sound transmission in this enclosure can be considered good.

If only a few individual reflections exceed this contour, steps are taken towards additional acoustic treatment of the surfaces from which the indicated reflections are coming. If a significant part of the oscillograph decay curve lies above the contour, this means that it is essential to carry out a considerable acoustic refurbishing of the enclosure.

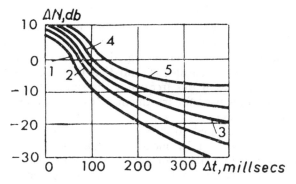

Fig. 10.18. Contours of perceptibility of interference: 1, 2, 3, 4, 5, correspond to perceptibility of 10, 30, 50, 70 and 90%.

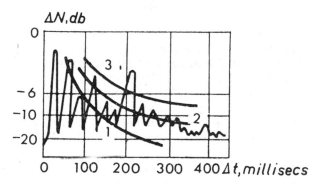

Fig. 10.19. Contours of 10, 30 and 50% perceptibility of interference (1, 2, 3) with an overlay of an oscillogram of the reflections of the sound impulse.

Fig. 10.19 shows the superimposition of the contours of perceptibility on the oscillogram. The enclosure corresponding to this example cannot be considered good acoustically as the second half of the oscillogram lies significantly above the curve of 10 per cent perceptibility. The individual reflections which exceed the contour of 50 per cent perceptibility of interference must be completely suppressed, to do which it is necessary to find the sources of these reflections.

361

Those elements of the interior design of the enclosure which result in the appearance of excessive reflections can be discovered by calculating the echogram of a series of first order reflections, carried out according to the formulae in 3.3 and by comparing this picture with the oscillogram.

The method of using models

Among the other pulse methods of investigating the acoustic properties of enclosures, the method of using models is of particular importance. It is based on the requirements for similarity between an enclosure and a model constructed to a given scale. For the investigation of a model of this kind it is necessary that the signal frequency should be increased in accordance with the scale factor, and that the materials at the bounding surfaces should have at this frequency the same properties of absorption as do the materials used for treatment of the enclosures themselves in the audio-frequency range.

The models of sound absorbing materials needed to carry out the experiments are chosen by measuring the coefficient of absorption of various materials in a model of a reverberation chamber at the ultrasonic frequencies which will be used in model tests. Moreover, because the absorption by the air is proportional to the square of the frequency, an appropriate correction must be introduced when making measurements of the coefficient of sound absorption of some model absorber.

As has been shown by investigations of a number of models of sound absorbing materials, a significant increase in the frequency of the measuring signal makes it necessary to define this correction very accurately, which is not always done. Moreover, for many such models at very high frequencies, apart from the usual losses caused by the structure of the material, there are also residual losses at the edges caused by the viscosity and heat conductivity of the material. These additional considerations can introduce significant distortions into the characteristics of an enclosure obtained through the use of models. To avoid such distortions it is recommended that the scale chosen for the model should be about 1/20, but not more than 1/40.

To carry out investigations on a model, it is necessary to have small (5–10 mm) radiators and receivers of ultrasonic vibrations with directional characteristics close to the characteristics of loudspeakers and microphones used in practice.

10.5 The correlation method of investigation

The correlation method is widely used in practice for investigating phenomena in which random influences mask the basic relationships. This method, which has a statistical basis, is intended for the study of the connections which exist between variables by defining the strength and the sign of the relationship (linear or non-linear).

362

The closeness of the connection between certain observed values of x_i and y_i is defined by use of the correlation coefficient set out in the form:

$$R_k = \frac{\frac{1}{n}\sum_i x_i y_i - \bar{X}\bar{Y}}{\sqrt{\frac{1}{n}\sum_i (x_i - \bar{x})^2}\sqrt{\frac{1}{n}\sum_i (y_i - \bar{y})^2}} \qquad 10.19$$

where $\bar{X}$ and $\bar{Y}$ are the arithmetical means of the observed values of x_i and y_i; n is the total number of observations and the two co-factors below the signs of the radicals, as follows from a comparison of them with equation 10.18, are the mean quadratic divergences.

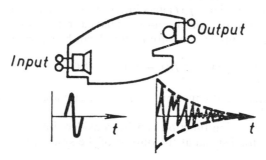

Fig. 10.20. An enclosure as a 'four-pole system' with signals at the input and the output.

The correlation coefficient lies within limits of -1 to $+1$ and the nearer it approaches to these values the stronger does the statistical link between the values of x_i and y_i become.

In investigating acoustic processes we determine the auto-correlation coefficient or the cross-correlation coefficient. The former can be used in solving problems about the coherence of signals. If the auto-correlation coefficient is zero, the signals are not coherent and the sum of their energies can be used to describe the sound field.

The correlation coefficient allows us to reveal distortions introduced via a certain 'four pole system' by comparing the processes at its output and input. As is shown in Fig. 10.20, an enclosure which is an oscillatory system can be considered as such a 'four pole system' to the input of which is fed one signal (for example an impulse), while another signal is received at the output.

In defining the cross-correlation coefficient we can either detach individual components from a complex signal or obtain the energy spectrum of this complex signal. This latter is important in particular for an analysis of the time sequence of change of a signal transmitted in an enclosed room. Thus the cross-correlation coefficient can, in its own way, be a measure of the evaluation of the acoustic properties of an enclosure.

The cross-correlation coefficient of the output and input signals by analogy with expression 10.19 is the name given to the function:

$$R_m = \frac{\overline{[f_1(t) - \overline{f_1(t)}][f_2(t) - \overline{f_2(t)}]}}{\sqrt{\overline{[f_1(t) - \overline{f_1(t)}]^2[f_2(t) - \overline{f_2(t)}]^2}}} \qquad 10.20$$

where $f_1(t)$ and $f_2(t)$ are functions which define the processes at the input and the output of the system and the line on the top replacing the plus sign indicates a mean of n observations.

In application to an enclosure $f_1(t)$, which is the input signal, represents the signal radiated by the source, while $f_2(t)$ represents the signal received at the output of the system, i.e. in the enclosure itself. If we recall that the system on which the signal $f_1(t)$ acts introduces changes due to the presence of a number of progressively more delayed reflections which are also diminishing in amplitude, the output signal can be shown in the form:

$$f_2(t) = f_1(t) + \beta_1 f_1(t - \tau_1) + \beta_2 f_1(t - \tau_2) + \dots, \qquad 10.21$$

where β_1, β_2 etc. are attenuation, and τ_1, τ_2, etc. are the times of delay of reflected signals.

The cross-correlation function can be considered as the mean power for a time T of the total signal at the output of the system, i.e. as

$$r_m(t) = \frac{1}{T}\int_{-\infty}^{t} f_1(\theta) \cdot f_2(\theta)\, \mathrm{d}\theta \qquad 10.22$$

Taking into account that, as a result of the integration process in hearing, the loudness level at a time t is determined not only by the value of the signal at that moment, but also by its earlier values (where $\theta < t$), the integration of the signal should be carried out from those values which correspond to the integrating time θ. However, if we recall that the reaction of an enclosure on a single impulse is expressed by a function of the type $e^{-(t-\theta)/T}$ (see Fig. 10.20) then the output signal $f_2(\theta)$ is represented as $f_2(\theta)\, e^{-(t-\theta)/T}$ and

$$r_m(t) = \frac{1}{T}\int_{-\infty}^{t} e^{-(t-\theta)/T} f_1(\theta) f_2(\theta)\, \mathrm{d}\theta \qquad 10.23$$

If we insert the value of $f_2(\theta)$ in accordance with equation 2.31 we can write that

$$r_m(t) = \frac{1}{T}\int_{-\infty}^{t} e^{-(t-\theta)/T} f_1^2(\theta)\, \mathrm{d}\theta + \frac{\beta_1}{T}\int_{-\infty}^{t} e^{-(t-\theta)/T} \cdot f_1(\theta)f_1(\theta - \tau_1)\, \mathrm{d}\theta +$$

$$+ \frac{\beta_2}{T}\int_{-\infty}^{t} e^{-(t-\theta)/T} f_1(\theta)f_1(\theta - \tau_2)\, \mathrm{d}\theta + \dots \qquad 10.24$$

364

In this equation the first term defines the character of change of the original signal, and all the other terms which can be written in the form of an equation:

$$r_i(t) = \frac{1}{T} \int_{-\infty}^{t} e^{-(t-\theta)/T} f_1(\theta) f_1(\theta - \tau_i)\, d\theta \qquad 10.25$$

define the auto-correlation function which changes in time, of the signal $f_1(\theta)$ for a given delay time $\tau = \tau_i$.

It follows from an analysis of expressions 10.24 and 10.25 that if the auto-correlation function $r_i(t) = 0$, the basic impulse signal in the enclosure changes according to exponential law, i.e. thanks to the elimination of the influence of accidental factors, processes in the enclosure conform to

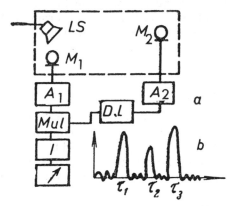

Fig. 10.21. Block diagram of a correlometer used for acoustic measurements (a), and curves of the function of mutual correlation (b).

established laws which depend upon an ideally diffuse field. If, on the other hand, the function $r_i(t)$ is close to unity, this indicates the presence of a close correlation between the original and reflected signals which can occur when the latter have a comparatively small delay time. Thus, reflected signals closely correlated with the basic signal are useful energy.

With a less close correlation when the auto-correlation function is not close enough to unity, the reflected signals, particularly if they are of high level, impair the quality of sound transmission.

The thoughts set out above enable us to find a practical method of using the correlation method for direct measurement. It is clear that it is necessary to have a sound source LS (Fig. 10.21) giving out a signal, and microphones M_1 and M_2 which receive signals $f_1(\theta)$ and $f_2(\theta - \tau_i)$. Next these signals must be multiplied by means of a special instrument AM and then passed on to an integrator at the output of which is placed a measuring instrument. By changing the delay time of the first signal by means of a delay line DL, it is possible to obtain at the output the various cross-correlation functions which have maximum values for τ, corresponding to time shifts of the reflected signals relative to the basic signals.

Thus, the correlation method enables a comparison to be made of signals reaching the microphone directly and indirectly, and the amplitudes and delay time of the reflected signals to be determined.

10.6 Measurement of the coefficient of sound absorption

The acoustic properties of materials used for the treatment of interior surfaces in the enclosure are defined by their coefficients of sound absorption. In defining acoustic conditions in an enclosure by the formulae of the statistical theory, it is necessary that these coefficients should be measured in a diffuse sound field. The reverberation method is usually used for the measurement of the diffuse coefficient of absorption. Alternatively the coefficient of the vertical incidence absorption is determined by analysis of standing waves formed in a tube.

Reverberation method

Measurements by this method are carried out in fairly large reverberation chambers whose walls are treated with reflecting materials. To create a diffuse field the walls are made non-parallel, and bands of random noise are used as a measuring symbol. Sometimes, in order to increase the diffuseness of the field in the chamber, scattering elements are introduced, made of reflecting materials.

To determine the coefficient of absorption in a chamber as accurately as possible the reverberation time is measured twice, once without the material under study and once with it in the chamber. The area of the material introduced is decided as a compromise: it should not be so large as to reduce reverberation time significantly, nor yet so small as to introduce edge effects.

On the basis of formula 2.43 the mean coefficients of absorption for the chamber after the first and second measurements respectively are:

$$\alpha_1 = \frac{0\cdot164V}{ST_1} \quad \text{and} \quad \alpha_2 = \frac{0\cdot164V}{ST_2}$$

The difference between these coefficients is expressed as:

$$\alpha_2 - \alpha_1 = \frac{0\cdot164V}{S}\left(\frac{1}{T_2} - \frac{1}{T_1}\right) \qquad 10.26$$

If α is the coefficient of the material being studied and S_0 is its area, the coefficient of absorption defined after the second measurement is:

$$\alpha_2 = \frac{\alpha_1 S + \alpha S_0 - \alpha_1 S_0}{S} \qquad 10.27$$

After this value for α_2 has been inserted in expression 10.26 it appears as:

$$\alpha_1 S + \alpha S_0 - \alpha_1 S_0 - \alpha_1 S = 0.164V \left(\frac{1}{T_2} - \frac{1}{T_1} \right)$$

From this the measured coefficient of absorption is:

$$\alpha = \frac{0.164V}{S_0} \left(\frac{1}{T_2} - \frac{1}{T_1} \right) + \alpha_1 \qquad 10.28$$

For a greater accuracy of it is necessary to take into account changes in humidity and temperature of the air during the measuring process. This may lead to errors, particularly in measurements at high frequencies. With this in mind, the coefficient of absorption which is being sought should, on the basis of formula 2.40 be determined by the equation:

$$\alpha = \frac{0.164V}{S_0} \left(\frac{1}{T_2} - \frac{1}{T_1} \right) - \frac{4V}{S_0} (m_2 - m_1) + \alpha_1 \qquad 10.29$$

where m_1 and m_2 are constants of the decay of sound in air respectively for the first and second measurements.

In using this method of measurement a number of contradictory factors which can reduce the accuracy of the measurements should be borne in mind.

To preserve a high degree of diffuseness of the field when the material being studied is introduced into the chamber, the material has to be placed on several walls of the chamber, which increases the adverse influence of diffraction phenomena (edge effects). Increases in the area of the sample with the aim of reducing the edge effect can result in a reduction of the measured reverberation time and to an increase in the error in determining the coefficient of absorption.

The standing wave method

The coefficient of absorption can be measured by using a pipe, at one end of which is placed the material being studied and at the other a sound source emitting a plane sound wave. The frequency range of the measurements is determined by the dimensions of the tube. The lower frequency must have a wavelength equal to the length of the tube and the upper frequency a wavelength of double its diameter. The ratio of the length of the tube to its diameter is therefore set at 20:1.

When the sinusoidal signal is switched on, standing waves are formed in the tubes. The amplitude sound waves reflected from the sample decreases as the coefficient of absorption of the sample increases, the sound pressure decreases at the anti-nodes and increases at the nodes.

The coefficient of absorption is determined from the formula:

$$\alpha = \frac{4R}{(R+1)^2} \qquad 10.30$$

where R is the ratio of the pressures measured at the anti-node and the node.

Measurement of these pressures is carried out by means of a small microphone or an acoustic probe. This latter is a microphone M (Fig. 10.22) situated outside the tube and moving on a trolley. Fastened to the receiving part of the microphone is a probe, the other end of which is inserted into the tube. As the microphone moves, this end passes alternately through nodes and anti-nodes of the standing wave. To determine the coefficient of absorption, it is essential to find the ratio of pressures at these points,

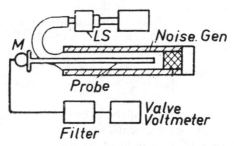

Fig. 10.22. Block diagram of an installation for measuring the coefficient of sound absorption using the standing wave method.

and this ratio is equal to the ratio of electrical voltages at the output of the microphone measured by a valve voltmeter VV for these positions of the probe.

Thus:

$$R = \frac{U_{\mathrm{max}}}{U_{\mathrm{min}}}$$

Attention should be drawn to the fact that as a result of the non-linear nature of the radiator LS, standing waves arise in the tube for all the harmonic components of the signal, and affect the accuracy of the measurements. The output of the microphone should therefore be fed to the measuring instrument through a narrow band filter F, tuned to the frequency of the generator SG.

So as to avoid the appearance of errors in the measuring process, a number of features of the measuring apparatus must be considered:

1.   The direct and reflected waves are attenuated by friction at the walls of the tube, and the attenuation increases as the distance from the material under test increases. It is therefore recommended that measurement of the pressures should be made at the maximum and minimum points nearest to the material.

2.   The presence of the probe in the tube distorts the sound field. Considering that measurement error increases as the thickness of the probe is increased, its diameter should be such that $d_1 \leqslant 0.013 \sqrt{d_2}$ where d_2 is

368

the diameter of the measuring tube. In this case measurement error will not exceed 1–2 per cent.

3.   Sound pressure at the minima is very small, and the point of minimum is itself very sharp. For these reasons measurements at the anti-nodal points should be made extremely carefully.

4.   The measuring tube is a distributed resonant system which is excited by the proximity of any of its natural modes of vibration to the frequency

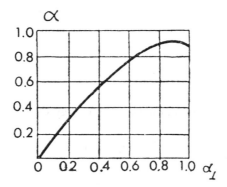

Fig. 10.23. Curve to find the random-incidence coefficient of sound absorption from the coefficient measured at vertical incidence.

of the measuring signal. The error caused by the distortion of the sound field which in this case is inevitable can be eliminated by damping the natural modes of vibration by embedding the tube in a sand box.

5.   Even the slightest air space between the material under test and the hard surface behind it makes it easier for particles of air in the pores to move. As a result of this they vibrate at a greater speed, which results in an increase in the absorption coefficient, particularly at low frequencies. Thus the testing of samples of materials should be carried out with the material at the same distance from a hard surface as it will be when it is used in practice. If the conditions of practical use are not known, it should be glued to the rear wall of the tube for the measurements.

The method just described is considerably simpler than the reverberation method, and gives sufficiently accurate results. However, the values of the coefficient of absorption obtained by this method hold true only for cases in which the sound waves impinge normally on the absorbing material. By using the graph in Fig. 10.23, plotted from formula 5.10, we can determine from the measured value of α_1 the coefficient of absorption for random incidence of the wave α.

10.7   Measurement of the natural sound insulation of partitions

Measurements of natural sound insulation of partitions are carried out both for airborne and impact sounds. They are made either directly in enclosures

369

divided by a wall, the sound insulation of which has to be measured, or under laboratory conditions, where the wall under test is placed between two contiguous, specially equipped chambers. In both cases it is important that the sound should penetrate from one enclosure to the other only through the partition under test.

Having in mind the use which can be made of the rule of analogy in acoustics, measurement of natural sound insulation can be carried out on models.

If the level of sound pressure in the enclosure where the sound source is situated is N_1, then, according to formula 6.43, the density of energy penetrating into the isolated enclosure is:

$$E_2 = \frac{E_{\text{threshold}}}{\alpha_2 S_2} \cdot S_0 10^{-0 \cdot 1 (N_1 - \tau_{db})} \qquad 10.31$$

and the noise level corresponding to this energy density is represented as:

$$N_2 = 10 \log \frac{E_2}{E_{\text{threshold}}} = 10 \log \frac{S_0}{\alpha_2 S_2} \cdot 10^{-0.1(N_1 - \tau_{db})}$$

$$= N_1 - \tau_{db} + 10 \log \frac{S_0}{\alpha_2 S_2} \qquad 10.32$$

From this the sound insulation is defined by the equation:

$$\tau_{db} = N_1 - N_2 + 10 \log \frac{S_0}{\alpha_2 S_2}$$

where S_2 and α_2 are the area of the inner surfaces and the coefficient of absorption of the enclosure; S_0 is the area of the partition being examined.

As the difference in levels $N_1 - N_2$ can be replaced by the ratio of the energies or of the square of the pressures, this last expression can be rewritten:

$$\tau_{db} = 10 \log \frac{E_1}{E_2} + 10 \log \frac{S_0}{\alpha_2 S_2} = 20 \log \frac{p_1}{p_2} + 10 \log \frac{S_0}{\alpha_2 S_2} \qquad 10.33$$

Although in the measurements of the sound insulation of partitions use is made of complex sound signals, e.g. multitone or noise, it is not possible fully to avoid variations of sound pressure in the enclosure. The sound pressures p_1 and p_2 in formula 10.33, must therefore be measured at several arbitrarily selected points, and their values replaced by root-mean square values:

$$P_{1_{\text{mean}}} = \sqrt{\frac{p_1{}^2 + p_2{}^2 + p_3{}^2 + \ldots + p_n{}^2}{n_1}} = \sqrt{\frac{1}{n_1} \sum_i p_{1_i}{}^2}$$

$$\text{and} \quad P_{2_{\text{mean}}} = \sqrt{\frac{1}{n_2} \sum_i p_{2_i}{}^2}$$

If we insert these values into equation 10.33, we obtain the formula:

$$\tau_{db} = 20 \log \frac{P_{1_{\text{mean}}}}{P_{2_{\text{mean}}}} + 10 \log \frac{S_0}{\alpha_2 S_2} + 10 \log \frac{n_2 \sum_i p_{1_i}^2}{n_1 \sum_i p_{2_i}^2} + 10 \log \frac{S_0}{\alpha_2 S_2} \quad 10.34$$

where p_{1_i} and p_{2_i} are sound pressures measured at various points in the first and second enclosures respectively; n_1 and n_2 are the number of such points.

Thus sound insulation is measured in an indirect way by direct measurement at the output of the microphone of voltages proportional to sound pressures at the points at which it is placed in front of the partition and behind it.

Equipment for measurements of this kind (Fig. 10.24) consists, in the transmitting part, of a noise generator G, a filter F, an amplifier A and a

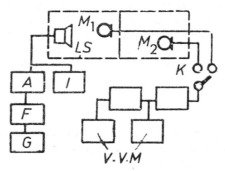

Fig. 10.24. Block diagram of an installation for measurements of sound insulation.

loudspeaker LS. The receiving part of the equipment consists of two identical microphones m_1 and m_2, a key k and a valve voltmeter VV, whose indications are proportional to the sound pressures.

Measurements are carried out in the following way.

The sound source is switched on in the first enclosure and is maintained at constant power by the device I. Measurement of the pressures at the outputs of the first and second microphones is carried out with the microphones at various points in enclosures 1 and 2. The number of points where the microphones are placed depends on the degree of accuracy required of the measurements, and can be found by comparing two or three calculations in sequence for an increasing number of microphone positions. These measurements are carried out first for one frequency band, and are then repeated for a number of other bands from 100 to 3200 Hz. For a complete calculation from formula 10.34 it is necessary to know the value of $\alpha_2 S_2$, which is found by measuring the reverberation time of the second enclosure and by calculating this value as $\alpha_2 S_2 = 0 \cdot 164(V_2/T_2)$.

In determining sound insulation properties under laboratory conditions, the area S_0 of the sample under test, and the volumes of both chambers,

371

remain constant. The interior surfaces of the chambers are treated with good reflecting materials, allowing a sufficiently diffuse sound field to be obtained. As the volumes of the chambers are comparatively small (40–100m³), we conclude on the basis of the wave theory that the sound fields in them can be considered diffuse only for the high frequency region of the range indicated above. So the number of points at which measurements should be made can be reduced for high frequencies. Measurements are still further simplified by the fact that in laboratory conditions the second part of formula 10.34 is a constant value.

When scale-models are used to determine the sound insulation properties of materials and constructions, in accordance with the rule for similarity the frequencies of the measuring signals are increased in the same scale as the rooms are reduced. Radiators and receivers must be small, and the other elements of the installation remain as before. The scale of models is usually 1:5, which reduces the volume of the chambers to 0·5–1m³.

In measuring sound insulation from impact noise, a standard impact-producing equipment is used as the source of vibration.

Levels of the impact sound behind the material under test are measured over a frequency range of one octave width by means of a microphone and a filter. The level of the impact sound penetrating into the enclosure may be shown to be:

$$N = N_1 - 10 \log \frac{A_0}{A}$$

where N_1 is the measured value of the mean level of sound pressure; A_0 is the total absorption in the transmitting chamber and A is the total absorption measured in the receiving chamber.

As the correlation method allows an amplitude comparison of the direct and indirect signals reaching a given point, it can also be used for the measurement of sound insulation. In this case part of the signal $f_1(\theta)$ going to the loudspeaker is fed via the delay line to one input of the correlator, while the second input receives the signal $f_2(\theta - \tau)$ from the microphone placed behind the material under test. The correlation functions at the output of the correlator give the relative value of amplitudes of the signals in front of the partition and behind it, from which its sound insulation is determined.

It is necessary to pay special attention to the accuracy of the measurement of sound insulation in view of the fact that an error of 1 dB in its calculation results in the need to increase the weight of the barrier by 15–20 per cent. Thus, accuracy of measurements in this case have an important economic basis.

Appendix 1

TABLE OF COEFFICIENTS OF ABSORPTION FOR SOME
MATERIALS AND CONSTRUCTIONS

Name of material	Coefficients of sound absorption at frequencies					
	125	250	500	1000	2000	4000
BUILDING MATERIALS						
1. Smooth concrete wall un-painted	0·01	0·012	0·015	0·019	0·023	0·035
2. Unplastered brick wall	0·024	0·025	0·032	0·042	0·049	0·070
3. Gypsum plaster smooth on a brick wall, unpainted	0·012	0·013	0·017	0·02	0·023	0·025
4. Smooth alabaster plaster on wooden battens	0·020	0·022	0·032	0·039	0·039	0·028
5. Rough lime plaster on wooden battens	0·025	0·045	0·060	0·085	0·043	0·058
6. Smooth lime plaster on wooden battens	0·024	0·027	0·030	0·037	0·019	0·034
7. Dry plaster tiles	0·02	0·05	0·06	0·08	0·04	0·06
8. Pine panel 19 mm thick	0·098	0·110	0·061	0·081	0·082	0·110
9. Plywood at 5 cm from wall	0·18	0·26	0·24	0·1	0·1	0·1
10. Plywood lying flat on wall	0·05	0·06	0·06	0·1	0·1	0·1
11. Parquet on asphalt	0·05	0·03	0·06	0·09	0·10	0·22
12. Lino 5 mm thick on firm base	0·02	0·025	0·03	0·035	0·04	0·04
13. Window glass	0·035	—	0·027	—	0·02	—
14. Mirror glass	0·035	0·025	0·019	0·012	0·07	0·04
DRAPES AND CARPETS						
1. Cotton cloth 360 g/m²	0·03	0·04	0·11	0·17	0·24	0·35
2. Cotton cloth 500 g/m²	0·04	0·07	0·13	0·22	0·33	0·35

Name of material	Coefficients of sound absorption at frequencies					
	125	250	500	1000	2000	4000
3. Velvet cloth 650 g/m²	0·05	0·12	0·35	0·45	0·38	0·36
4. As above at 10 cm from wall	0·06	0·27	0·44	0·50	0·40	0·35
5. As above at 20 cm from wall	0·08	0·29	0·44	0·50	0·40	0·35
6. Cotton cloth 500 g/m², draping ⅞ area	0·03	0·12	0·15	0·27	0·37	0·42
7. As above draping ¾ area	0·04	0·23	0·40	0·57	0·53	0·40
8. As above draping ½ area	0·07	0·37	0·49	0·81	0·65	0·54
9. Carpet 1 cm thick with pile on concrete	0·09	0·08	0·21	0·27	0·27	0·37
10. The same 0·3 cm thick on felt on concrete	0·11	0·14	0·37	0·43	0·27	0·25
11. As above on felt with pine board 2 cm lying on concrete	0·11	0·13	0·28	0·45	0·29	0·29
12. Rubber carpet 0·5 cm thick	0·04	0·04	0·08	0·12	0·13	0·10

SPECIAL MATERIALS AND CONSTRUCTIONS

Name of material	125	250	500	1000	2000	4000
1. Perforated arborite 2·5 cm thick, volume weight 150 kg/m³	—	—	0·73	0·72	0·84	0·80
2. Asbestite 3·5 cm thick 570 kg/m³	0·32	0·40	0·36	0·35	0·37	0·35
3. Asbosilicate 3·5 cm thick 250 kg/m³	0·60	0·73	0·80	0·82	0·81	0·70
4. Acoustic foam concrete 3 cm thick 940 kg/m³	—	—	0·39	0·42	0·42	0·39
5. Hair felt 25 mm	0·18	0·36	0·71	0·79	0·82	0·85
6. Same with a 50 mm air gap	0·35	0·62	0·88	0·92	0·78	0·84
7. Wool felt 2·5 cm	0·09	0·34	0·55	0·66	0·52	0·39
8. Mineral wool 15 cm	0·47	0·53	0·60	0·62	0·58	0·56
9. Mineral wool with fine wood wool 2·5 cm thick, 6·5 kg/m³	0·10	0·27	0·50	0·68	0·56	0·48
10. Same with perforated metal covering 10 apertures 1·5 mm diameter to 1 cm²	0·09	0·25	0·48	0·66	0·57	0·47
11. Mineral wool 35 mm thick covered by perforated acoustic plaster	0·28	0·37	0·40	0·38	0·39	—
12. Mineral wool 40 mm in calico cushion	0·32	0·40	0·53	0·55	0·61	0·66
13. Mineral wool 60 mm thick with perforated wood-fibre screen: h—4 mm, d—3·5 mm, D—15 mm, m—0·017 g/cm²	0·31	0·50	0·70	0·41	0·29	—
14. Mineral wool 60 mm thick with perforated wood-fibre screen D—25 mm, m—0·046 g/cm²	0·31	0·55	0·46	0·21	0·14	—

Name of material	Coefficients of sound absorption at frequencies					
	125	250	500	1000	2000	4000
15. Glass wool 9 cm	0·32	0·4	0·51	0·6	0·65	0·6
16. Glass wool 3 cm, lying close to wall	0·1	0·15	0·45	0·55	0·6	0·6
17. Glass wool 50 mm with perforated hardboard screen h—4 mm, d—3·5 mm, D—15 mm, m—0·017 g/cm^2	0·30	0·50	0·77	0·48	0·27	—
18. Same with screen D—25 mm, 0·32, m—0·046 g/cm^2	0·32	0·55	0·55	0·21	0·12	—
19. Bekesy's screens	0·8	0·81	0·73	0·58	0·46	0·45
20. Plywood semi-cylindrical panel; chord d—50 cm, height of section h—20 cm, b—35, 25, 20, 15, 20, 25, 35, 25 cm	0·25	0·3	0·33	0·22	0·2	0·21
21. The same, d—70 cm, h—25 cm, b—40, 30, 25, 20, 25, 30 . . . cm	0·32	0·35	0·3	0·25	0·2	0·23
22. The same d—90 cm, h—30 cm, b—45, 35, 25, 20, 25 . . . cm	0·37	0·35	0·32	0·28	0·22	0·22
23. The same d—115 cm, h—40 cm, b—50, 40, 30, 20, 30 . . . cm	0·41	0·4	0·33	0·25	0·2	0·22
24. As in 20 but with filling of mineral wool	0·3	0·42	0·35	0·23	0·19	0·2
25. As in 21 with filling of mineral wool	0·35	0·5	0·38	0·3	0·22	0·18
26. Woodfibre tiles 25 mm tight to wall	0·47	0·52	0·50	0·55	0·58	0·63
27. Woodfibre tiles in two layers, each layer 25 mm thick, tight to wall	0·33	0·42	0·47	0·41	0·42	0·36
28. Woodfibre tiles 12 mm thick tight to wall	0·06	0·15	0·28	0·30	0·33	0·31
29. The same 3 cm from wall	0·14	0·25	0·29	0·31	0·41	0·42
30. The same 5 cm from wall	0·22	0·30	0·34	0·32	0·41	0·42
31. Pemzolit tiles tight to wall	0·07	0·11	0·17	0·36	0·44	0·38
32. The same 5 cm from wall	0·12	0·33	0·43	0·41	0·34	0·42
33. Glass foam with unenclosed pores	0·1	0·36	0·38	0·36	0·45	0·55
34. Woodshaving tiles 20 mm (620 600 kg/m^3) unpainted 100 mm from wall	0·14	0·09	0·1	0·08	0·01	0·15
35. Acoustic fibrolite 35 mm thick (300 kg/m^3) unpainted (tiles of wood wool with cement) flat to wall	0·06	0·16	0·25	0·38	0·59	0·63
100 mm off wall	0·08	0·27	0·46	0·35	0·54	0·60
150 mm off wall	0·13	0·42	0·53	0·35	0·53	0·63

Name of material	Coefficients of sound absorption at frequencies					
	125	250	500	1000	2000	4000

36. Woodfibre tiles with slit perforations 18 mm thick (650 kg/m³). Dimensions of slits 2 times 56 mm, distance between slits 15 mm, and between rows 20 mm

	125	250	500	1000	2000	4000
flat to wall	0·03	0·20	0·56	0·78	0·66	0·39
50 mm off wall	0·15	0·42	0·64	0·80	0·65	0·31
100 mm off wall	0·27	0·53	0·57	0·74	0·67	0·39

37. Unperforated plastic 2 mm thick

	125	250	500	1000	2000	4000
50 mm off wall	0·62	0·27	0·12	0·05	—	—
50 mm off wall filled with PP-80	0·47	0·45	0·18	0·09	—	—
100 mm off wall	0·26	0·26	0·07	0·01	0·01	—
100 mm off wall, filled with PP-80	0·76	0·47	0·20	0·14	0·02	—
150 mm off wall	0·41	0·23	0·1	0·02	—	—
200 mm off wall, filled with 50 mm of PP-80, 150 mm air space	0·52	0·38	0·22	0·14	—	—

(Translator's note—there is no indication in the book as to what PP-80 is, save by implication that it is a form of mineral wool).

38. Mineral wool tiles PP-80, 25–30 mm thick.

	125	250	500	1000	2000	4000
flat to wall	0·08	0·4	0·64	0·89	0·95	0·81
50 mm off wall	0·21	0·47	0·72	0·98	0·97	0·79
two layers flat to wall	0·18	0·6	0·98	0·95	0·94	0·82

39. PP-80 tiles 50 mm thick

	125	250	500	1000	2000	4000
flat to wall	0·14	0·52	0·9	0·99	0·92	0·82
50 mm off wall	0·2	0·61	0·98	0·94	0·92	0·78
100 mm off wall	0·38	0·8	0·94	0·88	0·86	0·79

40. PP-80 tiles 100 mm thick

	125	250	500	1000	2000	4000
flat to wall	0·5	0·92	0·98	0·95	0·91	0·8
100 mm off wall	0·62	0·97	0·98	0·97	0·94	0·81

41. Foam plastic 50 mm thick

	125	250	500	1000	2000	4000
50 mm off wall, enclosed air space	0·14	0·28	0·18	0·09	0·12	0·25
50 mm off wall, filled with PP-80 50 mm thick	0·59	0·34	0·23	0·12	0·11	0·08
100 mm off wall, unenclosed air space	0·36	0·26	0·16	0·08	0·13	0·27

42. Panel of aceid 8 mm thick (asbocement tiles GOST 4248-52)

	125	250	500	1000	2000	4000
flat to wall	0·03	0·03	0·08	0·09	0·08	0·03
50 mm off wall	0·15	0·19	0·12	0·05	0·05	0·03
100 mm off wall	0·32	0·21	0·16	0·09	0·06	0·03
200 mm off wall	0·24	0·14	0·08	0·08	0·06	0·05
100 mm off wall, filler 100 mm	0·38	0·28	0·21	0·1	0·05	0·04

Name of material	Coefficients of sound absorption at frequencies					
	125	250	500	1000	2000	4000

43. Acoustic tiles PAS—nature of material unexplained.

44. Acoustic tiles PAO—nature of material unexplained.

45. Perforated plywood 4 mm thick, d—5 mm, D—35 mm

	125	250	500	1000	2000	4000
50 mm off wall, filler PP-80 50 mm thick	0·19	0·90	0·76	0·28	0·15	0·1
100 mm off wall, filler PP-80 50 mm thick	0·25	0·96	0·66	0·26	0·16	0·1
160 mm off wall, filler PP-80 100 mm thick	0·39	0·87	0·58	0·33	0·15	0·1
100 mm off wall, filler PP-80 100 mm thick	0·65	0·9	0·64	0·28	0·15	0·12
200 mm off wall, filler PP-80 100 mm thick	0·78	0·98	0·68	0·27	0·16	0·12
50 mm off wall, layer of mineral wadding 50 mm thick	0·23	0·9	0·72	0·35	0·18	0·15

46. Perforated plywood 4 mm thick, d—5 mm, D—65 mm

	125	250	500	1000	2000	4000
50 mm off wall, filler PP-80 50 mm thick	0·26	0·88	0·38	0·29	0·12	0·1
100 mm off wall, filler PP-80 50 mm thick	0·43	0·69	0·33	0·17	0·1	0·1
150 mm off wall, filler PP-80 100 mm thick	0·51	0·6	0·33	0·15	0·1	0·1
200 mm off wall, filler PP-80 100 mm thick	0·86	0·68	0·39	0·17	0·13	0·1

47. Perforated plywood 4 mm thick, d—5 mm, D—100 mm

	125	250	500	1000	2000	4000
50 mm off wall	0·06	0·42	0·2	0·07	0·07	0·06
50 mm off wall, filler PP-80 50 mm thick	0·37	0·68	0·31	0·15	0·1	0·09
100 mm off wall, filler PP-80 50 mm thick	0·4	0·7	0·3	0·12	0·1	0·05
100 mm off wall, filler PP-80 100 mm thick	0·77	0·64	0·3	0·15	0·15	0·1
200 mm off wall, filler PP-80 100 mm thick	0·8	0·58	0·27	0·14	0·12	0·1

48. Perforated plywood 4 mm thick, d—4 mm, D—40 mm

	125	250	500	1000	2000	4000
50 mm off wall	0·06	0·22	0·31	0·12	0·1	0·08
50 mm off wall, filler PP-80 50 mm thick	0·31	0·99	0·51	0·14	0·14	0·1
100 mm off wall, filler PP-80 50 mm thick	0·4	0·84	0·4	0·16	0·14	0·12
100 mm off wall, filler PP-80 100 mm thick	0·68	0·88	0·41	0·17	0·14	0·1
200 mm off wall, filler PP-80 100 mm thick	0·98	0·88	0·52	0·21	0·16	0·14

Name of material	Coefficients of sound absorption at frequencies					
	125	250	500	1000	2000	4000
49. Perforated plywood 4 mm thick, d—20 mm, D—60 mm						
50 mm off wall	0·06	0·08	0·07	0·16	0·1	0·08
50 mm off wall, filler PP-80 50 mm thick	0·23	0·85	0·99	0·54	0·31	0·28
100 mm off wall, filler PP-80 50 mm thick	0·27	0·84	0·96	0·36	0·32	0·26
100 mm off wall, filler PP-80 100 mm thick	0·50	0·89	0·97	0·6	0·32	0·26
200 mm off wall, filler PP-80 100 mm thick	0·78	0·98	0·95	0·53	0·32	0·27
50. Perforated plastic, d—5 mm, D—35 mm, with an overlay of glass-fibre fabric						
100 mm off wall	0·05	0·34	0·35	0·14	0·05	—
150 mm off wall	0·15	0·42	0·32	0·11	0·06	—
50 mm off wall, filler PP-80 50 mm thick	0·14	0·77	0·9	0·4	0·19	0·14
100 mm off wall, filler PP-80 50 mm thick	0·28	0·95	0·95	0·38	0·14	0·14
150 mm off wall, filler PP-80 50 mm thick	0·42	0·96	0·77	0·43	0·16	0·16
100 mm off wall, filler PP-80 100 mm thick	0·66	0·96	0·88	0·47	0·2	0·16
51. Perforated plastic, d—5 mm, D—65 mm, filler PP-80						
50 mm off wall	0·18	0·78	0·44	0·2	0·05	—
100 mm off wall	0·58	0·74	0·45	0·28	0·02	—
100 mm off wall, air layer 50 mm	0·23	0·71	0·39	0·22	0·05	—
150 mm off wall, air layer 100 mm	0·53	0·64	0·4	0·08	0·05	—
52. Perforated plastic, d—5 mm, D—35 mm, with double layer of gauze.						
50 mm off wall	—	0·09	0·36	0·22	0·04	0·02
150 mm off wall	0·06	0·47	0·3	0·11	0·04	—
200 mm off wall with overlay of cotton material r—59 g/sec	0·14	0·47	0·24	0·16	0·03	—
50 mm off wall	0·05	0·16	0·55	0·31	0·03	0·07
100 mm off wall	0·07	0·37	0·45	0·26	0·12	0·08
53. Perforated plastic, d—5 mm, D—65 mm, with two layers of gauze.						
50 mm off wall	0·06	0·32	0·35	0·12	0·07	—
150 mm off wall	0·22	0·41	0·24	0·13	0·07	—
200 mm off wall	0·25	0·40	0·19	0·12	0·07	—
with one layer of cotton material						

Name of material	Coefficients of sound absorption at frequencies					
	125	250	500	1000	2000	4000
50 mm off wall	0·04	0·33	0·4	0·16	0·05	0·03
100 mm off wall	0·07	0·47	0·31	0·12	0·06	0·04
54. Semicylindrical constructions chord 1000 mm, height of section 300 mm						
50 mm off wall	0·41	0·3	0·35	0·16	0·1	0·14
50 mm off wall, filler thickness 50 mm	0·49	0·44	0·39	0·19	0·13	0·22
chord 700 mm, height of section 200 mm						
50 mm off wall	0·51	0·21	0·32	0·18	0·08	0·13
50 mm off wall, filler thickness 50 mm	0·5	0·40	0·44	0·33	0·13	0·16
55. Corrugated constructions						
25 mm off wall	0·2	0·25	0·11	0·05	0·11	0·1
50 mm off wall	0·2	0·28	0·15	0·12	0·15	0·16
On woodshaving tiles 20 mm thick	0·29	0·28	0·17	0·13	0·15	0·17
56. VPE 100-3 (4)	0·32	0·35	0·19	0·13	0·11	0·10
57. VPE 100-10	0·34	0·19	0·10	0·09	0·12	0·11
58. SS-90·35	0·26	0·32	0·31	0·28	0·27	0·27
59. SS-70.32	0·30	0·34	0·35	0·32	0·28	0·26
60. SS-113.40	0·35	0·29	0·26	0·11	0·08	0·07
61. SS-50.23	0·30	0·30	0·32	0·30	0·27	0·25
62. SS-113.40—70.32	0·38	0·28	0·30	0·21	0·16	0·12
63. SS-113.40—50.25	0·32	0·32	0·31	0·32	0·13	0·12
64. PFC 100-4.4.15	0·48	0·74	0·66	0·68	0·50	0·42
65. PFC 100-4.4.20	0·75	0·80	0·65	0·53	0·39	0·26
66. PFC 100-4.4.40	0·47	0·47	0·36	0·28	0·25	0·27
67. PFC 100-4.6.25	0·52	0·54	0·54	0·50	0·41	0·33
68. PFC 100-4.7.30	0·45	0·51	0·55	0·48	0·34	0·21
69. PFC 50-4.6.15	0·17	0·42	0·36	0·25	0·15	0·13
70. PFC 50-4.6.25	0·20	0·46	0·58	0·52	0·42	0·31
71. PFC 50-7.7.30	0·19	0·36	0·45	0·43	0·30	0·24
72. PFCG 50-4.4.40	0·28	0·27	0·30	0 25	0·16	0·15
73. PFCS 100-45.2—10.20	0·17	0·20	0·26	0·24	0·27	0·26
74. PFCS 50-45.2—10.20	0·21	0·35	0·40	0·43	0·42	0·32
75. PFCSC 100-45.2—10.20	0·10	0·19	0·28	0·26	0·34	0·23
76. PCEC 100-4.6.25	0·19	0·24	0·25	0·22	0·18	0·18
77. PCESC 100-45.2—10.20	0·18	0·33	0·36	0·36	0·35	0·33

ABSORPTION OF OBJECTS

	125	250	500	1000	2000	4000
1. Chair with hard seat and back	0·02	0·02	0·03	0·035	0·038	0·038
2. Chair with soft seat and back	0·09	0·12	0·14	0·16	0·15	0·16
3. A listener	0·36	0·43	0·47	0·44	0·49	0·49

Note: the abbreviations for special sound absorbent materials mean the following:

VPE—veneered plywood, empty. 100 is distance from wall in millimetres, 3 (4) is thickness of veneer in mm.

SS—semicylindrical surface, 90 length of chord, 35 height of section.

PFC—Perforated filled construction.

PFCG—as above with gypsum plaster coating.

PFCS—as above with slit perforations.

PFCSC—as above with slit perforations and a glued on cloth overlay.

PCEC—Perforated construction, empty, with cloth overlay.

PCESC—as above, with slit perforations, empty, with cloth overlay.

For all perforated constructions 100 or 50 thickness of filling layer, 4.4.20 in order—thickness of veneer, diameter of circular perforation, and distance between perforations.

Appendix 2

Sound source	Distance from source in metres	Noise level, dB.
Heavy lorry transport, tramway	10	83
Light goods transport	10	67
Heavy traffic	3–5	88
Traffic in quiet street	10	60
Powerful street loudspeaker	10	85–90
Ambulance siren	15	90
Fire engine siren	10	100
Steam engine whistle	10–15	100–105
Heavy aircraft landing and taking off	—	120
Train in motion	—	110
Factory siren	20	105
Factory hooter	20	95
Hammer striking a steel sheet	5	110
Boilermaking shop	—	up to 100
Forge	—	up to 100
Loud conversation	5	70–80
Normal conversation	5	60–70
Typewriter	—	72
Door slam	5	75
Symphony orchestra	—	90
Piano	10	60–80
Chamber music	—	70
People passing in corridor	—	70–80
Office	—	70
Noisy vestibule	—	85
Ventilation noise	—	72
Movement on a staircase	—	62
Large shop or restaurant	—	61
Dwelling house with radio on quietly	—	50
Normal office	—	40
Living rooms	—	30–40
Libraries	—	25
Radio studio or film studio	—	25

Appendix 3

TABLE OF MEAN VALUES FOR SOUND INSULATION OF SOME WALLS AND CONSTRUCTIONS

Name of the material or construction	Thickness of the construction in cm	Weight kg/m²	Sound insulation dB
Hair felt, single layer	2·5	3·66	6
Same in two layers each 1·5 cm	3	5·65	9
Same in three layers	4·5	8·51	13
Same in four layers	6	10·3	17
Blanket of mineral wool covered by paper on both sides	1·3	1·5	16
Compressed card	0·5	3	16
Multi-ply card	2·0	12·0	20
Asbestos card	0·25	2·25	18
Solid pine board	3·0	19·5	12
Same, oak	4·5	33·5	27
Three-ply	0·32	2·54	19
Same	0·64	3·56	21
Sheet iron	0·2	15·6	33
Mirror glass	0·63	17·5	30
The same, 0·63 double with a space of 3·8 cm	—	—	40
The same with a gap of 19 cm	—	—	45
The same with a gap of 40 cm	—	—	48
Strawboard 9 cm thick, plastered on both sides	12	72	39
Clinker plaster wall panels 2 by 5 cm	13	120	40
Pumice concrete wall panels 2 by 6 cm	15	135	40
The same, 2 by 8·5 cm	20	185	43
Wall of pumice concrete	14	150	42
The same	23	250	50
The same of clinker filled concrete	14	150	42

Name of the material or construction	Thickness of the construction in cm	Weight kg/m²	Sound insulation dB
The same of ferroconcrete	10	240	49
The same of hollow concrete blocks	19	190	43
The same	29	270	50
Wall of brickwork half brick thick, without plaster	12	204	48
The same, one brick thick	25	425	53
The same, one and a half bricks thick	38	646	56
The same, two bricks thick	52	884	58
The same, two and a half bricks thick	64	1088	59
The same, three bricks thick	77	1340	63
The same, three and a half bricks thick	90	1560	65
A single partition of boards 2 cm thick, plastered on both sides	6	70	37
Single partition of boards 2·5 cm thick, plastered on both sides on felt	7	76	39
Double partition of beams 10 cm thick, covered on both sides in boards 2·5 cm thick and plastered on both sides	18	95	45
The same plastered on felt	18	96	47
Double partition of plywood sheets 3 mm thick with a 3 mm gap, with a gap of 2·5 mm filled with wadding	3	8	26
The same with a 5 mm gap	5·5	12	29
The same with a 6·5 mm gap	7	14	34
A special acoustic triple glazed window	—	—	50
Double window, well sealed, closed	—	—	25
Single window, badly closed	—	—	8·5
Ordinary door with a panel of 2·5 cm boards (two panels) with a frame 4·5 cm thick	—	—	18
Door with a frame 2·5 cm thick, and panels of 3 mm plywood	—	—	10
Door overpanelled in plywood, 90 by 200 cm	—	—	22
Heavy oak door 90 by 210 cm, tightly fitting	—	—	25
Double brick wall with an air gap	—	—	75

Appendix 4

Type of enclosure	Noise level, dB
Public offices	40
Theatres, auditoria	30
Hospitals	25
Cinema projection booths	20
Laboratories	45
Cinema auditoria	30–40
Sync. shooting stages	25–30
Music recording stages	20–25
Speech recording stages	25
Sound recording control rooms where aural control is exercised	40
Large or small television studio	30
Television presentation studio	30
Radio concert studio	25
Radio chamber music studios	25
Talks studios	25
Rehearsal rooms	40

Appendix 5

Type of studio	Floor area m³	Height m	Volume m³	Number of performers
Television talks studio	12–15	2·8–3·2	34–48	1–2
Film speech studios and radio studios	15–25	3·2–3·5	48–90	1–2
TV presentation studios	50–80	4–5	200–400	10–15
Radio chamber music studio	50–80	4–4·5	200–360	10–15
Small radio concert studio	150–200	6–7	900–1400	25–40
Small music film studio	150	6·5–7	1000	25–30
Orchestra recording studio	450–600	9·0–9·5	4000–5700	75–100
Large radio concert studio	400–450	10–11	4000–5000	115–140
Large studio for recording orchestra and choir	720	10	7200	120
Small TV studio	100	5–6	600	20
	150	6	900	35
	200	7	1400	50
Large TV studio	300	8	2400	100
	450	10	4500	200
	600	11	6600	250
	1000	15	15000	400
Sync. shooting stage	500–2000	10–18	5000–35000	—

Appendix 6

TABLE TO FIND THE VALUES OF $-\ln(1-\alpha)$ FROM THE VALUES OF THE MEAN
COEFFICIENT OF SOUND ABSORPTION

$-\ln(1-\alpha)$	α	$-\ln(1-\alpha)$	α	$-\ln(1-\alpha)$	α
0·01	0·01	0·26	0·229	0·50	0·393
0·02	0·02	0·27	0·237	0·51	0·400
0·03	0·03	0·28	0·244	0·52	0·406
0·04	0·039	0·29	0·252	0·53	0·413
0·05	0·049	0·30	0·259	0·54	0·418
0·06	0·058	0·31	0·267	0·55	0·423
0·07	0·068	0·32	0·274	0·56	0·430
0·08	0·077	0·33	0·281	0·57	0·435
0·09	0·086	0·34	0·288	0·58	0·440
0·10	0·095	0·35	0·295	0·59	0·447
0·11	0·104	0·36	0·302	0·60	0·452
0·12	0·113	0·37	0·309	0·61	0·457
0·13	0·122	0·38	0·316	0·62	0·462
0·14	0·131	0·39	0·323	0·63	0·468
0·15	0·139	0·40	0·330	0·64	0·473
0·16	0·148	0·41	0·336	0·65	0·479
0·17	0·156	0·42	0·343	0·66	0·484
0·18	0·165	0·43	0·349	0·67	0·488
0·19	0·173	0·44	0·356	0·68	0·494
0·20	0·181	0·45	0·362	0·69	0·499
0·22	0·197	0·46	0·369	0·70	0·505
0·23	0·206	0·47	0·375		
0·24	0·213	0·48	0·381		
0·25	0·221	0·49	0·387		

Bibliography

BIBLIOGRAPHY

1. Andreev, N. N. and Lyamshev, L. M. (editors). *Problemy sovremennoi akustiki* (Problems of modern acoustics), pub. by Academy of Sciences of the USSR, 1962.
2. Alekseev, S. P., Vorobyov, S. I., Zharinov, B. D. *Zvukoisolyatsiya v stroitel'stve* (Sound Isolation in building), pub. Gosstroiizdat, 1949.
3. Belov, A. I. *Akusticheskiye ozmereniya* (Acoustic measurements), pub. VETA, 1941.
4. Beranek, L. L. *Akusticheskiye ozmereniya* (Acoustic measurements), pub. Izdatel'stvo Inostrannoi Literatury, 1962.
5. Bekesy, G. *Über die Schallverzerzungen in der Nahe vor absorbierenden Flüchen und ihre Bedeutung für die Raumakustik.* Z. f. techn. Phys., Bd. 14, 1933.
6. Breus, Yu. V. *Kakim dolzhen byt' ob"yom atelye dlya zapisi muzyki* (On the desirable volume of music recording studios), pub. Tekhnika kino i televideniya (TK i T) 1963, 7.
7. Belov, A. I., Fainstein, N. D. *Eksperimental'noye issledovanie zaglusheniya zvuka v ventilyatsionnykh kanalakh* (Experimental research into sound deadening in ventilation channels), pub. in Zhurn. tekhn. fiz. IX, 16, 1939.
8. Brekhovskikh, L. M. *Rasprostranenie zvuka v sloyakh* (The distribution of sound in layers), pub. by Academy of Sciences of the USSR, 1958.
9. Brekhovskikh, L. M. *Predely primenimosti nekotorikh priblizhennykh metodov, upotreblyaemykh v arkhitekturnoi akustike* (The permissible limits of use of certain approximate methods employed in architectural acoustics), pub. Usp. fiz. nauk, XXXII, 1947.
10. Bolt, R. *Studio acoustics,* Radio-Electronic Eng. Edition of Radio News, I, 1946.
11. Watson, F. R. *Arkhitekturnaya akustika* (Architectural acoustics), pub. Izdatel'stvo inostrannoi literatury, 1948.
12. Velizhanina, K. A. *Zvukopoglotiteli s perforirovannoi panel'yu* (Sound absorbers with perforated panels), pub. in Akust. zhurn. of the Academy of Sciences of the USSR, VII, 2, 1961.
13. Vermeulen, P. *Stereoreverberation,* Phil. Tech. Review, 17, 1956.
14. Gorelik, G. S. *Kolebaniya i Volny* (Vibrations and waves), pub. by Fizmatizdat, 1959.
15. Gardashyan, V. M. *Issledovanie akusticheskikh svoistv pomeshchenij na modelyakh* (The use of models to study the acoustic properties of rooms), pub. in Trudy NIFKI, 4 (14), 1957.

16. Goron, I. E. *Radioveshchanie* (Radio broadcasting), pub. by Svyazizdat, 1944.
17. Genzel', G. S. and Zaezdnyi, A. M. *Osnovy akustiki* (Basic acoustics), pub. by Izdatel'stvo Morskovo Transporta, 1952.
18. Ganus, K. *Arkhitekturnaya Akustika* (Architectural acoustics), pub. Gosstroiizdat, 1963.
19. Gershman, S. G. *Koeffitsient korrelyatsii kak kriterij akusticheskovo kachestva zakrytovo pomeshcheniya* (The correlation coefficient as a criterion for determining the acoustic quality of an enclosure), pub. in Zhurn. tekhn. fiz., XXI, 2, 1959.
20. Golikov, E. E. *K. voprosu o novykh koeffitsientakh kachestvennoi otsenki akustiki pomeshchenij* (On some new coefficients of the qualitative evaluation of the acoustics of enclosures), pub. in Akust. Zhurn. of the Academy of Science of the USSR, II, 3, 1956.
21. Golikov, E. E. *Obshchij i chastnyj kriterij akusticheskovo kachestva pomeshchenij* (A general and particular criterion of the acoustic quality of an enclosure), pub. in Akust. Zhurn. of the Academy of Sciences of the USSR, III, 2, 1957.
22. Dreisen, I. G. *Elektroakustika v shiroveshchanii* (Electroacoustics in broadcasting), pub. Svyazizdat, 1932.
23. Dreisen, I. G. *Kurs Elektroakustiki, chast'* 1, 2 (A course in electro-acoustics, parts 1 and 2), pub. by Svyazizdat, 1940.
24. Dreisen, I. G. *Elektroakustika i zvukovoye veshchaniye* (Electroacoustics and sound broadcasting), Svyazizdat, 1961.
25. Davis, A. *Sovremennaya akustika* (Modern Acoustics), pub. ONTI, 1938.
26. Dunin-Barkovskij, I. V. and Smirnov, N. V. *Teoriya veroyatnostej i matematicheskaya statistika v tekhnike* (The theory of probabilities and mathematical statistics in technology), pub. Gostekhteoretizdat, 1955.
27. Zaborov, B. I. *Teoriya zvukoizolyatsii ograzhdayushchikh konstruktsij* (The theory of sound isolation of enclosing structures), pub. Gosstroiizdat, 1962.
28. Zaborov, V. I. *O zvukoizolyatsii dlya dvoinovo ograzhdeniya s poristym uprugim promezhotuchym sloem* (On sound isolation for a double wall with a porous sprung intermediary layer), pub. in Akust. zhurn. of the Academy of Sciences of the USSR, IX, 2, 1963.
29. Zarkov, N. and Velchev, N. *Issledovanie akusticheskikh kachestv zakrytykh pomeshchenij s pomoshch'yu korotkikh zvukovykh impul'sov* (The use of brief sound impulses in the study of the acoustic properties of an enclosure), pub. in Izvest. NIIKR., Sofia, Bulgaria, IV, 1, 1964.
30. Zarkov, N. and Velchev, N. *Issledovanie akusticheskikh kachestv zakrytykh pomeshchenij s pomoshch'yu modeli* (The use of a model in the study of the acoustic qualities of enclosures), pub. in Izvest. NIIKR., Sofia, Bulgaria, IV, 2, 1964.
31. *Instruktsiya po zvukoizolyatsii pomeshchenij zhilykh i obshchestvennykh zdanij* (SN 39–58), (Instructions for the sound isolation of rooms in dwelling houses and public buildings) (Building norms 39–58) pub. Gosstroiizdat, 1959.
32. Ioffe, V. K. *Elektroakustika* (Electroacoustics), pub. by Svyazizdat, 1954.
33. Ioffe, V. K. and Yampolskij, A. *Raschotnye grafiki i tablitsy po elektroakustike* (Calculating graphs and tables for electroacoustics), pub. by Gosenergoizdat, 1954.
34. Ingerslev, F. *Akustika v sovremennoi stroitel'noi praktike* (Acoustics in modern building practice), pub. Gosstroiizdat, 1957.
35. *Instruction and application apparatus*, Bruel and Kjaer, Copenhagen, 1956.
36. Knudsen, V. O. *Arkhitekturnaya akustika* (Architectural acoustics), pub. by ONTI, 1936.
37. Conturi, L. *Akustika v stroitel'stve* (Acoustics in building), pub. by Gosstroiizdat, 1960.
38. Krasil'nikov, V. *Zvukovye volny* (Sound waves), pub. by Fitzmatgiz, 1956.
39. Kacherovich, A. N. *Akustika kinostudij i kinoteatrov* (The acoustics of film studios and cinemas), pub. Goskinoizdat, 1949.
40. Kacherovich, A. N. *Akusticheskoe proektirovanie kinostudij i kinoteatrov* (The acoustic design of film studios and cinemas), pub. Goskinoizdat, 1952.
41. Kacherovich, A. N. and Khomutov, E. E. *Akustika i Arkhitektura kinoteatrov* (The acoustics and architecture of cinemas), pub. Iskusstvo, 1961.

42. Kacherovich, A. N. *Ob otrazheniyakh zvukovoi energii v pomeshcheniyakh* (On the reflections of sound energy within enclosures), pub. in Trudy NIKFI, 12 (22), 1957.
43. Kacherovich, A. N. *Ob uluchshenij kachestva zvuchaniya fonogrammy v kinoteatre* (On the improvement of sound track reproduction quality in cinemas), pub. TK i T, 3, 1963.
44. Kacherovich, A. N. *Ob obyome atele dlya zapisi muzyki* (On the volume of a music recording studio), pub. by TK i T, 5, 1964.
45. Kuhl, W. *Über Versuche zur Ermittlung der günstigsten Nachhallzeit grossen Musik-studion*, Acustica, 4, 1954.
46. Lifshits, S. Ya. *Akustika zdanij i ikh izolyatsii ot shuma i sotryasenij* (The acoustics of buildings and means of isolating them from sound and from concussion), pub. by Nauch. tekhn. izd., 1931.
47. Lifshits, S. Ya. *Kurs arkhitekturnoi akustiki* (A course in architectural acoustics), ONTI, 1937.
48. Lebedeva, I. V. *Issledovanie reverberatsionnoi kamery MGU* (A study of the Moscow State University's reverberation chamber), in Akust. zhurn. of the Academy of Sciences of the USSR, VI, 3, 1960.
49. Lebedeva, I. V. *K voprosu o metodike izmereniya koeffitsienta zvukopogloshcheniya v reverberatsionnoi kamere* (On the methods of measurement of the coefficient of sound absorption in a reverberation chamber), in Akust. zhurn. of the Academy of Sciences of the USSR, VIII, 3, 1962.
50. Morse, P. *Kolebaniya i zvuk* (Vibrations and sound), pub. Gostekhizdat, 1949.
51. Morse, P. and Bolt, R. *Zvukovye volny v pomeshcheniyakh* (Sound waves in enclosures), in Usp. fiz. nauk, XXX, 1–2, 1946.
52. Myasnikov, L. L. *Akusticheskiye izmereniya* (Acoustic measurements), pub. ONTI, 1937.
53. Mexfild, J. and Albershein, W. Acoust. Soc. Amer., 19, 1947.
54. Molodaya, N. T. *Akusticheskoye proetirovanie radioveshchatel'nykh i televizionnykh studij* (The acoustic design of radio and television studios), pub. Svyazizdat, 1964.
55. Mankovsky, V. S. *K voprosu ob akusticheskikh usloviyakh v zritelnykh zalakh kinoteatrov* (On acoustic conditions in cinema auditoria), pub. in Trudy Leningrads-kovo Instituta kinoinzhenerov (LIKI), III, 1955.
56. Mankovsky, V. S. *Akusticheskiye osobennosti pervichnykh pomeshchenij, prednas-nachennykh dlya stereofonicheskikh peredachakh* (The acoustic properties of primary enclosures designed for stereophonic transmissions), pub. in Trudy LIKI, 1961.
57. Mankovsky, V. S. *K voprosu o vliyanii akustiki pomeshchenij na kachestvo stereo-fonicheskoi peredachi* (On the influence of room acoustics on the quality of stereo-phonic transmissions), pub. by TK i T, 4, 1962.
58. Mankovsky, V. S. *Teoreticheskij raschot nekotorykh velichin, vliyayushchikh na peredachu prostranstvennykh kharakteristik kashushchevosya istochnika zvuka* (The theoretical calculation of certain values which have an influence on the transmission of the spatial characteristics of the seeming sound source), pub. in Trudy LIKI, X, 1964.
59. Mankovsky, V. S. *K voprosu o vospriyatii prostranstva pri stereofonicheskikh pere-dachakh* (On the perception of space in stereophonic transmissions), pub. in Trudy LIKI, X, 1964.
60. Molodaya, N. T. and Papernov, L. Z. *Apparatura studijnykh traktov i sistem zvukoy-sileniya* (The apparatus of studio channels and sound amplification systems), pub. Svyazizdat, 1963.
61. Nikol'skij, V. N. (ed.) *Voprosy zvukoizolyatsii v arkhitekturnoi akustike* (Questions of sound isolation in architectural acoustics), pub. Gosstroiizdat, 1959.
62. Nikol'skij, V. N. and Zaborov, V. I. *Zvukoizolyatsiya krupnopanel'nykh zdanij* (Sound isolation in buildings of large panel construction), pub. Stroiizdat, 1964.
63. *Normy i tekhnicheskie usloviya proektirovaniya zdanij kinoteatrov* (Norms and technical conditions for the design of cinema buildings (Building norms SN 30–58), pub. Stroiizdat, 1958.

64. Papernov, L. Z. *Ozvuchanie otkrytykh prostranstv* (The irradiation of open spaces with sound), pub. Svyazizdat, 1963.
65. Panfilov, N. D. *Eksperimental'noye issledovanie akusticheskikh uslovij v pomeshcheniyakh impul'snym metodom* (The use of the impulse method in experimental studies of acoustic conditions in enclosures), pub. in Trudy NIKFI, 12 (22), 1957.
66. Raleigh, *Teoriya zvuka t.* 1 (Sound theory, vol. 1), pub. by Gostekhteoretizdat, 1940.
67. Rzhevskij, S. N. *Obzor rabot po rezonansnym zvukopoglotitelyam* (A review of works on resonance sound absorbers), pub. in Usp. fiz. nauk, XXX, 1-2, 1946.
68. Rzhevskij, S. N. *Slukh i rech, v svete sovremennykh fizicheskikh issleodvanij* (Hearing and speech in the light of modern studies in physics), pub. by ONTI, 1936.
69. Rabinovich, A. A. and Sukharevskij, Yu. M. *Radioveshchatel'nye studii i mikrofony* (Radio broadcasting studios and microphones), pub. Svyazizdat, 1939.
70. Rozenberg, L. D. *Summarnaya reverberatsiya pri zapis i i vosproizvedenii zvuka* (Summary reverberation in sound-recording and reproduction), pub. in Zhurn. tekhn. fiz., 2, 1931.
71. Rozenberg, L. D. *Metod raschota zvukovykh polej, obrazovannykh raspredelyonnymi sistemami izluchatelej* (The method of calculation of sound fields formed by distributed systems of radiators), pub. in Zhurn. tekhn. fiz., 12, 1940.
72. Rozenberg, L. D. *Metod raschota zvukovykh polej, obrasovannykh raspredelyonnymi sistemami izluchatelej, rabotayushchikh v zakrytykh pomeshcheniyakh* (The method of calculation of sound fields formed by distributed systems of radiators working in enclosures), pub. in Zhurn. takhn. fiz., 12, 1942.
73. Stuchik, E. *Osnovy akustiki*, tt. 1, 2 (Basic acoustics, vols. 1 and 2), pub. by Izdatel'stvo inostrannoi literatury, 1959.
74. Sabine, W. C. *Collected papers on acoustics*, Harvard University Press, 1923.
75. Sapozhkov, M. A. *K voprosu ob opredelenii optimal'noi reverberatsii v svyazannykh pomeshcheniyakh* (On the determination of optimum reverberation in linked enclosures), pub. in Zhurn. tekhn. fiz., IV, 8, 1934.
76. Stoletov, A. G. *Vvedenie v akustiku i optiku* (Introduction to acoustics and optics), pub. 1895.
77. *Stroitel'nye normy i pravila t.* 2 (Building norms and rules, vol. 2), Moscow, 1960.
78. Tager, P. G. (ed.). *Akustika kinematografii* (Acoustics of cinematography), pub. in Trudy NIKFI, 12 (22), 1957.
79. Timofeev, A. K. and Osipov, G. L. *Zvukoizolyatsiya ograzhdayushchikh konstruktsij zhilykh domov* (Sound isolation of the walls of dwelling houses), pub. by NII stroit. fiz., 1959.
80. *Teletsentry, retranslyatsionnye stantsii i radiodoma. Normy proektirovaniya* (Television centres, retransmitting stations, and radio studio centres. Design norms), TU—560—60, pub. by Svyazizdat, 1961.
81. Furduev, V. V. *Elektroakustika* (Electroacoustics), pub. Gostekhizdat, 1948.
82. Furduev, V. V. *Akusticheskiye osnovy veshchaniya* (Acoustic bases of broadcasting), pub. Svyazizdat, 1960.
83. Furduev, V. V. (ed.). *Arkhitekturnaya akustika, sbornik* (A handbook of architectural acoustics), pub. Stroiizdat, 1961.
84. Furduev, V. V. *Metod akusticheskovo proektirovaniya auditorij, oborudovannykh gromkogovoritelyami* (Method of acoustic design of auditoria equipped with loudspeakers), pub. in Doklady (Reports) of the Academy of Sciences of the USSR, 15, 1937.
85. Furduev, V. V. and Chen-Tun. *Izmerenie diffuznosti zvukovovo polya v pomeshcheniyakh metodom napravlennovo mikrofona* (The use of the method of the directional microphone in the measurement of diffuseness of the sound field in enclosures), pub. in Akust. zhurn. of the Academy of Sciences of the USSR, V, 1, 1960.
86. Furduev, V. V. *Korrelyatsionnyj kriterij optimuma reverberatsii* (The correlation criterion of optimum reverberation), pub. in Akust. zhurn of the Academy of Sciences of the USSR, III, 1, 1957.

87. Furduev, V. V. *Ambiofonicheskaya reverberatsiya* (Ambiophonic reverberation), pub. in Akust. zhurn. of the Academy of Sciences of the USSR, VII, 2, 1961.
88. Haas, H. H. *Über die Einfluss eines Einfachechos aus die Hörsamkeit von Sprache.* Acustica, I, 49, 1951.
89. Khokhlov, A. D. and others. *Rukovodstvo k laboratornym rabotam po elektroakustike* (A guide to laboratory work on electroacoustics), pub. LIKI, 1957.
90. Zwikker, K. and Kosten, K. *Zvukopogloshchayushchiye materialy* (Sound absorbent materials), pub. by Izdatel'stvo inostrannoi literatury, 1952.
91. Tseller, V. *Tekhnika bor'by s shumami* (The technique of overcoming noise), pub. Gosstroiizdat, 1958.
92. Chigrinskij, G. A. *Kartina otrazhenij i ee primenenie v arkhitekturnoi akustike* (The reflection pattern and its application in architectural acoustics), pub. in Reports of the Academy of Sciences of the USSR, XXIII, 7, 1939.
93. Chigrinskij, G. A. *Kartina otrazhenij i reverberatsiya nezamknutykh prostranstv* (The reflection pattern and reverberation of non-enclosed spaces), pub. in Zhurn. tekhn. fiz., IX, 16, 1939.
94. Yudin, E. Ya. *Glushenie shuma ventilyatsionnykh ustanovok* (Sound baffling in ventilation installations), pub. Gosstroiizdat, 1958.

Index